Lagrangian Analysis and Quantum Mechanics

Lagrangian Analysis and Quantum Mechanics
A Mathematical Structure Related to
Asymptotic Expansions and the Maslov Index

Jean Leray

English translation by Carolyn Schroeder

The MIT Press
Cambridge, Massachusetts
London, England

This book was set in Monophoto Times Roman by Asco Trade Typesetting Ltd., Hong Kong, and printed and bound by Murray Printing Company in the United States of America.

Library of Congress Cataloging in Publication Data

Leray, Jean, 1906–
Lagrangian analysis and quantum mechanics.

Bibliography: p.
Includes indexes.
1. Differential equations, Partial—Asymptotic theory. 2. Lagrangian functions. 3. Maslov index. 4. Quantum theory. I. Title.
QA377.L414 1982 515.3′53 81-18581
ISBN 0-262-12087-9 AACR2

To Hans Lewy

Contents

Preface xi
Index of Symbols xiii
Index of Concepts xvii

I. The Fourier Transform and Symplectic Group

Introduction 1

§1. Differential Operators, The Metaplectic and Symplectic Groups 1

0. Introduction 1
1. The Metaplectic Group $\mathrm{Mp}(l)$ 1
2. The Subgroup $\mathrm{Sp}_2(l)$ of $\mathrm{Mp}(l)$ 9
3. Differential Operators with Polynomial Coefficients 20

§2. Maslov Indices; Indices of Inertia; Lagrangian Manifolds and Their Orientations 25

0. Introduction 25
1. Choice of Hermitian Structures on $\mathrm{Z}(l)$ 26
2. The Lagrangian Grassmannian $\Lambda(l)$ of $Z(l)$ 27
3. The Covering Groups of $\mathrm{Sp}(l)$ and the Covering Spaces of $\Lambda(l)$ 31
4. Indices of Inertia 37
5. The Maslov Index m on $\Lambda^2_\infty(l)$ 42
6. The Jump of the Maslov Index $m(\lambda_\infty, \lambda'_\infty)$ at a Point (λ, λ') Where $\dim \lambda \cap \lambda' = 1$ 47
7. The Maslov Index on $\mathrm{Sp}_\infty(l)$; the Mixed Inertia 51
8. Maslov Indices on $\Lambda_q(l)$ and $\mathrm{Sp}_q(l)$ 53
9. Lagrangian Manifolds 55
10. q-Orientation ($q = 1, 2, 3, \ldots, \infty$) 56

§3. Symplectic Spaces 58

0. Introduction 58
1. Symplectic Space Z 58
2. The Frames of Z 60
3. The q-Frames of Z 61
4. q-Symplectic Geometries 65

Conclusion 65

II. Lagrangian Functions; Lagrangian Differential Operators

Introduction 67

§1. Formal Analysis 68

0. Summary 68
1. The Algebra $\mathscr{C}(X)$ of Asymptotic Equivalence Classes 68
2. Formal Numbers; Formal Functions 73
3. Integration of Elements of $\mathscr{F}_0(X)$ 80
4. Transformation of Formal Functions by Elements of $\mathrm{Sp}_2(l)$ 86
5. Norm and Scalar product of Formal Functions with Compact Support 91
6. Formal Differential Operators 97
7. Formal Distributions 102

§2. Lagrangian Analysis 104

0. Summary 104
1. Lagrangian Operators 105
2. Lagrangian Functions on $\check{V}$ 109
3. Lagrangian Functions on V 115
4. The Group $\mathrm{Sp}_2(Z)$ 123
5. Lagrangian Distributions 123

§3. Homogeneous Lagrangian Systems in One Unknown 124

0. Summary 124
1. Lagrangian Manifolds on Which Lagrangian Solutions of $aU = 0$ Are Defined 124
2. Review of E. Cartan's Theory of Pfaffian Forms 125
3. Lagrangian Manifolds in the Symplectic Space Z and in Its Hypersurfaces 129
4. Calculation of aU 135
5. Resolution of the Lagrangian Equation $aU = 0$ 139
6. Solutions of the Lagrangian Equation $aU = 0 \bmod(1/v^2)$ with Positive Lagrangian Amplitude: Maslov's Quantization 143
7. Solution of Some Lagrangian Systems in One Unknown 145

8. Lagrangian Distributions That Are Solutions of a Homogeneous Lagrangian System 151

Conclusion 151

§4. Homogeneous Lagrangian Systems in Several Unknowns 152

1. Calculation of $\Sigma_{m=1}^{\mu} a_n^m U_m$ 152

2. Resolution of the Lagrangian System $aU = 0$ in Which the Zeros of det a_0^0 Are Simple Zeros 156

3. A Special Lagrangian System $aU = 0$ in Which the Zeros of det a_0^0 Are Multiple Zeros 159

III. Schrödinger and Klein-Gordon Equations for One-Electron Atoms in a Magnetic Field

Introduction 163

§1. A Hamiltonian H to Which Theorem 7.1 (Chapter II, §3) Applies Easily; the Energy Levels of One-Electron Atoms with the Zeeman Effect 166

1. Four Functions Whose Pairs Are All in Involution on $\mathbf{E}^3 \oplus \mathbf{E}^3$ Except for One 166

2. Choice of a Hamiltonian H 170

3. The Quantized Tori $T(l, m, n)$ Characterizing Solutions, Defined mod$(1/v)$ on Compact Manifolds, of the Lagrangian System $aU = (a_{L^2} - L_0^2)U = (a_M - M_0)U = 0 \bmod (1/v^2)$ 174

4. Examples: The Schrödinger and Klein-Gordon Operators 179

§2. The Lagrangian Equestion $aU = 0 \bmod(1/v^2)$ (a Associated to H, U Having Lagrangian Amplitude $\geqslant 0$ Defined on a Compact V) 184

0. Introduction 184

1. Solutions of the Equation $aU = 0 \bmod(1/v^2)$ with Lagrangian Amplitude $\geqslant 0$ Defined on the Tori $V[L_0, M_0]$ 185

2. Compact Lagrangian Manifolds V, Other Than the Tori $V[L_0, M_0]$, on Which Solutions of the Equation $aU = 0 \bmod(1/v^2)$ with Lagrangian Amplitude $\geqslant 0$ Exist 190

3. Example: The Schrödinger–Klein-Gordon Operator 204

Conclusion 207

§3. The Lagrangian System
$aU = (a_M - \text{const.})U = (a_{L^2} - \text{const.})U = 0$
When a Is the Schrödinger–Klein-Gordon Operator 207

0. Introduction 207

1. Commutivity of the Operators a, a_{L^2}, and a_M Associated to the Hamiltonians H (§1, Section 2), L^2, and M (§1, Section 1) 207

2. Case of an Operator a Commuting with a_{L^2} and a_M 210

3. A Special Case 221

4. The Schrödinger–Klein-Gordon Case 226

Conclusion 230

§4. The Schrödinger–Klein-Gordon Equation 230

0. Introduction 230

1. Study of Problem (0.1) without Assumption (0.4) 231

2. The Schrödinger–Klein-Gordon Case 234

Conclusion 237

IV. Dirac Equation with the Zeeman Effect

Introduction 238

§1. A Lagrangian Problem in Two Unknowns 238

1. Choice of Operators Commuting mod$(1/v^3)$ 238

2. Resolution of a Lagrangian Problem in Two Unknowns 240

§2. The Dirac Equation 248

0. Summary 248

1. Reduction of the Dirac Equation in Lagrangian Analysis 248

2. The Reduced Dirac Equation for a One-Electron Atom in a Constant Magnetic Field 254

3. The Energy Levels 258

4. Crude Interpretation of the Spin in Lagrangian Analysis 262

5. The Probability of the Presence of the Electron 264

Conclusion 266

Bibliography 269

Preface

Only in the simplest cases do physicists use exact solutions, $u(x)$, of problems involving temporally evolving systems. Usually they use asymptotic solutions of the type

$$u(\nu, x) = \alpha(\nu, x)e^{\nu\varphi(x)}, \tag{1}$$

where

- the *phase* φ is a real-valued function of $x \in X = \mathbf{R}^l$;
- the *amplitude* α is a formal series in $1/\nu$,

$$\alpha(\nu, x) = \sum_{r=0}^{\infty} \frac{1}{\nu^r}\alpha_r(x),$$

whose coefficients α_r are complex-valued functions of x;
- the *frequency* ν is purely imaginary.

The differential equation governing the evolution,

$$a\left(\nu, x, \frac{1}{\nu}\frac{\partial}{\partial x}\right)u(\nu, x) = 0, \tag{2}$$

is satisfied in the sense that the left-hand side reduces to the product of $e^{\nu\varphi}$ and a formal series in $1/\nu$ whose first terms or all of whose terms vanish. The construction of these asymptotic solutions is well known and called the WKB method:

- The phase φ has to satisfy a first-order differential equation that is nonlinear if the operator a is not of first order.
- The amplitude α is computed by integrations along the characteristics of the first-order equation that defines φ.

In quantum mechanics, for example, computations are first made as if

$$\nu = \frac{i}{\hbar} = \frac{2\pi i}{h}, \qquad \text{where } h \text{ is Planck's constant,}$$

were a parameter tending to $i\infty$; afterwards ν receives its numerical value ν_0.

Physicists use asymptotic solutions to deal with problems involving equilibrium and periodicity conditions, for example, to replace problems of wave optics with problems of geometrical optics. But φ has a jump and α has singularities on the envelope of characteristics that define φ: for example, in geometrical optics, α has singularities on the caustics, which are the images of the sources of light; nevertheless geometrical optics holds beyond the caustics.

V. P. Maslov introduced an index (whose definition was clarified by *I. V. Arnold*) that described these phase jumps, and he showed by a convenient use of the Fourier transform that these amplitude singularities are only apparent singularities. But he had to impose some "quantum conditions." These assume that ν has some purely imaginary numerical value ν_0, in contradiction with the previous assumption about ν, namely, that ν is a parameter tending to $i\infty$. The assumption that ν tends to $i\infty$ is necessary for the Fourier transform to be pointwise, which is essential for Maslov's treatment. *A procedure, avoiding that contradiction and guided by purely mathematical motivations, that makes use of the Fourier transform, expressions of the type* (1), *Maslov's quantum conditions, and the datum of a number* ν_0 *does exist*, but no longer tends to define a function or a class of functions by its asymptotic expansion. It leads to a *new mathematical structure*, ***lagrangian analysis***, which requires the datum of a constant ν_0 and is based on symplectic geometry. Its interest can appear only a posteriori *and could be quantum mechanics.* Indeed this structure allows a new interpretation of the *Schrödinger*, *Klein-Gordon*, and *Dirac* equations provided

$$\nu_0 = \frac{i}{\hbar} = \frac{2\pi i}{h}, \qquad \text{where } h \text{ is Planck's constant.}$$

Therefore the real number $2\pi i/\nu_0$ whose choice defines this new mathematical structure can be called *Planck's constant.*

The introductions, summaries, and conclusions of the chapters and parts constitute an abstract of the exposition.

Historical note. In Moscow in 1967 *I. V. Arnold* asked me my thoughts on Maslov's work [10, 11]. The present book is an answer to that question.

It has benefited greatly from the invaluable knowledge of *J. Lascoux.*

It introduces ν_0 for defining lagrangian functions on V (chapter II, §2, section 3) in the same manner as Planck introduced $\hbar$ for describing the spectrum of the blackbody. Thus the book could be entitled

The Introduction of Planck's Constant into Mathematics.

January 1978
Collège de France
Paris 05

Index of Symbols

$\mathbf{A}$	I,§1,definition 1.2*
$\mathbf{C}$	field of the complex numbers; $\dot{\mathbf{C}} = \mathbf{C}\backslash\{0\}$
$\mathbf{E}^3$	3-dimensional euclidean space
$\mathbf{I} = i[1, \infty[$	II,§1,1
$\mathbf{N}$	set of the natural numbers (i.e., integers $\geqslant 0$)
$\mathbf{R}$	field of the real numbers
$\mathbf{R}_+$	set of the real positive numbers
$\mathbf{S}^n$	n-dimensional sphere
$\mathbf{Z}$	ring of the integers
A	element of $\mathbf{A}$: I; II
B	bounded set: I; II
A, B, C	functions of M, coefficients of the Schrödinger–Klein-Gordon operator: III,§1,example 4
E	neutral element of a group: I; II
	atomic energy level: III; IV
F	any function: II
	the function in III,§1,(4.23)
G	group: I; II
	function: III; IV
H	hamiltonian: II,§3,1; II,§3,definition 6.1
Hess	hessian: I,§1,definition 2.3
I_k, J_k	elements of $\mathbf{E}^3$: III,§1,1
Inert	index of inertia: I,§1,2; I,§2,4; I,§2,definition 7.2

*Each chapter (I, II, III, IV) is divided into parts (§1, §2, §3, ...), which in turn are divided into sections (0, 1, 2, ...). References to elements of sections (for example, theorems, equations, definitions) in the same chapter, part, and section are by one or two numbers: in the latter case the first number refers to the section and is followed by a period. References to elements of sections in another chapter, part, and/or section are by a string of numbers separated by commas. For example, a reference *in chapter I, §2, section 3* to the one theorem in this chapter, part, and section is simply theorem 3; to the one theorem in this chapter and part but section 4 is theorem 4; to the one theorem in this chapter but §3 of section 4 is §3,theorem 4; and to the one theorem in chapter II, §3, section 4 is II,§3,theorem 4. Similarly, a reference *in chapter I, §2, section 3* to the first definition (of more than one) in chapter II, §3, section 4 is II,§3,definition 4.1.

$J^{(k)}$	matrices: II,§4,3; IV,§1,1
K	characteristic curve: II,§3,definition 3.1; III,§1,(2.14)
	function: III,§1,4
L, M, P, Q, R	functions: III,§1,1
N	function of (L, M): III,§1,(2.9); N_L and N_M are its derivatives
R (I, II), R_0 (III, IV)	frame in symplectic geometry: I,§3,2; I,§3,3
$\mathrm{Sp}(l)$	symplectic group: I,§1,definition 1.1
$\mathrm{Sp}_2(l)$	the covering group, of order 2, of the symplectic group: I,§1,definition 2.1
S	element of $\mathrm{Sp}_2(l)$
T	torus: III,§1,3
U	lagrangian function on V: II,§2,3
$\check{U}_R$	formal functions on $\check{V}$: II,§2,2
$U(l)$	unitary group: I,§2,2
V	lagrangian manifold: I,§2,9; I,§3,1
W	hypersurface of Z
$W(l)$	subset of $U(l)$: I,§2,lemma 2.1
X, Z	spaces: I,§1,1; I,§3,1
a	differential, formal or lagrangian operator: I,§1,definition 3.1; II,§1,definition 6.2; II,§2, definition 1.1
arg	argument
d	differentiation
$d^l x$	Lebesgue measure
det	determinant
e	2.71828...
	neutral element of a group: I
i	$\sqrt{-1}$
i_ξ	interior product: II,§3,2
j	quantum number: IV,§1,example 2
l	dimension of X: I; II (dimension of $X = 3$ in III, IV)

l, m, n	quantum numbers: III; IV
m, m_R	Maslov index: I,§1,definition 1.2; I,§1,(2.15); I,§2,definition 5.3; I,§2,theorem 5; I,§3,theorem 1; I,§3,3
s	element of $\mathrm{Sp}(l)$: I; II
	function: III; IV
t	function: II,§3,(3.10); II,§3,(3.13); III,§1,(2.6)
u	element of $U(l)$: I
${}^t u$	transpose of u: I,§2,2
u	formal number or function: II,§1,2
w	element of $W(l)$
x, y	elements of X
z	elements of Z
$\mathscr{A}$	I,§1,1
$\mathscr{B}, \mathscr{C},$	II,§1,1
$\mathscr{F}, \mathscr{F}'$	space of formal or lagrangian functions or distribution: II,§1,2; II,§1,7; II,§2,2; II,§2,3; II,§2,5
$\mathscr{H}$	Hilbert space: I,§1,1
$\mathscr{L}$	Lie derivative: II,§3,definition 3.2
$\mathscr{N}$	neighborhood
$\mathscr{S}, \mathscr{S}'$	Schwartz spaces: I,§1,1
Γ	arc: I,§2
	curve: III,§1,(2.5)
Δ	laplacian (Δ_0 is the spherical laplacian): III,§3,(2.4)
Λ	lagrangian grassmannian: I,§2,2; I,§3,1
$\wedge$	exterior product
Π	projection: II,§1,theorem 2.1
Σ	apparent contour, Σ_{Sp}: I,§1,definition 1.3
	Σ_V: I,§2,9
	Σ_R: I,§3,2
Φ, Ψ, Θ	Euler angles: III,§1,(1.11); III,§1,(1.12)
Ω	open set in Z: II,§1,6; II,§2,1

	function: III,§1,(2.8)
Ω^6	open set in $\mathbf{E}^3 \oplus \mathbf{E}^3$: III,§1,1
α	amplitude: II,§1,2
β_0	lagrangian amplitude: II,§2,theorem 2.2
γ	arc or homotopy class
η, η_V	invariant measure of V: II,§3,definition 3.2
κ	characteristic vector: II,§3,definition 3.1
λ	element of Λ: I; II
λ, μ	functions: III,§1,(2.10)
ν	element of $\mathbf{I}$: II,§1,1
ν_0	$i/\hbar = 2\pi i/h$ ($h \in \mathbf{R}_+$): II,§2,3; II,§3,6
π	3.14159...
π_j	jth homotopy group: I,§2,3
$\sigma[\cdot], \sigma_j$	Pauli matrices: IV,§1,(1.6); IV,§1,(1.7)
χ	$\eta/d^l x$
φ	phase: I,§2,9; I,§3,2; II,§1,2
ψ	lagrangian phase: I,§3,1
ω, ϖ	pfaffian forms
ω_j	III,§1,(1.7)

Atomic Symbols: III; IV; passim (see III,§1,4, Notations)

E	energy
c	speed of light
h	Planck's constant
$\mathscr{A}_j$	potential vector
$\mathscr{H}$	magnetic field
α	1/137
β	Bohr magneton
ε	charge
μ	mass

Index of Concepts

amplitude	α
asymptotic class	$\mathscr{C}$
characteristic curve	K
characteristic vector	κ
energy	E
Euler angles	Φ, Ψ, Θ
formal number, functions	$u, \check{U}_R$
frames	$R; (I_1, I_2, I_3); (J_1; J_2; J_3)$
groups	$\mathrm{Sp}(l); U(l)$
hamiltonian	H
hessian	Hess
homotopy	π_j
index of inertia	Inert
interior product	i_ξ
lagrangian amplitude	β_0
lagrangian function	U
lagrangian manifold	V
lagrangian operator	a
lagrangian phase	ψ
Lie derivative	$\mathscr{L}$
Maslov index	m
matrix	$J^{(k)}, \sigma$
operator	a
Planck's constant	$\nu_0 = i/\hbar$
quantum numbers	l, m, n, j
spaces	$X, Z, \mathscr{F}. \mathscr{F}', \mathscr{H}, \mathscr{S}, \mathscr{S}'$
symplectic space	Z

Lagrangian Analysis and Quantum Mechanics

I The Fourier Transform and Symplectic Group

Introduction

Chapter I explains the connection between two very classical notions: the Fourier transform and the symplectic group.

It will make possible the study of asymptotic solutions of partial differential equations in chapter II.

§1. Differential Operators, the Metaplectic and Symplectic Groups

0. Introduction

Historical account. The metaplectic group was defined by *I. Segal* [14]; his study was taken up by *D. Shale* [15]. *V. C. Buslaev* [3, 11] showed that it made Maslov's theory independent of the choice of coordinates. *A. Weil* [18] studied it on an arbitrary field in order to extend *C. Siegel's* work in number theory.

Summary. We take up the study of the metaplectic group in order to specify its action on $\mathscr{S}(\mathbf{R}^l)$, $\mathscr{H}(\mathbf{R}^l)$, and $\mathscr{S}'(\mathbf{R}^l)$ (see theorem 2) and its action on differential operators (see theorem 3.1).

1. The Metaplectic Group $\mathrm{Mp}(l)$

Let X be the vector space $\mathbf{R}^l$ ($l > 1$) provided with *Lebesgue measure* d^lx. Let X^* be its dual, and let $\langle p, x \rangle$ be the value obtained by acting $p \in X^*$ on $x \in X$.

Spaces of functions and distributions on X. The *Hilbert space* $\mathscr{H}(X)$ consists of functions $f\colon X \to \mathbf{C}$ satisfying

$$|f| = \left(\int_X |f(x)|^2\, d^lx\right)^{1/2} < \infty.$$

The *Schwartz* space $\mathscr{S}(X)$ [13] consists of infinitely differentiable, rapidly decreasing functions $f\colon X \to \mathbf{C}$. That is, for all pairs of l-indices (q, r)

$$|f|_{q,r} = \operatorname*{Sup}_X \left|x^q \left(\frac{\partial}{\partial x}\right)^r f(x)\right| < \infty.$$

The topology of $\mathscr{S}(X)$ is defined by a *countable* fundamental system of neighborhoods of 0, each depending on a pair of l-indices (q, r) and a rational number $\varepsilon > 0$ as follows:

$$\mathcal{N}(q, r, \varepsilon) = \{f \mid |f|_{q,r} \leqslant \varepsilon\}.$$

The bounded sets B of $\mathcal{S}(X)$ are thus all subsets of bounded sets of $\mathcal{S}(X)$ of the following form:

$$B(\{b_{q,r}\}) = \{f \mid |f|_{q,r} \leqslant b_{q,r}\ \forall q, r\}, \qquad q, r \in \mathbf{N}^l, \quad b_{q,r} \in \dot{\mathbf{R}}_+.$$

The *Schwartz space* $\mathcal{S}'(X)$ is the dual of $\mathcal{S}(X)$ [13]; its elements are the *tempered distributions*: such an element f' is a continuous linear functional

$$\mathcal{S}(X) \to \mathbf{C}.$$

The value of f' on f will be denoted by $\int_X f'(x) f(x)\, d^l x$, although the value of f' at x is not in general defined. The bound of f' on a bounded set B in $\mathcal{S}(X)$ is denoted by

$$|f'|_B = \operatorname*{Sup}_{f \in B} \left| \int_X f'(x) f(x)\, d^l x \right|.$$

The continuity of f' is equivalent to the condition that f' is *bounded*: $|f'|_B < \infty\ \forall B$. The topology of $\mathcal{S}'(X)$ is defined by a fundamental system of neighborhoods of 0, each depending on a bounded set B of $\mathcal{S}(X)$ and a number $\varepsilon > 0$, as follows:

$$\mathcal{N}'(B, \varepsilon) = \{f' \mid |f'|_B \leqslant \varepsilon\}.$$

Unlike the above, this topology cannot be given by a countable fundamental system of neighborhoods of zero.

Let us recall the following theorems. $\mathcal{H}(X)$ can be identified with a subspace of $\mathcal{S}'(X)$:

$$\mathcal{S}(X) \subset \mathcal{H}(X) \subset \mathcal{S}'(X).$$

The *Fourier transform* is a continuous automorphism of $\mathcal{S}'(X)$ whose restrictions to $\mathcal{H}(X)$ and $\mathcal{S}(X)$ are, respectively, a unitary automorphism and a continuous automorphism.

$\mathcal{S}(X)$ is dense in $\mathcal{S}'(X)$.

For the proof of the last theorem, see L. Schwartz [13]: chapter VII, §4, the commentary on theorem IV, and chapter III, §3, theorem XV; alternatively, see chapter VI, §4, theorem IV, theorem XI and its commentary.

Differential operators associated with elements of $Z(l) = X \oplus X^*$. Let v be an imaginary number with argument $\pi/2$: $v/i > 0$.

Let a^0 be a linear function, $a^0: Z(l) \to \mathbf{R}$. Let $a^0(z) = a^0(x, p)$ be its value at $z = x + p\,[z \in Z(l), x \in X, p \in X^*]$. The operator

$$a = a^0\left(x, \frac{1}{v}\frac{\partial}{\partial x}\right)$$

is a *self-adjoint* endomorphism of $\mathscr{S}'(X)$: the adjoint of a, which is an endomorphism of $\mathscr{S}(X)$, is the restriction of a to $\mathscr{S}(X)$. The operators a and the functions a^0 are, respectively, elements of two vector spaces $\mathscr{A}$ and $\mathscr{A}^0$. These spaces are both of dimension $2l$ and are naturally isomorphic:

$$\mathscr{A}^0 \ni a^0 \mapsto a \in \mathscr{A}.$$

We say that a is the differential operator associated to $a^0 \in \mathscr{A}^0$. By (1.2), $\mathscr{A}^0$, which is the dual of $Z(l)$, will be identified with $Z(l)$.

The commutator of a and $b \in \mathscr{A}$ is

$$[a, b] = ab - ba \in \mathbf{C};$$

$c \in \mathbf{C}$ denotes the endomorphism of $\mathscr{S}'(X)$:

$$c: f \mapsto cf \qquad \forall f \in \mathscr{S}'(X).$$

In order to study this commutator, we give $Z(l)$ the *symplectic structure* $[\cdot,\cdot]$ defined by

$$[z, z'] = \langle p, x' \rangle - \langle p', x \rangle,$$

where $z = x + p$, $z' = x' + p'$, x and $x' \in X$, and p and $p' \in X^*$.

Each function $a^0 \in \mathscr{A}^0$ is defined by a unique element a^1 in $Z(l)$ such that

$$a^0(z) = [a^1, z]. \tag{1.1}$$

This gives a natural isomorphism

$$Z(l) \ni a^1 \mapsto a^0 \in \mathscr{A}^0. \tag{1.2}$$

The commutator of a and $b \in \mathscr{A}$ is clearly

$$[a, b] = \frac{1}{\nu}[a^1, b^1], \tag{1.3}$$

where the right-hand side is defined by the symplectic structure.

An automorphism S of $\mathscr{S}'(X)$ transforms each $a \in \mathscr{A}$ into an operator $b = SaS^{-1}$, defined by the condition

$$bSf = Saf \qquad \forall f \in \mathscr{S}'(X).$$

$b \neq 0$ if $a \neq 0$. In general, $b \notin \mathscr{A}$.

Definition 1.1. $G(l)$ is the group of continuous automorphisms S of $\mathscr{S}'(X)$ that transform $\mathscr{A}$ into itself in the sense that

$$SaS^{-1} \in \mathscr{A} \qquad \forall a \in \mathscr{A}. \tag{1.4}$$

$G(l)$ is clearly a semigroup. If $S \in G(l)$,

$$a \mapsto SaS^{-1} \tag{1.5}$$

is clearly an automorphism of $\mathscr{A}$. Therefore $S^{-1} \in G(l)$, and $G(l)$ is a group.

Under the natural isomorphism $Z(l) \to \mathscr{A}$, the automorphism (1.5) of $\mathscr{A}$ becomes an automorphism of the vector space $Z(l)$:

$$s : a^1 \mapsto sa^1. \tag{1.6}$$

Since S commutes with the automorphisms of $\mathscr{S}'(X)$ given by $c \in \dot{\mathbf{C}}$, and since $[a, b] \in \mathbf{C}$, we have

$$[SaS^{-1}, SbS^{-1}] = [a, b],$$

or, considering (1.3) and the equivalence of (1.5) and (1.6),

$$[sa^1, sb^1] = [a^1, b^1].$$

Therefore s is an automorphism of the symplectic space $Z(l)$.

The group of automorphisms of the symplectic space $Z(l)$ is called the *symplectic group* and is denoted $\mathrm{Sp}(l)$:

$$s \in \mathrm{Sp}(l).$$

By (1.1),

$$[sa^1, z] = [a^1, s^{-1}z] = (a^0 \circ s^{-1})(z).$$

In summary:

LEMMA 1.1. Under the natural isomorphisms of $\mathscr{A}$, $Z(l)$, and $\mathscr{A}^0$, the automorphism

$$a \mapsto SaS^{-1}$$

of $\mathscr{A}$, which is defined for all $S \in G(l)$, becomes

- an automorphism s of $Z(l)$, $s: a^1 \mapsto sa^1$, $s \in \mathrm{Sp}(l)$,
- an automorphism of $\mathscr{A}^0$ given by $a^0 \mapsto a^0 \circ s^{-1}$.

The function $S \mapsto s$ is a natural morphism

$$G(l) \to \mathrm{Sp}(l). \tag{1.7}$$

LEMMA 1.2. The kernel of the morphism (1.7) is a subgroup of $G(l)$ consisting of automorphisms of $\mathscr{S}'(X)$ of the form

$f \to cf$, where $f \in \mathscr{S}'(X)$ and $c \in \dot{\mathbf{C}}$ (complex plane minus the origin).

Remark. This subgroup will be written as $\dot{\mathbf{C}}$.

Proof. All $c \in \dot{\mathbf{C}}$ commute with all $a \in \mathscr{A}$ and thus belong to the kernel.

Conversely, let S be an element of the kernel. Therefore S is an automorphism of $\mathscr{S}'(X)$ commuting with all $a \in \mathscr{A}$. Let $p \in X^*$. We have

$$\left(\frac{1}{v}\frac{\partial}{\partial x} + p\right)e^{-v\langle p,x\rangle} = 0.$$

Therefore, since S and $(1/v)(\partial/\partial x) + p$ commute,

$$\left(\frac{1}{v}\frac{\partial}{\partial x} + p\right)Se^{-v\langle p,x\rangle} = 0.$$

By integration of this system of differential equations,

$$Se^{-v\langle p,x\rangle} = c(p)e^{-v\langle p,x\rangle}, \text{ where } c: X^* \to \mathbf{C}.$$

Taking the derivative with respect to p, we see that the gradient of c, c_p, exists and satisfies

$$-vS[xe^{-v\langle p,x\rangle}] = -vxSe^{-v\langle p,x\rangle} + c_p e^{-v\langle p,x\rangle};$$

equivalently, since S and multiplication by x commute,

$$c_p = 0.$$

$c(p)$ is independent of p and will be denoted c. Let F be the *Fourier* transform and let $g = F^{-1}f \in \mathscr{S}(X)$. By the definition of F,

$$f(x) = \left(\frac{|\nu|}{2\pi i}\right)^{l/2} \int_X e^{-\nu\langle p,x\rangle} g(p) d^l p. \tag{1.8}$$

Since $Se^{-\nu\langle p,x\rangle} = ce^{-\nu\langle p,x\rangle}$, we obtain

$$Sf = cf \qquad \forall f \in \mathscr{S}(X).$$

Now $\mathscr{S}(X)$ is dense in $\mathscr{S}'(X)$. Therefore $S = c \in \dot{\mathbf{C}}$. This proves the lemma.

Some other subgroups of $G(l)$ will be needed in proving that the map $G(l) \to \mathrm{Sp}(l)$ is an epimorphism. They are

i. the finite group generated by the Fourier transforms in one of the coordinates (some base of the vector space X having been fixed);
ii. the group consisting of automorphisms of $\mathscr{S}'(X)$ of the form

$$f \mapsto e^{\nu Q} f,$$

where Q is a *real quadratic form* mapping $X \to \mathbf{R}$;
iii. the group consisting of automorphisms of $\mathscr{S}'(X)$ of the form

$f' \mapsto f$, where $f(x) = \sqrt{\det T}\, f'(Tx)$, T an automorphism of X.

Each of these groups has a restriction to $\mathscr{S}(X)$ that gives a group of automorphisms of $\mathscr{S}(X)$ and a restriction to $\mathscr{H}(X)$ that gives a group of unitary (that is, isometric and invertible) transformations of $\mathscr{H}(X)$. The following definition uses these properties.

Definition 1.2. Let $\mathbf{A}$ be the collection of elements A each consisting of

1°) a quadratic form $X \oplus X \to \mathbf{R}$, whose value at $(x, x') \in X \oplus X$ is

$$A(x, x') = \tfrac{1}{2}\langle Px, x\rangle - \langle Lx, x'\rangle + \tfrac{1}{2}\langle Qx', x'\rangle, \tag{1.9}$$

where, if ${}^tP$ denotes the transpose of P,

$$P = {}^tP : X \to X^*, \qquad L : X \to X^*, \qquad Q = {}^tQ : X \to X^*,$$

$\det L \neq 0$;

2°) a choice of $\arg \det L = \pi m(A)$, $m(A) \in \mathbf{Z}$, which allows us to define

$\Delta(A) = \sqrt{\det L}$ by $\arg \Delta(A) = (\pi/2) m(A)$.

Remark. $\det L$ is calculated using coordinates in X^* dual to the coordinates in X and is independent of coordinates chosen such that $dx^1 \wedge \cdots \wedge dx^l = d^l x$.

Remark. $m(A)$ will be identified with the Maslov index by 2,(2.15) and §2,8,(8.6).

To each A we associate S_A, an endomorphism of $\mathscr{S}(X)$ defined by

$$(S_A f')(x) = \left[\frac{|\nu|}{2\pi i}\right]^{l/2} \Delta(A) \int_X e^{\nu A(x,x')} f'(x') d^l x',$$
$$\text{where } f' \in \mathscr{S}(X), \quad \arg[i]^{l/2} = \pi l/4. \tag{1.10}$$

Clearly S_A is a product of elements belonging to the groups (i), (ii), and (iii). Therefore S_A is an *automorphism* of $\mathscr{S}(X)$ that extends by continuity to a *unitary automorphism of* $\mathscr{H}(X)$ and to an *automorphism* of $\mathscr{S}'(X)$. These three automorphisms will be denoted S_A; $S_A \in G(l)$.

The image s_A of S_A in $\mathrm{Sp}(l)$ is characterized as follows (where A_x is the gradient of A with respect to x):

$$(x, p) = s_A(x', p') \text{ is equivalent to}$$
$$p = A_x(x, x'), \qquad p' = -A_{x'}(x, x'). \tag{1.11}$$

Proof of (1.11). Let $f' \in \mathscr{S}(X)$. $\partial(S_A f')/\partial x$ and $S_A(\partial f'/\partial x)$ are calculated by differentiation of (1.10) and integration by parts; the result of these calculations gives the following relations among differential operators of $\mathscr{A}$:

$$\frac{1}{\nu}\frac{\partial}{\partial x} - Px = -S_A({}^t L x) S_A^{-1}, \qquad S_A\left(\frac{1}{\nu}\frac{\partial}{\partial x} + Qx\right) S_A^{-1} = Lx;$$

writing

$$(x, p) = s_A(x', p'),$$

these relations mean

$$p - Px = -{}^t L x', \qquad p' + Qx' = Lx \qquad \forall x' \in X, \quad p' \in X^*.$$

This is proposition (1.11).

Definition 1.3. We shall write Σ_{Sp} for the set of $s \in \mathrm{Sp}(l)$ such that x and x' are not independent on the $2l$-dimensional plane in $Z(l) \oplus Z(l)$ determined by the equation

$$(x, p) = s(x', p').$$

Let us recall the well-known theorem that *the set of* s_A *characterized by* (1.11) *is* $\mathrm{Sp}(l)\backslash\Sigma_{\mathrm{Sp}}$.

Proof. Clearly $s_A \notin \Sigma_{\mathrm{Sp}}$. Conversely, let $s \in \mathrm{Sp}(l)$. On the $2l$-dimensional plane in $Z(l) \oplus Z(l)$ determined by the equation

$$(x, p) = s(x', p')$$

we have, since s is symplectic,

$$\langle p, dx\rangle - \langle dp, x\rangle = \langle p', dx'\rangle - \langle dp', x'\rangle.$$

Therefore

$$\tfrac{1}{2}d[\langle p, x\rangle - \langle p', x'\rangle] = \langle p, dx\rangle - \langle p', dx'\rangle.$$

We assume $s \notin \Sigma_{\mathrm{Sp}}$. Then x and x' are independent on the above $2l$-dimensional plane. On this plane we define

$$A(x, x') = \tfrac{1}{2}\langle p, x\rangle - \tfrac{1}{2}\langle p', x'\rangle. \qquad (1.12)$$

We therefore have

$$dA = \langle p, dx\rangle - \langle p', dx'\rangle, \text{ that is,}$$

$$p = A_x, \qquad p' = -A_{x'}.$$

x and A_x have to be independent. Hence $\det_{jk}(A_{x^j x'^k}) \neq 0$. Therefore $s = s_A$, which completes the proof.

The s_A clearly generate $\mathrm{Sp}(l)$. Thus:

LEMMA 1.3. The natural morphism $G(l) \to \mathrm{Sp}(l)$ is an epimorphism.

By lemma 1.2, $G(l)$ is a *Lie group* and

$$G(l)/\dot{\mathbf{C}} = \mathrm{Sp}(l). \qquad (1.13)$$

[$\dot{\mathbf{C}}$ is the center of $G(l)$ because the center of $\mathrm{Sp}(l)$ is just the identity element.]

Definition 1.4. The *metaplectic group* $\mathrm{Mp}(l)$ is the subgroup of $G(l)$

consisting of those elements whose restriction to $\mathscr{H}(X)$ is a unitary automorphism of $\mathscr{H}(X)$.

We have $S_A \in \mathrm{Mp}(l)$ $\forall A$. Now the s_A generate $\mathrm{Sp}(l)$, so the natural morphism

$$\mathrm{Mp}(l) \to \mathrm{Sp}(l)$$

is an epimorphism. By (1.13), all elements of $G(l)$ can be written uniquely in the form

$$cS, \qquad \text{where } S \in \mathrm{Mp}(l), \quad c > 0.$$

Writing $\mathbf{R}_+$ for the multiplicative group of real numbers > 0, we obtain

$$G(l) = \mathbf{R}_+ \times \mathrm{Mp}(l). \tag{1.14}$$

The study of $G(l)$ therefore reduces to that of $\mathrm{Mp}(l)$, which has the following properties:

THEOREM 1. *$\mathrm{Mp}(l)$ is a group of automorphisms of $\mathscr{S}'(X)$ whose restrictions to $\mathscr{H}(X)$ are unitary automorphisms.*

1°) *Let $\mathbf{S}^1$ be the multiplicative group of complex numbers of modulus* 1. *Then*

$$\mathrm{Mp}(l)/\mathbf{S}^1 = \mathrm{Sp}(l). \tag{1.15}$$

2°) *Let Σ_{Mp} be the hypersurface of* $\mathrm{Mp}(l)$ *that projects onto Σ_{Sp}. Every element of* $\mathrm{Mp}(l)\backslash\Sigma_{\mathrm{Mp}}$ *can be written as cS_A, where $c \in \mathbf{S}^1$ and S_A is given by an expression of the form* (1.10).

3°) *The restriction of every $S \in$* $\mathrm{Mp}(l)$ *to $\mathscr{S}(X)$ is an automorphism of $\mathscr{S}(X)$.*

Proof of 1°): (1.13) and (1.14); $\mathbf{S}^1$ is identified with a subgroup of $\mathrm{Mp}(l)$.

Proof of 2°). Let $S \in \mathrm{Mp}(l)\backslash\Sigma_{\mathrm{Mp}}$. Then the image of S in $\mathrm{Sp}(l)$ is some element s_A, $A \in \mathbf{A}$; $SS_A^{-1} \in \mathbf{S}^1$ *by* (1.15).

Proof of 3°). By 2°), $S = cS_{A_1} \cdots S_{A_k}$. Now the restrictions of c, $S_{A_1}, \ldots, S_{A_k}$ to $\mathscr{S}(X)$ are automorphisms of $\mathscr{S}(X)$.

2. The Subgroup $\mathrm{Sp}_2(l)$ of $\mathrm{Mp}(l)$

Definition 2.1. We denote by $\mathrm{Sp}_2(l)$ the subgroup of $\mathrm{Mp}(l)$ that is generated by the S_A.

The purpose of this section is to prove that $\mathrm{Sp}_2(l)$ is a covering group of $\mathrm{Sp}(l)$ of order 2.

In order to prove this, we calculate inverses and compositions of the elements S_A.

Definition 2.2. Given $A \in \mathbf{A}$, we define $A^* \in \mathbf{A}$ as follows:

$$A^*(x, x') = -A(x', x), \qquad \Delta(A^*) = i^l\overline{\Delta(A)},$$

$$m(A^*) = l - m(A).$$

LEMMA 2.1. $S_A^{-1} = S_{A^*}$; thus $s_A^{-1} = s_{A^*}$.

Proof. This amounts to proving the equivalence of the following two conditions for any f and $f' \in \mathcal{S}(X)$:

$$f(x) = \left(\frac{|\nu|}{2\pi i}\right)^{l/2} \Delta(A) \int_X e^{\nu A(x,x')} f'(x')d^l x',$$

$$f'(x') = \left(\frac{|\nu| i}{2\pi}\right)^{l/2} \overline{\Delta(A)} \int_X e^{-\nu A(x,x')} f(x) d^l x.$$

Using the expression for A given by (1.9), this is the same as the equivalence of the following two conditions:

$$f(x) = \int_X e^{-\nu\langle LX, X'\rangle} f'(x') d^l x',$$

$$f'(x') = \left(\frac{|\nu|}{2\pi}\right)^l |\det L| \int_X e^{\nu\langle LX, X'\rangle} f(x) d^l x.$$

The equivalence is deduced from the Fourier inversion formula; the lemma follows.

To compute compositions of the S_A, we will find an explicit expression for $S_A(e^{\nu\varphi'})$, where φ' is a second-degree polynomial. This is made possible by the following definition.

Definition 2.3. Choose linear coordinates in X such that $d^l x = dx^1 \wedge \cdots \wedge dx^l$ and choose the dual coordinates in X^*. The following notions are independent of this choice.

Let φ be a real function, twice differentiable:

$$\varphi : X \to \mathbf{R}.$$

$\text{Hess}_x(\varphi)$ denotes the *hessian* of φ, the determinant of its second derivatives. Alternatively this is the determinant of the quadratic form

$$X \ni dx \mapsto \langle d\varphi_x, dx \rangle \in \mathbf{R}.$$

$\text{Inert}_x(\varphi)$ denotes the *index of inertia* of this form. It is defined[1] when $\text{Hess}(\varphi) \neq 0$. Clearly

$$\text{Inert}(-\varphi) = l - \text{Inert}(\varphi),$$

$$\arg \text{Hess}(\varphi) = \pi \, \text{Inert}(\varphi) \mod 2\pi.$$

This formula makes possible the *definition*

$$\arg \text{Hess}(\varphi) = \pi \, \text{Inert}(\varphi). \tag{2.1}$$

Thus, for example,

$$[\text{Hess}(\varphi)]^{1/2} = |\text{Hess}(\varphi)|^{1/2} i^{\text{Inert}(\varphi)}. \tag{2.2}$$

If φ is a real quadratic form,

$$\varphi : X \ni x \mapsto \tfrac{1}{2}\langle Rx, x \rangle, \quad \text{where } R = {}^{t}R : X \to X^*,$$

then $\text{Hess}(\varphi)$ and $\text{Inert}(\varphi)$ will be denoted $\text{Hess}(R)$ and $\text{Inert}(R)$. $\text{Hess}(R)$ is the determinant of the symmetric matrix R. $\text{Inert}(R)$ is the number of negative eigenvalues of R. Clearly

$$\text{Inert}(R) = \text{Inert}(R^{-1}), \qquad [\text{Hess}(R)]^{1/2}[\text{Hess}(-R^{-1})]^{1/2} = i^l. \tag{2.3}$$

LEMMA 2.2. Let φ' be a real second-degree polynomial. Let $A \in \mathbf{A}$ be such that $\text{Hess}_{x'}(\varphi'(x') + A(x, x')) \neq 0$. Denote by $\varphi(x)$ the critical value of the polynomial

$$X \ni x' \mapsto A(x, x') + \varphi'(x');$$

φ is a second-degree polynomial. We have

$$S_A(e^{\nu\varphi'}) = \Delta(A)[\text{Hess}_{x'}(\varphi' + A)]^{-1/2} e^{\nu\varphi}. \tag{2.4}$$

Remark 2.1. This lemma assumes $\nu/i > 0$. Up to this point, it was sufficient to assume ν/i real and nonzero.

Proof. We know that

[1] It is the number of negative eigenvalues of the linear symmetric operator $dx \mapsto d\varphi_x$.

$$\int_{-\infty}^{\infty} \exp\left[-\frac{x^2}{2}\right] dx = \sqrt{2\pi}.$$

Therefore if $c \in \mathbf{C}$ and $|\arg \mu| < \pi/2$,

$$\int_{-\infty}^{\infty} \exp\left[-\frac{\mu}{2}(x + c)^2\right] dx = \frac{\sqrt{2\pi}}{\sqrt{\mu}}, \qquad |\arg \sqrt{\mu}| < \frac{\pi}{4}.$$

We then have, for any $p \in \mathbf{C}$,

$$\int_{-\infty}^{\infty} \exp\left[-\nu p x - \frac{\mu}{2}(x + c)^2\right] dx$$

$$= e^{\nu\varphi} \int_{-\infty}^{\infty} \exp\left\{-\frac{1}{2\mu}[\nu p + \mu(x + c)]^2\right\} dx = \frac{\sqrt{2\pi}}{\sqrt{\mu}} e^{\nu\varphi},$$

where φ is the critical value of the function

$$x \mapsto \varphi'(x) - px, \quad \text{where } \varphi' = -\frac{\mu}{2\nu}(x + c)^2.$$

The Fourier transform F is the automorphism of $\mathscr{S}$ defined by

$$(Ff')(p) = \left(\frac{|\nu|}{2\pi i}\right)^{l/2} \int_X e^{-\nu\langle p, x'\rangle} f'(x') d^l x' \qquad \forall f' \in \mathscr{S}(X). \tag{2.5}$$

We then have, for $l = 1$, $|\arg \mu| < \pi/2$,

$$Fe^{\nu\varphi'} = \frac{\sqrt{|\nu|}}{\sqrt{\mu}\sqrt{i}} e^{\nu\varphi}; \qquad \sqrt{i} = e^{\pi i/4}.$$

Since F is a continuous automorphism of $\mathscr{S}'(X)$, the preceding formula remains valid for $\mu = -\varepsilon\nu$, $\varepsilon \in \dot{\mathbf{R}}$; then

$$\frac{\sqrt{\mu}\sqrt{i}}{\sqrt{|\nu|}} = \begin{cases} \sqrt{\varepsilon} & \text{if } \varepsilon > 0 \\ i\sqrt{|\varepsilon|} & \text{if } \varepsilon < 0. \end{cases}$$

In other words, when $l = 1$, the following result holds: Let $\varphi' : X \to \mathbf{R}$ be a real second-degree polynomial such that $\operatorname{Hess} \varphi' \neq 0$; let $\varphi(p)$ be the critical value of the polynomial

$$x \mapsto \varphi'(x) - \langle p, x\rangle;$$

we have

$$Fe^{\nu\varphi'} = [\text{Hess}\,\varphi']^{-1/2}e^{\nu\varphi}. \tag{2.6}$$

Let us show that, since relation (2.6) holds for $l = 1$, it holds for all $l \geqslant 1$. It suffices to choose the coordinates x^j in X such that

$$\varphi'(x) = \sum_{j=1}^{l} \varphi'_j(x^j).$$

Now using the definitions (1.9) of A, (1.10) of S_A, and (2.5) of F, we have in the case $P = Q = 0$,

$$(S_A f')(x) = \Delta(A)(Ff')(Lx).$$

Then (2.6) establishes (2.4) in this case. From the definitions of A and S_A, the general case is clearly equivalent to this one.

Before taking compositions of the S_A,we consider compositions of the s_A:

LEMMA 2.3. 1°) Let A and $A' \in \mathbf{A}$. The condition

$$s_A s_{A'} \notin \Sigma_{\text{Sp}} \tag{2.7}$$

is equivalent to the condition

$$\text{Hess}_{x'}[A(x, x') + A'(x', x'')] \neq 0 \text{ (the Hessian is constant).} \tag{2.8}$$

2°) This condition is equivalent by lemma 2.1 to the existence of $A'' \in \mathbf{A}$ such that

$$s_A s_{A'} s_{A''} = e \text{ [identity element of Sp}(l)]. \tag{2.9}$$

A'' is defined by the condition that the critical value of the polynomial

$$x' \mapsto A(x, x') + A'(x', x'') + A''(x'', x)$$

be zero.

3°) Just as (1.9) defines A by P, Q, L, let A' and A'' be defined by P', Q', L' and P'', Q'', L''. The condition (2.8) for the existence of A'' is expressed as

$P' + Q$ is invertible.

A'' can be defined by the formulas

$$\begin{aligned} P'' + Q' &= L'(P' + Q)^{-1}\,{}^tL', \qquad P + Q'' = {}^tL(P' + Q)^{-1}L, \\ L'' &= -{}^tL(P' + Q)^{-1}\,{}^tL'. \end{aligned} \tag{2.10}$$

Remark 2.2. Writing $A + A' + A''$ for $A(x, x') + A'(x', x'') + A''(x'', x)$, we have

$$\operatorname{Inert}_x(A + A' + A'') = \operatorname{Inert}_{x'}(A + A' + A'') = \operatorname{Inert}_{x''}(A + A' + A''), \tag{2.11}$$

$$\operatorname{Hess}_{x'}(A + A' + A'') = \frac{\Delta^2(A)\Delta^2(A')}{\Delta^2(A''^*)}. \tag{2.12}$$

Proof of 1°). By (1.11), the relations

$$(x, p) = s_A(x', p'), \qquad (x', p') = s_{A'}(x'', p'')$$

may be written

$$p = A_x(x, x'), \qquad p' = -A_{x'}(x, x') = A'_{x'}(x', x''),$$

$$p'' = -A'_{x''}(x', x'').$$

It results from the elimination of p' and x' in these relations that

$$(x, p) = s_A s_{A'}(x'', p'').$$

The condition (2.7) that $s_A s_{A'} \notin \Sigma_{\mathrm{Sp}}$ is then equivalent to each of the following conditions:

- The elimination of p' and x' in the preceding step leaves x and x'' independent.
- The relation

$$A_{x'}(x, x') + A'_{x'}(x', x'') = 0$$

leaves x and x'' independent.

- For any x and x'', there exists an x' satisfying this relation.

Now in (1.9), $\det L \neq 0$. Therefore (2.7) is equivalent to (2.8).

Proof of 2°). Assumption (2.9) means that any two of the following three relations implies the third:

$$(x, p) = s_A(x', p'), \qquad (x', p') = s_{A'}(x'', p''), \qquad (x'', p'') = s_{A''}(x, p).$$

Then by (1.11), each of the next three relations implies the other two:

$$(A + A' + A'')_x = 0, \qquad (A + A' + A'')_{x'} = 0, \qquad (A + A' + A'')_{x''} = 0, \tag{2.13}$$

where

$$A + A' + A'' = A(x, x') + A'(x', x'') + A''(x'', x).$$

Now by Euler's formula, these three relations imply

$$A + A' + A'' = 0.$$

Therefore

$$(A + A' + A'')_{x'} = 0, \text{ that is, } (A + A')_{x'} = 0, \text{ implies } A + A' + A'' = 0.$$

Proof of 3°). We have

$$\operatorname{Hess}_{x'}(A + A' + A'') = \operatorname{Hess}(P' + Q),$$

which gives the first statement. For the other, the three pairwise equivalent relations (2.13) can be written

$$(P + Q'')x - {}^{t}Lx' - L''x'' = 0,$$

$$-Lx + (P' + Q)x' - {}^{t}L'x'' = 0,$$

$$-{}^{t}L''x - L'x' + (P'' + Q')x'' = 0.$$

(2.10) clearly expresses the equivalence of these three relations.

Proof of Remark 2.2. By (2.10), the symmetric matrices

$$P'' + Q', \qquad (P' + Q)^{-1}, \qquad P + Q''$$

can be transformed one into the other. They therefore have the same inertia. This is (2.11).

By $(2.10)_3$,

$$\operatorname{Hess}(P' + Q) = (\det L)(\det L')/(-1)^{l} \det L''.$$

By definition 2.2, this is (2.12).

Definition 2.4. Given

$$s_A, s_{A'}, s_{A''} \in \operatorname{Sp}(l) \backslash \Sigma_{\operatorname{Sp}} \text{ such that } s_A\, s_{A'}\, s_{A''} = e,$$

we define

$$\operatorname{Inert}(s_A, s_{A'}, s_{A''}) = \operatorname{Inert}_x(A + A' + A'') \quad [\text{see (2.11)}]. \tag{2.14}$$

We define

$$\operatorname{Inert}(S_A, S_{A'}, S_{A''}) = \operatorname{Inert}(s_A, s_{A'}, s_{A''}).$$

Moreover, we define the *Maslov index* of S_A, $m(S_A) \in \mathbf{Z}_4$, by

$$m(S_A) = m(A) \mod 4. \tag{2.15}$$

§2,8 will connect this with the index that V. I. Maslov actually introduced.

Lemma 2.1 and (2.15) have these obvious consequences:

$$\operatorname{Inert}(s_{A''}^{-1}, s_{A'}^{-1}, s_A^{-1}) = l - \operatorname{Inert}(s_A, s_{A'}, s_{A''}), \tag{2.16}$$

$$m(S_A^{-1}) = l - m(S_A), \qquad m(-S_A) = m(S_A) + 2 \mod 4.$$

We can at last study compositions of the S_A.

LEMMA 2.4. Consider a triple A, A', A'' of elements of $\mathbf{A}$ such that

$$s_A s_{A'} s_{A''} = e. \tag{2.17}$$

Then

$$S_A S_{A'} S_{A''} = \pm E \quad [E \text{ is the identity element of } \operatorname{Mp}(l)]. \tag{2.18}$$

We have

$$S_A S_{A'} S_{A''} = E \tag{2.19}$$

if and only if

$$\operatorname{Inert}(S_A, S_{A'}, S_{A''}) = m(S_A) - m(S_{A'}^{-1}) + m(S_{A''}) \mod 4. \tag{2.20}$$

Remark. Condition (2.17), which is equivalent to (2.18), implies (2.20) mod 2.

Proof. Let $y \in X$. Formula (1.10) holds if f' is replaced by the Dirac measure with support y, given by

$$\delta'(x) = \delta(x - y).$$

We obtain

$$(S_{A'}\delta')(x) = \left(\frac{|\nu|}{2\pi i}\right)^{l/2} \Delta(A') e^{\nu A'(x,y)},$$

from which follows, by lemmas 2.2 and 2.3,2°),

$$(S_A S_{A'} \delta')(x) = \left(\frac{|v|}{2\pi i}\right)^{l/2} \Delta(A)\Delta(A')\{\mathrm{Hess}_{x'}[A(x, x') + A'(x', y)]\}^{-1/2} e^{-vA''(y,x)}.$$

Multiplying this by $f'(y)d^l y$, where $f' \in \mathscr{S}(X)$, and integrating, we get

$$S_A S_{A'} f' = \frac{\Delta(A)\Delta(A')}{\Delta(A''*)} [\mathrm{Hess}_{x'}(A + A' + A'')]^{-1/2} S_{A''*} f',$$

which gives, by lemma 2.1 and formula (2.12),

$$S_A S_{A'} S_{A''} = \pm E.$$

Now specify the sign. By definition 2.4,

$$\arg[\mathrm{Hess}_{x'}(A + A' + A'')]^{1/2} = \frac{\pi}{2}\mathrm{Inert}(S_A, S_{A'}, S_{A''}) \mod 2\pi.$$

By definition 1.2, (2.16), and lemma 2.1,

$$\arg \Delta(A) = \frac{\pi}{2} m(S_A), \qquad \arg \Delta(A') = \frac{\pi}{2} m(S_{A'}) = \frac{\pi}{2}[l - m(S_{A'}^{-1})],$$

$$\arg \Delta(A''*) = \frac{\pi}{2} m(S_{A''}^{-1}) = \frac{\pi}{2}[l - m(S_{A''})] \mod 2\pi.$$

Therefore

$$\arg(\pm 1) = \frac{\pi}{2}[\mathrm{Inert}(S_A, S_{A'}, S_{A''}) - m(S_A) + m(S_{A'}^{-1}) - m(S_{A''})] \mod 2\pi,$$

which proves the lemma.

Recall that $\mathrm{Sp}_2(l)$ denotes the group generated by the S_A.

LEMMA 2.5. Every element of $\mathrm{Sp}_2(l)$ is a product of two of the S_A.

Proof. By lemma 2.1, every element of $\mathrm{Sp}_2(l)$ is a product of the S_A. It then suffices to prove that given $U, V, W \in \mathbf{A}$, there exist B and C in $\mathbf{A}$ such that

$$S_U S_V S_W = S_B S_C. \tag{2.21}$$

Now, by lemmas 2.3,1°) and 2.4, for every $W \in \mathbf{A}$ and every T a generic element of $\mathbf{A}$, $S_W S_T$ belongs to $\{S_A\}$ and is generic. Therefore, for T generic,

$$S_V S_T \in \{S_A\}, \qquad S_U S_V S_T \in \{S_A\}, \qquad S_T^{-1} S_W \in \{S_A\},$$

which gives (2.21) with

$$S_B = S_U S_V S_T \in \{S_A\}, \qquad S_C = S_T^{-1} S_W \in \{S_A\}.$$

The restriction to $\mathrm{Sp}_2(l)$ of the natural morphism $\mathrm{Mp}(l) \to \mathrm{Sp}(l)$ is clearly a natural morphism:

$$\mathrm{Sp}_2(l) \to \mathrm{Sp}(l).$$

LEMMA 2.6. The kernel of this morphism is the subgroup

$$S^0 = \{E, -E\}.$$

Therefore

$$\mathrm{Sp}_2(l)/S^0 = \mathrm{Sp}(l).$$

Proof. By the preceding lemma, the kernel of this morphism is the collection of the $S_A S_{A'} (A, A' \in \mathbf{A})$ such that $s_A s_{A'} = e$. From this, by lemma 2.1,

$$s_{A'} = s_{A^*}.$$

Therefore, by (1.11),

$$A'(x, x') = A^*(x, x') \qquad \forall x, x' \in X.$$

Consequently by definition 1.2.

$$\Delta(A') = \pm \Delta(A^*),$$

and

$$S_{A'} = \pm S_{A^*}; \text{ therefore } S_A S_{A'} = \pm E.$$

LEMMA 2.7. The group $\mathrm{Sp}_2(l)$ is connected.

Proof. Given $k \in \mathbf{Z}_4$ (additive group of integers mod 4), let D_k be the collection of S_A such that

$$m(A) = k, \text{ or equivalently, } i^{-k}\Delta(A) > 0.$$

The collection of quadratic forms A satisfying $\Delta^2(A) > 0$ [or $\Delta^2(A) < 0$] is connected. Each D_k is thus a connected set in $\mathrm{Sp}_2(l)$.

Given $k \in \mathbf{Z}_4$, let S_A and $S_{A'}$ be such that

$m(S_A) - m(S_{A'}^{-1}) = -k \mod 4;$

$P' + Q$ has one eigenvalue equal to zero and $l - 1$ eigenvalues > 0. Let B and B' be elements of $\mathbf{A}$ near A and A' and such that

$\text{Hess}_{x'}(B + B') \neq 0.$

$\text{Inert}_{x'}(B + B')$ takes the values 0 and 1. Since m is locally constant,

$m(S_B) = m(S_A), \qquad m(S_{B'}^{-1}) = m(S_{A'}^{-1}).$

We define $B'' \in \mathbf{A}$ by $S_B S_{B'} S_{B''} = E$. By (2.20), $m(S_{B''})$ takes the values k and $k + 1$ in any neighborhood of the element $(S_A S_{A'})^{-1}$ of $\text{Sp}_2(l)$. This element thus belongs to $\overline{D_k} \cap \overline{D_{k+1}}$:

$\overline{D_k} \cap \overline{D_{k+1}} \neq \varnothing,$

which gives the lemma.

The above lemmas prove the following theorem. Part 1 of the theorem reduces the study of $\text{Mp}(l)$ to that of $\text{Sp}_2(l)$. Its equivalent can be found in the work of D. Shale and A. Weil, but the proof we have given has established various other results that will be indispensible to us. One of these is part 3 of the theorem. This will be used in §2,8.

THEOREM 2. 1°) *The elements S_A of* $\text{Mp}(l)$ *that are defined by* (1.10) *generate a subgroup* $\text{Sp}_2(l)$ *of* $\text{Mp}(l)$. $\text{Sp}_2(l)$ *is a covering group* (see Steenrod [17], 1.6, 14.1) *of the group* $\text{Sp}(l)$ *of order* 2. *It is a group of automorphisms of $\mathscr{S}(X)$ that extend to unitary automorphisms of $\mathscr{H}(X)$ and to automorphisms of $\mathscr{S}'(X)$.*

2°) *The formulas* (2.11) *and* (2.14) *define the inertia of every triple s, s', s'' of elements of* $\text{Sp}(l)\backslash\Sigma_{\text{Sp}}$ *such that*

$ss's'' = e$ [*identity element of* $\text{Sp}(l)$].

The inertia is a locally constant function (discontinuous on Σ_{Sp}) with values in $\{0, 1, \ldots, l\}$ satisfying

$$\begin{aligned}\text{Inert}(s, s', s'') &= \text{Inert}(s'', s, s') = \cdots \\ &= 1 - \text{Inert}(s''^{-1}, s'^{-1}, s^{-1}).\end{aligned}$$

Let Σ_{Sp_2} be the hypersurface of $\text{Sp}_2(l)$ *that is mapped onto Σ_{Sp} in* $\text{Sp}(l)$ *under the natural projection. The elements S_A defined by* (1.10) *are the elements of* $\text{Sp}_2(l)\backslash\Sigma_{\text{Sp}_2}$. *Let S, S', S'' be a triple of such elements satisfying*

$SS'S'' = E$ [*identity element of* $\mathrm{Sp}_2(l)$].

Let s, s', s'' *be the images of these elements under the natural projection onto* $\mathrm{Sp}(l)$. *We define*

$$\mathrm{Inert}(S, S', S'') = \mathrm{Inert}(s, s', s'').$$

3°) *Formula* (2.15) *and definition* 1.2 *define the Maslov index* m *on* $\mathrm{Sp}_2(l)\backslash\Sigma_{\mathrm{Sp}_2}$. *It is a locally constant function (discontinuous on* Σ_{Sp_2}*) with values in* Z_4. *It satisfies*

$$m(S^{-1}) = l - m(S), \qquad m(-S) = m(S) + 2 \mod 4,$$

$$\mathrm{Inert}(S, S', S'') = m(S) - m(S'^{-1}) + m(S'') \mod 4.$$

Remark 2.3. We shall see later that m is characterized by the last formula and the property of being locally constant.

Remark 2.4. $\mathrm{Sp}_2(l)$ contains the three subgroups of $G(l)$ defined in section 1 by (i) Fourier transformation, (ii) quadratic forms, and (iii) automorphisms of X.

Proof. Let S be an element of one of the three subgroups. It is easy to find $A \in \mathbf{A}$ such that

$$SS_A = S_{A'}, \quad \text{where } A' \in \mathbf{A}.$$

Remark 2.5. It can be shown that every $S \in \mathrm{Sp}_2(l)$ is of the form

$$S = S_1 S_2 S_3 S_4,$$

where $S_3 \in$ (i), that is, S_3 is a *Fourier* transformation in at most l coordinates; S_1 and $S_4 \in$ (ii), that is, they are of the form $f' \mapsto e^{\nu Q} f'$, where Q is a *real quadratic form*; and $S_2 \in$ (iii), that is, S_2 has the form $f' \mapsto \sqrt{\det T}\, f' \circ T$, where T is an *automorphism* of X.

3. Differential Operators with Polynomial Coefficients

By definition 1.1, the elements of $\mathrm{Sp}_2(l)$ transform differential operators with polynomial coefficients into operators of the same type. Section 3 describes this transformation more explicitly.

Let a^+ and a^- be two polynomials in $1/\nu$, x, and p:

$$a^+(\nu, x, p) = \sum_\alpha a_\alpha^+(\nu, x) p^\alpha, \qquad a^-(\nu, p, x) = \sum_\alpha p^\alpha a_\alpha^-(\nu, x)$$

(α a multi-index). We consider the two differential operators

$$a^{+}\left(v, x, \frac{1}{v}\frac{\partial}{\partial x}\right): f \mapsto \sum_{\alpha} a_{\alpha}^{+}(v, \cdot)\left(\frac{1}{v}\frac{\partial}{\partial x}\right)^{\alpha} f(\cdot), \tag{3.1}$$

$$a^{-}\left(v, \frac{1}{v}\frac{\partial}{\partial x}, x\right): f \mapsto \sum_{\alpha}\left(\frac{1}{v}\frac{\partial}{\partial x}\right)^{\alpha}[a_{\alpha}^{-}(v, \cdot) f(\cdot)]. \tag{3.2}$$

LEMMA 3.1. These two operators are identical, that is,

$$a^{+}\left(v, x, \frac{1}{v}\frac{\partial}{\partial x}\right) = a^{-}\left(v, \frac{1}{v}\frac{\partial}{\partial x}, x\right), \tag{3.3}$$

if and only if there exists a polynomial a^0 in $1/v$, x, and p such that

$$\begin{aligned} a^{+}(v, x, p) &= \left[\exp\frac{1}{2v}\left\langle\frac{\partial}{\partial x}, \frac{\partial}{\partial p}\right\rangle\right] a^{\circ}(v, x, p). \\ a^{-}(v, p, x) &= \left[\exp -\frac{1}{2v}\left\langle\frac{\partial}{\partial x}, \frac{\partial}{\partial p}\right\rangle\right] a^{\circ}(v, x, p). \end{aligned} \tag{3.4}$$

The notation is the following:

$$\left\langle\frac{\partial}{\partial x}, \frac{\partial}{\partial p}\right\rangle = \sum_{j=1}^{l} \frac{\partial^2}{\partial x^j \partial p_j} \quad (x^j \text{ and } p_j \text{ dual coordinates in } X \text{ and } X^*);$$

$$\exp \lambda \left\langle\frac{\partial}{\partial x}, \frac{\partial}{\partial p}\right\rangle = \sum_{k=0}^{\infty} \frac{1}{k!}\lambda^k \left\langle\frac{\partial}{\partial x}, \frac{\partial}{\partial p}\right\rangle^k.$$

Proof. Relation (3.3) defines a bijection $a^{-} \mapsto a^{+}$ such that, for all $p \in X^*$,

$$\begin{aligned} a^{+}(v, x, p) &= e^{-v\langle p,x\rangle} a^{+}\left(v, x, \frac{1}{v}\frac{\partial}{\partial x}\right) e^{v\langle p,x\rangle} \\ &= e^{-v\langle p,x\rangle} a^{-}\left(v, \frac{1}{v}\frac{\partial}{\partial x}, x\right) e^{v\langle p,x\rangle} \\ &= \sum_{\alpha}\left(p + \frac{1}{v}\frac{\partial}{\partial x}\right)^{\alpha} a_{\alpha}^{-}(v, x) = \left[\exp\frac{1}{v}\left\langle\frac{\partial}{\partial x}, \frac{\partial}{\partial p}\right\rangle\right] a^{-}(v, p, x), \end{aligned}$$

since, by Taylor's formula, for every polynomial $P: X^* \to \mathbf{C}$ and every function $f: X \to \mathbf{C}$,

$$P\left(p + \frac{1}{v}\frac{\partial}{\partial x}\right) f(x) = \sum_{\beta} \frac{1}{\beta!}\left(\frac{\partial}{\partial p}\right)^{\beta} P(p)\left(\frac{1}{v}\frac{\partial}{\partial x}\right)^{\beta} f(x)$$

$$= \left[\exp\frac{1}{v}\left\langle\frac{\partial}{\partial x},\frac{\partial}{\partial p}\right\rangle\right][P(p)f(x)].$$

The bijection $a^- \mapsto a^+$ can then be defined by the relation

$$a^+(v, x, p) = \left[\exp\frac{1}{v}\left\langle\frac{\partial}{\partial x},\frac{\partial}{\partial p}\right\rangle\right]a^-(v, p, x).$$

This is what the lemma asserts.

Definition 3.1. Let a be a differential operator that can be expressed as in (3.1) and (3.2). It is defined by the polynomial a^0 in $(1/v, x, p)$ that satisfies (3.4). We say that *a is the differential operator associated to the polynomial a^0*.

Theorem 3.1 will describe the transform SaS^{-1} of a by $S \in \mathrm{Sp}_2(l)$; Lemma 1.1 has already dealt with the case in which a° is linear in (x, p). The proof of this theorem will use the following properties.

LEMMA 3.2. If a and b are the operators associated to the polynomials a^0 and b^0, then the operator

$$c = ab$$

is associated to the polynomial c^0, where

$$c^0(v, x, p) = \left\{\left[\exp\left(\frac{1}{2v}\left\langle\frac{\partial}{\partial y},\frac{\partial}{\partial p}\right\rangle - \frac{1}{2v}\left\langle\frac{\partial}{\partial x},\frac{\partial}{\partial q}\right\rangle\right)\right] \cdot [a^0(v, x, p)b^0(v, y, q)]\right\}_{\substack{y=x\\ q=p}}. \tag{3.5}$$

Proof. If $b^0(v, x, p)$ only depends on p, then the polynomial c^0 associated to $c = ab$ is

$$\begin{aligned} c^0(v, x, p) &= \left[\exp -\frac{1}{2v}\left\langle\frac{\partial}{\partial x},\frac{\partial}{\partial p}\right\rangle\right][a^+(v, x, p)b^0(p)] \\ &= \left\{\left[\exp -\frac{1}{2v}\left\langle\frac{\partial}{\partial x},\frac{\partial}{\partial p}+\frac{\partial}{\partial q}\right\rangle\right][a^+(v, x, p)b^0(q)]\right\}_{q=p} \\ &= \left\{\left[\exp -\frac{1}{2v}\left\langle\frac{\partial}{\partial x},\frac{\partial}{\partial p}\right\rangle\right][a^0(v, x, p)b^0(q)\right\}_{q=p}. \end{aligned}$$

Similarly, if $b^0(v, x, p)$ only depends on x, then the polynomial associated to $c = ab$ is

$$c^0(v, x, p) = \left\{\left[\exp \frac{1}{2v}\left\langle\frac{\partial}{\partial y}, \frac{\partial}{\partial p}\right\rangle\right][a^0(v, x, p)b^0(y)]\right\}_{y=x}.$$

Thus if $b^+(x, p) = b'(x)b''(p)$, then the polynomial associated to $c = ab$ is

$$\begin{aligned} c^0(v, x, p) &= \left[\exp -\frac{1}{2v}\left\langle\frac{\partial}{\partial x}, \frac{\partial}{\partial q}\right\rangle\right]\left\{\left[\exp \frac{1}{2v}\left\langle\frac{\partial}{\partial y}, \frac{\partial}{\partial p}\right\rangle\right]\right. \\ &\qquad \left.\cdot [a^0(v, x, p)b'(y)]\right\}_{y=x} b''(q)\big|_{q=p} \\ &= \left\{\left[\exp -\frac{1}{2v}\left\langle\frac{\partial}{\partial x}, \frac{\partial}{\partial q}\right\rangle - \frac{1}{2v}\left\langle\frac{\partial}{\partial y}, \frac{\partial}{\partial q}\right\rangle + \frac{1}{2v}\left\langle\frac{\partial}{\partial y}, \frac{\partial}{\partial p}\right\rangle\right]\right. \\ &\qquad \left.\cdot [a^0(v, x, p)b'(y)b''(q)]\right\}_{\substack{y=x \\ q=p}}. \end{aligned}$$

This is (3.5) since, by (3.4),

$$\left[\exp -\frac{1}{2v}\left\langle\frac{\partial}{\partial x}, \frac{\partial}{\partial q}\right\rangle\right][b'(y)b''(q)] = b^0(y, q).$$

This implies lemma 3.2, which has the following obvious consequence:

LEMMA 3.3. The operator

$c = \frac{1}{2}(ab + ba)$

is associated to the polynomial

$$\begin{aligned} c^0(v, x, p) = &\left\{\cosh\left[\frac{1}{2v}\left\langle\frac{\partial}{\partial y}, \frac{\partial}{\partial p}\right\rangle - \frac{1}{2v}\left\langle\frac{\partial}{\partial x}, \frac{\partial}{\partial q}\right\rangle\right]\right. \\ &\left.\cdot [a^0(v, x, p)b^0(v, y, q)]\right\}_{\substack{y=x \\ q=p}}. \end{aligned}$$

If b is linear in (y, q), then

$$\left[\left\langle\frac{\partial}{\partial y}, \frac{\partial}{\partial p}\right\rangle - \left\langle\frac{\partial}{\partial x}, \frac{\partial}{\partial q}\right\rangle\right]^2 [a^0(v, x, p)b^0(v, y, q)] = 0,$$

from which follows

$\cosh[\cdots]a^0b^0 = a^0b^0;$

therefore we have the following lemma.

LEMMA 3.4. If b is linear in (x, p), then the operator associated to a^0b^0 is $\frac{1}{2}(ab + ba)$.

This lemma enables us to prove the following theorem.

THEOREM 3.1. *The transform SaS^{-1} of a by S is the differential operator associated to the polynomial $a^0 \circ s^{-1}$ [$S \in \mathrm{Sp}_2(l)$; s is the image of S in* $\mathrm{Sp}(l)$].

Proof. Let b be a differential operator associated to a polynomial b^0 that is linear or affine in (x, p); lemma 1.1 shows that theorem 3.1 holds for b. To prove the theorem by induction on the degree of a^0 in (x, p), it suffices to prove that, if the theorem holds for a^0, then it holds for a^0b^0.

Since the theorem holds for a^0 and b^0, the operators associated to the polynomials

$$a^0b^0 \text{ and } (a^0b^0) \circ s^{-1} = (a^0 \circ s^{-1})(b^0 \circ s^{-1})$$

are, respectively, by lemma 3.4,

$$\tfrac{1}{2}(ab + ba);$$
$$\tfrac{1}{2}(SaS^{-1}SbS^{-1} + SbS^{-1}SaS^{-1}) = \tfrac{1}{2}S(ab + ba)S^{-1}.$$

The theorem thus holds for a^0b^0, which completes the proof.

We supplement this by a theorem about adjoint operators.

Definition 3.2. Recall that $\mathscr{H}(X)$ has a scalar product:

$$(f \mid g) = \int_X f(x)\overline{g(x)}\, d^l x \qquad \forall f, g \in \mathscr{H}(X),$$

where $\overline{g(x)}$ is the complex conjugate of $g(x)$. Two differential operators a and b are said to be *adjoint* if

$$(af \mid g) = (f \mid bg) \qquad \forall f, g \in \mathscr{H}(X). \tag{3.6}$$

THEOREM 3.2. *Two differential operators a and b associated to two polynomials a^0 and b^0 are adjoint if and only if*

$$b^0(v, x, p) = \overline{a^0(v, x, p)} \qquad \forall v \in i\mathbf{R}, \quad x \in X, \quad p \in X^*. \tag{3.7}$$

Proof. It is clear that (3.6) is equivalent to

$$b^-(v, p, x) = \overline{a^+(v, x, p)},$$

that is to say, since v is pure imaginary, to

$$\left[\exp -\frac{1}{2\nu}\left\langle \frac{\partial}{\partial x}, \frac{\partial}{\partial p}\right\rangle\right] b^0(\nu, x, p) = \left[\exp -\frac{1}{2\nu}\left\langle \frac{\partial}{\partial x}, \frac{\partial}{\partial p}\right\rangle\right] \overline{a^0(\nu, x, p)},$$

and hence to (3.7).

Theorems 3.1 and 3.2 obviously have the following corollary.

COROLLARY 3.1. *If a^* is the adjoint of a, then $\forall S \in \mathrm{Sp}_2(l)$, Sa^*S^{-1} is the adjoint of SaS^{-1}.*

Theorem 3.2 clearly has the following corollary, which will be important later.

COROLLARY 3.2. *The operator a associated to a polynomial a^0 is self-adjoint if and only if the polynomial a^0 is real valued $\forall \nu \in i\mathbf{R}, x \in X, p \in X^*$.*

§2. Maslov Indices; Indices of Inertia; Lagrangian Manifolds and Their Orientations

0. Introduction

Historical account. Following *V. C. Buslaev* [3], [11], §1 has defined a Maslov index mod 4 on $\mathrm{Sp}_2(l)$ by (2.15) and has connected it by (2.20) to an index of inertia that is a function of a pair of elements of $\mathrm{Sp}(l)$.

On the other hand, *V. I. Arnold* [1], [11] defined another Maslov index on the covering space of the lagrangian grassmannian $\Lambda(l)$ of $Z(l)$; this index is connected to the preceding one and to a second index of inertia that is a function of a triple of points of $\Lambda(l)$. *J. M. Souriau* [16] has given a variant of the definition of the Maslov index that is considered in this section.

Summary. Chapter I, §3, and chapter II use these two Maslov indices and a third index of inertia, which is a function of an element of $\mathrm{Sp}(l)$ and a point of $\Lambda(l)$.

We review and modify the various definitions of these indices (Arnold's, section 5; Maslov's, section 6; Buslaev's, section 7) so as to clarify their properties (sections 4–8). In §3 those properties that will be used in chapter II are set forth.

First of all we must recall and supplement the topological properties of $\mathrm{Sp}(l)$ and $\Lambda(l)$ (theorem 3). To study these properties we follow Arnold in employing a hermitian structure on $Z(l)$ (sections 1 and 2).

1. Choice of Hermitian Structures on $Z(l)$

Let $(\cdot|\cdot)$ be the scalar product defining a hermitian structure on $Z(l)$; clearly

$$\operatorname{Im}(z \mid z') = -\operatorname{Im}(z' \mid z)$$

is a symplectic structure on $Z(l)$. Now in §1,1, a symplectic structure $[\cdot,\cdot]$ was defined on $Z(l)$.

LEMMA 1.1. Restriction to X defines a homeomorphism between the set of hermitian structures $(\cdot|\cdot)$ on $Z(l)$ such that

$$\operatorname{Im}(z|z') = [z, z'], \qquad iX = X^*, \tag{1.1}$$

and the set of euclidean structures on X.

Proof. (i) The restriction to X of a hermitian structure on $Z(l)$ satisfying (1.1) is euclidean since

$$[x, x'] = 0 \qquad \forall x, x' \in X.$$

Observe that, by (1.1),

$$(z \mid z') = [iz, z'] + i[z, z'], \tag{1.2}$$

and in particular

$$(x \mid x') = [ix, x'] \qquad \forall x, x' \in X. \tag{1.3}$$

Hence

$$ix = \frac{1}{2}\frac{\partial |x|^2}{\partial x} \in X^*. \tag{1.4}$$

(ii) A given $(\cdot \mid \cdot)$ on X defines

- by (1.4), the restriction of i to X,

$$i_1 \colon X \to X^*;$$

- the restriction of i to X^*,

$$i_2 \colon X^* \to X,$$

because $i_2 = -i_1^{-1}$ since $i^2 = -1$;

- hence the automorphism i of $Z(l)$,

$$i(x, p) = (i_2 p, i_1 x); \tag{1.5}$$

- finally, by (1.2), the hermitian structure on $Z(l)$.

The restriction to X of hermitian structures on $Z(l)$ satisfying (1.1) is thus an *injective* mapping of the set of such structures into the set of euclidean structures on X.

(iii) It is *bijective*. Indeed, the automorphism i of $Z(l)$ defined by the given $(\cdot \mid \cdot)$ on X, that is, by (1.5), satisfies

$$i^2 = -1, \qquad [iz, z'] = [iz', z],$$

since ${}^t i_1 = i_1, {}^t i_2 = i_2$; the function $(\cdot \mid \cdot)$, which (1.2) defines on $Z(l)$, is clearly linear in $z \in \mathbf{C}^l$, and

- satisfies $(z', z) = \overline{(z', z)}$,
- hence is sesquilinear,
- satisfies $|x + iy|^2 = |x|^2 + |y|^2$,
- and indeed defines a hermitian structure.

Lemma 1.1 has as the following corollary.

LEMMA 1.2. The set of hermitian structures on $Z(l)$ satisfying (1.1) is an open convex cone. It is therefore *connected*.

Remark 1. We choose arbitrarily one of these hermitian structures on $Z(l)$, which we shall use to define topological notions (the Maslov indices). By the preceding lemma, these notions will not depend on this choice.

2. The Lagrangian Grassmannian $\Lambda(l)$ of $Z(l)$

Definition 2.1 A subspace of $Z(l)$ is called *isotropic* when the restriction of $[\cdot, \cdot]$ to this subspace is identically zero, that is, by (1.1), when the restriction of the hermitian structure on $Z(l)$ is a euclidean structure on this subspace.

Every orthonormal frame of an isotropic subspace of dimension k is thus composed of vectors orthogonal in $Z(l)$; hence $k \leqslant l$.

Definition 2.2. The isotropic subspaces of maximal dimension l are called *lagrangian subspaces*; the collection of lagrangian subspaces $\Lambda(l)$ is called the *lagrangian grassmannian*:

X and $X^* \in \Lambda(l)$.

Let $\lambda \in \Lambda(l)$ and let r be an orthonormal frame of λ. It is a frame of $Z(l)$: the elements of $Z(l)$ (respectively λ) are linear combinations with complex (respectively real) coefficients of the vectors that make up r.

Let $U(l)$ denote the group of unitary automorphisms u of $Z(l)$ (that is, $uu^* = e$, where $u^* = {}^t\bar{u}$, and e is the identity). By (1.1),

$$U(l) \subset \mathrm{Sp}(l).$$

Further let $\lambda' \in \Lambda(l)$ and let r' be an orthonormal frame of λ'. There is a unique element u in $U(l)$ such that

$$r = ur'$$

from which follows

$$\lambda = u\lambda'.$$

The group $U(l)$ thus acts *transitively* on $\Lambda(l)$. The same holds a fortiori for $\mathrm{Sp}(l)$; whence 1°) of the lemma below, where $\mathrm{St}(l)$ [respectively, $O(l)$] denotes the *stabilizer* of X^* in $\mathrm{Sp}(l)$ [respectively, $U(l)$], that is, the subgroup of s such that $sX^* = X^*$.

Now $O(l)$ is clearly the *orthogonal group*. Lemma 2.3 characterizes $\mathrm{St}(l)$; part 2 shows why the stabilizer of X^* interests us more than that of X.

LEMMA 2.1. 1° We have

$$\Lambda(l) = \mathrm{Sp}(l)/\mathrm{St}(l) = U(l)/O(l). \tag{2.1}$$

2°) Let $W(l)$ be the set of symmetric elements w in $U(l)$, that is, the set of elements w such that ${}^t w = w$; thus $w \in W(l)$ means $w = {}^t w = \bar{w}^{-1}$. The diagram

$$\begin{array}{ccc} & U(l) \ni u \mapsto u\,{}^t u = w \in W(l) & \\ & \downarrow & \\ U(l)/O(l) = \Lambda(l) \ni \lambda = & uX^* & \end{array} \tag{2.2}$$

defines a natural *homeomorphism*

$$\Lambda(l) \ni \lambda \mapsto w(\lambda) \in W(l). \tag{2.3}$$

Then

$$z \in \lambda \text{ is equivalent to } z + w(\lambda)\bar{z} = 0. \tag{2.4}$$

Let

$$z = x + iy, \quad \text{where } x \text{ and } y \in X.$$

Assume

$$1 \notin \mathrm{sp}(w(\lambda),$$

where $\mathrm{sp}(w)$ is the spectrum of w, a 0-chain of the unit circle $\mathbf{S}^1$. Then

$$z \in \lambda \text{ is equivalent to } y = i\frac{e + w(\lambda)}{e - w(\lambda)}x, \tag{2.5}$$

where $i(e + w(\lambda)]/[e - w(\lambda)]$ is a real symmetric matrix (that is, equal to its transpose).

3°) $\dim(\lambda \cap \lambda')$ is the multiplicity of 1 in $\mathrm{sp}(w(\lambda)w^{-1}(\lambda'))$.

Remark 2.1. Part 3 is preparation for the topological definition of the Maslov index (section 5).

The proof of lemma 2.1 is based on the following lemma. Writing $u \in U(l)$ in terms of its eigenvectors and eigenvalues, the proof of lemma 2.2 is clear.

LEMMA 2.2. 1°) Let $u \in U(l)$. A necessary and sufficient condition for $u \in W(l)$ is that all of its eigenvectors can be chosen to be *real*.

2°) Every surjective mapping

$$F : \mathbf{S}^1 \to \mathbf{S}^1 \; (\mathbf{S}^1 \text{ is the unit circle in } \mathbf{C})$$

defines a surjective mapping

$$W(l) \ni w \mapsto F(w) \in W(l).$$

Proof of lemma 2.1,2°). The diagram (2.2) defines a mapping (2.3) since, if u and $uv \in U(l)$ have the same image in $\Lambda(l) = U(l)/O(l)$, then $v \in O(l)$, and so

$$uv\,{}^{t}(uv) = uv\,{}^{t}v\,{}^{t}u = u\,{}^{t}u.$$

By lemma 2.2,2°), given $w \in W(l)$, there exists some $u \in W(l)$ such that $w = u^2$. Then $w = u\,{}^{t}u$, and so the map (2.3) is surjective.

Since $\lambda = uX^*$, where $u \in U(l)$, the condition $z \in \lambda$ means $u^{-1}z \in X^*$, or $\mathrm{Re}(u^{-1}z) = 0$, or $u^{-1}z + \bar{u}^{-1}\bar{z} = 0$, or $z + w\bar{z} = 0$. The map (2.3) is therefore injective.

Proof of lemma 2.1,3°). Let $w = w(\lambda)$, $w' = w(\lambda')$. Then $\lambda \cap \lambda'$ is given by the equations

$$z + w\bar{z} = 0, \qquad z + w'\bar{z} = 0;$$

that is,

$$\lambda \cap \lambda': \quad w^{-1}z = w'^{-1}z, \qquad z = -w\bar{z}.$$

Let T be the analytic subspace of $Z(l)$ given by the equation

$$T: w^{-1}z = w'^{-1}z.$$

Then $\dim_c T = k$, where k is the multiplicity of 1 in $\mathrm{sp}(ww'^{-1})$. The equation of $\lambda \cap \lambda'$ in T is

$$z + w\bar{z} = 0.$$

By lemma 2.2,2°), there exists a $u \in W(l)$ such that

$$-w = \bar{u}^2 = u^{-1}\bar{u}.$$

Thus the equation of $\lambda \cap \lambda'$ in T may be written

$$uz = \bar{u}\bar{z}.$$

The isomorphism

$$T \ni z \mapsto uz \in \mathbf{C}^k$$

therefore maps $\lambda \cap \lambda'$ onto the real part $\mathbf{R}^k$ of $\mathbf{C}^k$, and so

$$\dim_r \lambda \cap \lambda' = k.$$

LEMMA 2.3. The stabilizer $\mathrm{St}(l)$ of X^* in $\mathrm{Sp}(l)$ has the following properties:

1°) The elements s of $\mathrm{St}(l)$ are characterized as follows:

$$s(x', p') = (x, p)$$

is equivalent to

$$x = s_1 x', \qquad p = {}^t s_1^{-1}(p' + s_2 x'), \tag{2.6}$$

where s_1 is an arbitrary automorphism of X and $s_2 = {}^t s_2$ is an arbitrary symmetric morphism $X \to X^*$.

2°) An element s of $\mathrm{St}(l)$ is the projection of two elements S of $\mathrm{Sp}_2(l)$ defined by

$$(Sf)(x) = \sqrt{\det s_1^{-1}}\,[e^{(\nu/2)\langle x', s_2 x'\rangle} f(x')]_{x' = s_1^{-1}x}\,. \tag{2.7}$$

Remark 2.2. We denote by $\mathrm{St}_2(l)$ the subgroup of $\mathrm{Sp}_2(l)$ whose projection onto $\mathrm{Sp}(l)$ is $\mathrm{St}(l)$. By Remark 2.5 in §1, $\mathrm{St}_2(l)$ is the set of $S \in \mathrm{Sp}_2(l)$ that act pointwise on $\mathscr{S}(X)$: the value of Sf at a point x of X depends only on the behavior of f at a point x' of X (in fact on the value of f at x').

Proof of 1°). The elements of the stabilizer of X^* in the group of automorphisms of the vector space $Z(l)$ are the mappings $(x', p') \mapsto (x, p)$ defined by

$$x = s_1 x', \qquad p = s_*(p' + s_2 x'),$$

where s_1 and s_* are automorphisms of X and X^* and s_2 is a morphism $X \to X^*$. These elements belong to $\mathrm{Sp}(l)$ when

$$s_* = {}^t s_1^{-1}, \qquad s_2 = {}^t s_2.$$

Proof of 2°). Formula (2.7) defines an automorphism S of $\mathscr{S}'(X)$ that belongs to $\mathrm{Sp}_2(l)$ by Remark 2.4 in §1. Clearly

$$x \cdot (Sf) = S[f \cdot s_1 x'], \qquad \frac{1}{\nu}\frac{\partial}{\partial x}(Sf) = S\left[{}^t s_1^{-1}\left(\frac{1}{\nu}\frac{\partial}{\partial x'} + s_2 x'\right) f\right].$$

Hence, for any a in $\mathscr{A}$ (§1,1),

$$a^0\left(x, \frac{1}{\nu}\frac{\partial}{\partial x}\right)(Sf) = Sa^0\left(s_1 x', {}^t s_1^{-1}\left(\frac{1}{\nu}\frac{\partial}{\partial x'} + s_2 x'\right)\right) f,$$

that is, by (2.6),

$S^{-1} a S$ is associated to $a^0 \circ s$,

and so s is the natural image in $\mathrm{Sp}(l)$ of $\pm S \in \mathrm{Sp}_2(l)$.

3. The Covering Groups of $\mathrm{Sp}(l)$ and the Covering Spaces of $\Lambda(l)$

The properties of these covering groups and spaces[(2)] follow from properties of $\pi_1[\mathrm{Sp}(l)]$ and $\pi_1[\Lambda(l)]$, which are obtained by studying $\pi_1[U(l)]$. Here π_k denotes the kth homotopy group, (see Steenrod [17]; we note that N. Steenrod uses the expression *symplectic groups* in a different sense than we do.)

[2] See Steenrod [17], 1.6, 14.1.

LEMMA 3.1. 1°) The inclusion $O(l) \subset \mathrm{St}(l)$ induces an isomorphism

$$\pi_k[O(l)] \simeq \pi_k[\mathrm{St}(l)] \qquad \forall k \in \mathbf{N}.$$

2°) The inclusion $U(l) \subset \mathrm{Sp}(l)$ induces an isomorphism

$$\pi_k[U(l)] \simeq \pi_k[\mathrm{Sp}(l)] \qquad \forall k \in \mathbf{N}.$$

3°) The morphism

$$\pi_1[U(l)] \ni \gamma \mapsto \frac{1}{2\pi i}\int_\gamma \frac{d(\det u)}{\det u} \in \mathbf{Z} \tag{3.1}$$

is a natural isomorphism:

$$\pi_1[U(l)] \simeq \mathbf{Z}.$$

Proof of 1°). The elements s of $\mathrm{St}(l)$ are characterized by (2.6); those for which $s_2 = 0$ form a subgroup $GL(l)$ of $\mathrm{St}(l)$. The inclusions

$$O(l) \subset GL(l) \subset \mathrm{St}(l)$$

induce natural morphisms

$$\pi_k[O(l)] \xrightarrow{i} \pi_k[GL(l)] \to \pi_k[\mathrm{St}(l)].$$

The second morphism is an isomorphism, since

$$\mathrm{St}(l) = GL(l) \times \mathbf{R}^n, \quad \text{where } n = l(l+1)/2.$$

It has to be shown that i is an isomorphism. Now $GL(l)$ acts transitively on the set Q_+ of positive definite quadratic forms on X, and $O(l)$ is the stabilizer of one of them. Hence

$$GL(l)/O(l) = Q_+, \quad \text{where } Q_+ \text{ is convex.}$$

The exactness of the homotopy sequence of this fibration (see Steenrod [17], 17.3, 17.4) proves that i is indeed an isomorphism.

Proof of 2°). The inclusions

$$U(l) \subset \mathrm{Sp}(l), \qquad \mathrm{St}(l) \cap U(l) = O(l) \subset \mathrm{St}(l)$$

define a mapping (see Steenrod [17], 17.5) of the fibration

$$U(l)/O(l) = \Lambda(l) \quad \text{into} \quad \mathrm{Sp}(l)/\mathrm{St}(l) = \Lambda(l);$$

its restriction to $\Lambda(l)$ is the identity. This mapping induces a morphism of the homotopy sequences of these two fibrations (see Steenrod [17], 17.3, 17.11, 17.5):

$$\begin{array}{ccccccccc}
\pi_{k+1}[\Lambda(l)] & \xrightarrow{\Delta'} & \pi_k[O(l)] & \xrightarrow{i'} & \pi_k[U(l)] & \xrightarrow{p'} & \pi_k[\Lambda(l)] & \cdots & \pi_0[O(l)] \\
\downarrow i_0 & & \downarrow i_0 & & \downarrow i_1 & & \downarrow i_0 & & \downarrow i_0 \\
\pi_{k+1}[\Lambda(l)] & \xrightarrow{\Delta} & \pi_k[\mathrm{St}(l)] & \xrightarrow{i} & \pi_k[\mathrm{Sp}(l)] & \xrightarrow{p} & \pi_k[\Lambda(l)] & \cdots & \pi_0[\mathrm{St}(l)].
\end{array}$$

This diagram, in which the lines are exact, is thus commutative. Since the mappings i_0 are isomorphisms, it follows that the mappings i_1 are necessarily isomorphisms.

Proof of 3°). (Steenrod [17], 25.2, proves part of this by other means.) We denote by det (that is, determinant) the epimorphism

$$U(l) \ni u \mapsto \det u \in \mathbf{S}^1 \subset \dot{\mathbf{C}} \tag{3.2}$$

and by $SU(l)$ its kernel. Since $u \in SU(l)$ when $\det u = 1$, we have

$$U(l)/SU(l) = \mathbf{S}^1;$$

(3.2) is the natural projection of $U(l)$ onto $\mathbf{S}^1$. The homotopy sequence of this fibration contains the following, which is thus exact:

$$\pi_1[SU(l)] \xrightarrow{i} \pi_1[U(l)] \xrightarrow{p} \pi_1[\mathbf{S}^1] \xrightarrow{\Delta} \pi_0[SU(l)]; \tag{3.3}$$

here p is induced by the morphism (3.2). Since $SU(l)$ is connected, $\pi_0[SU(l)]$ is trivial. Let us compute $\pi_1[SU(l)]$.

$SU(l)$ acts transitively on the sphere

$$\mathbf{S}^{2l-1}: |z| = 1.$$

The stabilizer of the vector $(1, 0, \ldots, 0)$ in $\mathbf{C}^l$ is $SU(l-1)$; thus

$$SU(l)/SU(l-1) = \mathbf{S}^{2l-1}.$$

The homotopy sequence of this fibration contains the following, which is thus exact:

$$\pi_2[\mathbf{S}^{2l-1}] \xrightarrow{\Delta'} \pi_1[SU(l-1)] \xrightarrow{i'} \pi_1[SU(l)] \xrightarrow{p'} \pi_1[\mathbf{S}^{2l-1}]$$

where $\pi_1[\mathbf{S}^{2l-1}]$ and $\pi_2[\mathbf{S}^{2l-1}]$ are trivial for $l \geqslant 2$ (see Steenrod [17], 21.2). Thus i' is an isomorphism. Now $\pi_1[SU(1)]$ is trivial, since $SU(1)$ is trivial, and so

$$\pi_1[SU(l)] \text{ is trivial.} \tag{3.4}$$

Since $\pi_1[SU(l)]$ and $\pi_0[SU(l)]$ are trivial in the exact sequence (3.3), p is an isomorphism. Now

$$\pi_1[\mathbf{S}^1] \ni \Gamma \mapsto \frac{1}{2\pi i}\int_\Gamma \frac{dz}{z} \in \mathbf{Z}$$

is an isomorphism.

The composition of p, which is induced by (3.2), with this isomorphism is an isomorphism $\pi_1[U(l)] \to \mathbf{Z}$, which clearly is defined by (3.1).

LEMMA 3.2. 1°) The composition of the natural isomorphism $\pi_1[\Lambda(l)] \simeq \pi_1[W(l)]$ [cf. (2.3)] and the morphism

$$\pi_1[W(l)] \ni \gamma \mapsto \frac{1}{2\pi i}\int_\gamma \frac{d(\det w)}{\det w} \in \mathbf{Z} \tag{3.5}$$

is a natural isomorphism (Arnold [1]):

$$\pi_1[\Lambda(l)] \simeq \mathbf{Z}.$$

2°) The fibration $\mathrm{Sp}(l)/\mathrm{St}(l) = \Lambda(l)$ defines a monomorphism

$$p : \mathbf{Z} \simeq \pi_1[\mathrm{Sp}(l)] \to \pi_1(\Lambda(l)] \simeq \mathbf{Z}, \tag{3.6}$$

which is multiplication by 2 on $\mathbf{Z}$.

Proof of 1°). The homeomorphism (2.3) allows us to define

$$\det\lambda = \det w \in \mathbf{S}^1; \tag{3.7}$$

the mapping

$$\Lambda(l) \ni \lambda \mapsto \det\lambda \in \mathbf{S}^1 \tag{3.8}$$

is clearly an epimorphism. By (2.2) we have

$$\det\lambda = \det^2 u, \text{ if } \lambda = uX^*.$$

Hence for all $u \in U(l)$

$$\det(u\lambda) = \det^2 u \cdot \det\lambda. \tag{3.9}$$

The mapping (3.8) thus defines a fibration on which $U(l)$ permutes the fibers. The fibration is

$$\Lambda(l)/S\Lambda(l) = \mathbf{S}^1,$$

where $S\Lambda(l)$ denotes the variety in $\Lambda(l)$ defined by the equation

$$S\Lambda(l): \det w = 1.$$

The exactness of the homotopy sequence of this fibration proves the exactness of the sequence

$$\pi_1[S\Lambda(l)] \xrightarrow{i} \pi_1[\Lambda(l)] \xrightarrow{p} \pi_1[\mathbf{S}^1]. \qquad (3.10)$$

Since

$$p: \pi_1[\Lambda(l)] \to \pi_1[\mathbf{S}^1]$$

is induced by

$$\det: W(l) \to \mathbf{S}^1,$$

p is an epimorphism. Let us compute $\pi_1[S\Lambda(l)]$.

$SU(l)$ acts transitively on $S\Lambda(l)$ and $X^* \in S\Lambda(l)$. The stabilizer of X^* in $SU(l)$ is $SO(l)$, the connected component of the identity element of $O(l)$, and so we have the fibration

$$SU(l)/SO(l) = S\Lambda(l).$$

The exactness of its homotopy sequence implies the exactness of

$$\pi_1[SU(l)] \xrightarrow{p'} \pi_1[S\Lambda(l)] \xrightarrow{\Delta'} \pi_0[SO(l)],$$

where $\pi_1[SU(l)]$ and $\pi_0[SO(l)]$ are trivial, by (3.4) and the fact that $SO(l)$ is connected. Thus $\pi_1[S\Lambda(l)]$ is trivial, and so p in (3.10) is an isomorphism

Now p is induced by the mapping (3.8). Taking its composition with the isomorphism

$$\pi_1[\mathbf{S}^1] \ni \Gamma \mapsto \frac{1}{2\pi i} \int_\Gamma \frac{dz}{z} \in \mathbf{Z},$$

we obtain an isomorphism $\pi_1[\Lambda(l)] \to \mathbf{Z}$, defined by (3.5).

Proof of 2°). In (3.6) the isomorphism $\mathbf{Z} \simeq \pi_1[\mathrm{Sp}(l)]$ is the composition of the isomorphisms defined by parts 2 and 3 of lemma 3.1. The isomorphism $\pi_1[\Lambda(l)] \simeq \mathbf{Z}$ is the composition of the isomorphism (3.5) and the one that induces the homeomorphism of $\Lambda(l)$ and $W(l)$. Proving part 2 of lemma 3.2 is thus the same as proving the following:

The fibration $U(l)/O(l) = W(l)$ induces a morphism

$$p:\mathbf{Z} \simeq \pi_1[U(l)] \to \pi_1[W(l)] \simeq \mathbf{Z} \tag{3.11}$$

that is multiplication by 2 on $\mathbf{Z}$.

This fibration is defined by the mapping

$U(l) \ni u \mapsto w = u\,{}^t u$, which satisfies $\det w = (\det u)^2$.

Since the isomorphisms entering into (3.11) are defined by (3.1) and (3.5), we have

$$p:\mathbf{Z} \ni \frac{1}{2\pi i}\int_\gamma \frac{d(\det u)}{\det u} \mapsto \frac{1}{2\pi i}\int_\gamma \frac{d(\det u)^2}{(\det u)^2} \in \mathbf{Z}.$$

This morphism $p:\mathbf{Z} \to \mathbf{Z}$ is evidently multiplication by 2.

The two preceding lemmas have the following theorem as an immediate consequence. It is clearly independent of the hermitian structure on $Z(l)$ used above.

Definition 3. α and β *are the generators of* $\pi_1[\mathrm{Sp}(l)]$ and $\pi_1[\Lambda(l)]$ whose natural images in $\mathbf{Z}$ are 1.

THEOREM 3. 1°) $\mathrm{Sp}(l)$ *has a unique covering group,* $\mathrm{Sp}_q(l)$, *of order* q ($q = 1, 2, \ldots, \infty$) [*namely: the number of points having the same projection onto* $\mathrm{Sp}(l)$ *is* q]; α *acts on* $\mathrm{Sp}_q(l)$; α^r *does not act as the identity on* $\mathrm{Sp}_q(l)$ *unless* $r = 0 \mod q$.

2°) $\Lambda(l)$ *has a unique covering space,* $\Lambda_q(l)$, *of order* q; β *acts on* $\Lambda_q(l)$; β^r *does not act as the identity on* $\Lambda_q(l)$ *unless* $r = 0 \mod q$.

3°) $\mathrm{Sp}_q(l)$ *acts transitively* on $\Lambda_{2q}(l)$:

$$(\alpha S_q)\lambda_{2q} = S_q(\beta^2\lambda_{2q}) = \beta^2(S_q\lambda_{2q}), \text{ where } \lambda_{2q} \in \Lambda_{2q}(l), \quad S_q \in \mathrm{Sp}_q(l). \tag{3.12}$$

Example 3.1. $\Lambda_2(l)$ is the set of oriented (in the euclidean sense) lagrangian subspaces of $Z(l)$; $\mathrm{Sp}(l)$ acts on $\Lambda_2(l)$.

Example 3.2. $\mathrm{Sp}_2(l)$ acts on $\Lambda_4(l)$. *This result is essential* for the theory of asymptotic expansions (chapter II).

Notation 3. Denote by s the projection of $S \in \mathrm{Sp}_q(l)$ onto $\mathrm{Sp}(l)$; by λ the projection of $\lambda_q \in \Lambda_q(l)$ onto $\Lambda(l)$; by λ_2 the projection of $\lambda_{2q} \in \Lambda_{2q}(l)$ onto $\Lambda_2(l)$; by e the identity element of $\mathrm{Sp}(l)$; and by E the identity element of $\mathrm{Sp}_q(l)$.

Let us choose an element X_∞^* of $\Lambda_\infty(l)$ projecting onto X^* in $\Lambda(l)$; X_q^* denotes its projection onto $\Lambda_q(l)$.

4. Indices of Inertia

Definition of the index of inertia Inert$(\lambda, \lambda', \lambda'')$ *of a triple* $(\lambda, \lambda', \lambda'')$ *of elements of* $\Lambda(l)$ *pairwise transverse.* We have

$$\lambda \oplus \lambda' = \lambda' \oplus \lambda'' = \lambda'' \oplus \lambda = Z(l).$$

The conditions

$$z \in \lambda, \qquad z' \in \lambda', \qquad z'' \in \lambda'', \qquad z + z' + z'' = 0 \tag{4.1}$$

therefore define three isomorphisms

$$\begin{array}{ccc} & z \in \lambda & \\ \nearrow & & \searrow \\ \lambda'' \ni z'' & \longleftarrow & z' \in \lambda' \end{array} \tag{4.2}$$

whose product is the identity. By (4.1)

$$[z, z'] = [z', z''] = [z'', z]; \tag{4.3}$$

this number is the value of a quadratic form at $z \in \lambda$ (at $z' \in \lambda'$ or at $z'' \in \lambda''$). The isomorphisms (4.2) transform each one of these forms into the others. Thus they all have *the same index of inertia*, which will be denoted

Inert$(\lambda, \lambda', \lambda'')$.

LEMMA. The quadratic form

$$\lambda \ni z \mapsto [z, z'] \in \mathbf{R}$$

is nondegenerate (that is, has no zero eigenvalues).

Proof. Take a second triple

$$\zeta \in \lambda, \qquad \zeta' \in \lambda', \qquad \zeta'' \in \lambda'' \text{ such that } \zeta + \zeta' + \zeta'' = 0.$$

Since λ, λ', and λ'' are lagrangian, the bilinear form

$$(z, \zeta) \mapsto [z, \zeta'] = [\zeta', z''] = [z'', \zeta] = [\zeta, z'] = [z', \zeta''] = [\zeta'', z]$$

is symmetric and, hence, is the polar form of the quadratic form

$$z \mapsto [z, z'].$$

If this were degenerate, then there would exist $z \neq 0$ such that

$$z \in \lambda, \qquad [z, \zeta'] = 0 \quad \forall \zeta' \in \lambda',$$

that is,

$$z \in \lambda \cap \lambda',$$

contrary to hypothesis.

Thus Inert$(\cdot,\cdot,\cdot)$ *has the following properties:*

$$\text{Inert}(\lambda, \lambda', \lambda'') = \text{Inert}(\lambda', \lambda'', \lambda) = l - \text{Inert}(\lambda, \lambda'', \lambda'). \tag{4.4}$$

Inert$(\cdot,\cdot,\cdot)$, which is defined when its arguments are pairwise transverse, is locally constant on its domain of definition.

$$\text{Inert}(s\lambda, s\lambda', s\lambda'') = \text{Inert}(\lambda, \lambda', \lambda'') \qquad \forall s \in \text{Sp}(l). \tag{4.5}$$

Formulas (2.11) and (2.14) of I,§1 defined Inert(s, s', s'') for

$$s, s', s'' \in \text{Sp}(l)\backslash\Sigma_{\text{Sp}}, \qquad s\, s'\, s'' = e. \tag{4.6}$$

The following relation exists between these two indices of inertia:

THEOREM 4. *Under assumption* (4.6),

$$\text{Inert}(s, s', s'') = \text{Inert}(sX^*, X^*, s''^{-1}X^*). \tag{4.7}$$

Remark 4.1. The condition $s \notin \Sigma_{\text{Sp}}$ is equivalent to the following: sX^* and X^* are transverse.

Proof. We return to the notation of I,§1,2 (specifically that of lemma 2.3), setting

$$s = s_A, \qquad s' = s_{A'}, \qquad s'' = s_{A''};$$

$s_A : (x', p') \mapsto (x, p)$ means $p = Px - {}^tLx', \qquad p' = Lx - Qx'$;

$s_{A''} : (x, p) \mapsto (x'', p'')$ means $p'' = P''x'' - {}^tL''x, \qquad p = L''x'' - Q''x$.

The equations of the subspaces

$$\lambda = sX^*, \qquad \lambda' = X^*, \qquad \lambda'' = s''^{-1}X^*$$

are then

$$\lambda : p = Px, \qquad \lambda' : x' = 0, \qquad \lambda'' : p'' = -Q''x''.$$

Condition (4.1),

$$z = (x, p) \in \lambda, \qquad z'' = (x'', p'') \in \lambda'', \qquad z + z'' \in \lambda',$$

may be written

$$z = (x, Px), \qquad z'' = (-x, Q''x),$$

from which follows

$$[z'', z] = \langle (P + Q'')x, x \rangle.$$

Hence, by definition (2.14) in §1,

$$\begin{aligned}\operatorname{Inert}(\lambda, \lambda', \lambda'') &= \operatorname{Inert}(P + Q'') = \operatorname{Inert}_x(A + A' + A'') \\ &= \operatorname{Inert}(s, s', s'').\end{aligned}$$

The two lemmas below express the index of inertia Inert in terms of the Kronecker index KI. After we have defined the Maslov index m in terms of the Kronecker index in section 5, lemma 6.1 will deduce the relation between the indices of inertia and the Maslov index from these lemmas.

LEMMA 4.1. $\operatorname{Sp}(l)$ acts transitively on the set of pairs of transverse elements of $\Lambda(l)$.

Remark 4.2. Therefore every triple of pairwise transverse elements of $\Lambda(l)$ has the form

$$(s\lambda, sX, sX^*).$$

We recall that

$$\operatorname{Inert}(s\lambda, sX, sX^*) = \operatorname{Inert}(\lambda, X, X^*).$$

Proof. Since $\operatorname{Sp}(l)$ acts transitively on $\Lambda(l)$, it suffices to show that the stabilizer $\operatorname{St}(l)$ of X^* acts transitively on the set of elements of $\Lambda(l)$ transverse to X^*. The element s of $\operatorname{St}(l)$ defined by (2.6) transforms

$$X : p' = 0 \text{ into } sX : p = {}^t s_1^{-1} s_2 s_1^{-1} x,$$

where $s_2 : X \to X^*$ is symmetric. Now the condition that the equation $p = s_3 x$ define an element of $\Lambda(l)$ is clearly that $s_3 : X \to X^*$ be symmetric.

Notation 4. Recall that the spectrum of $u \in U(l)$, $\operatorname{sp}(u)$, is a 0-chain in $\mathbf{S}^1$. Let

$$\mathrm{sp}(\lambda) = \mathrm{sp}(w), \tag{4.8}$$

where w is the image of λ under the natural homeomorphism (2.3).

Denote by $(\exp i\theta)$ the point in $\mathbf{S}^1$ with the coordinate $\exp i\theta$ $(\theta \in \mathbf{R})$ and by $l(\exp i\theta)$ the 0-chain consisting of this point with multiplicity l $(l \in \mathbf{Z})$. We have, for example,

$$\begin{aligned} \mathrm{sp}(X^*) &= \mathrm{sp}(e), & \mathrm{sp}(X) &= \mathrm{sp}(-e), \\ \mathrm{sp}(e) &= l(1), & \mathrm{sp}(-e) &= l(-1). \end{aligned} \tag{4.9}$$

Let σ denote a 1-chain in $\mathbf{S}^1$, $|\sigma|$ its support, $\partial\sigma$ its boundary, and KI the *Kronecker index* (see Lefschetz [9] or any theory of chain intersection).

Recall that the integer $\mathrm{KI}[\sigma_1, \sigma_0]$

- is defined when σ_1 and σ_0 are a 1-chain and a 0-chain in $\mathbf{S}^1$ such that $|\sigma_0| \cap |\partial\sigma_1| = \varnothing$,
- is zero if $|\sigma_0| \cap |\sigma_1| = \varnothing$,
- is linear in σ_0 and σ_1,
- is equal to 1 if σ_1 is a positively oriented arc and σ_0 is an interior point of this arc,
- satisfies $\mathrm{KI}[\sigma_1, \partial\sigma_1'] = -\mathrm{KI}[\sigma_1', \partial\sigma_1]$.

LEMMA 4.2. Let σ and σ^* be two 1-chains in $\mathbf{S}^1$ such that

$$\partial\sigma = \mathrm{sp}(\lambda) - \mathrm{sp}(X^*), \qquad \partial\sigma^* = \mathrm{sp}(\lambda) - \mathrm{sp}(X), \tag{4.10}$$

$$\sigma - \sigma^* \text{ belongs to the half-circle in } \mathbf{S}^1, \text{ where } \mathrm{Im}(z) \geqslant 0. \tag{4.11}$$

Then

$$\mathrm{Inert}(\lambda, X, X^*) = \mathrm{KI}[\sigma,(-1)] - \mathrm{KI}[\sigma^*, (1)]. \tag{4.12}$$

Remark 4.3. Lemma 5.1 will use this decomposition of Inert.

Proof. Let

$$z = x + iy \in \lambda, \qquad z' \in X, \qquad z'' \in X^*$$

be such that

$$z + z' + z'' = 0, \quad \text{where } x, y \in X.$$

Clearly

$$z' = -x, \qquad z'' = -iy,$$

$$[z', z''] = \operatorname{Im}(z' \mid z'') = -(x \mid y).$$

Let w be the natural image of λ under (2.3). By (2.5)

$$y = i\frac{e + w}{e - w}x \quad (e \text{ is the identity}).$$

Since w is unitary and symmetric, $i(e + w)/(e - w)$ is real and symmetric. Hence $\operatorname{Inert}(\lambda, X, X^*)$ is the index of inertia of the quadratic form

$$X \ni x \mapsto -\left(x \,\middle|\, i\frac{e + w}{e - w}x\right) \in \mathbf{R},$$

that is, the number of positive eigenvalues of the real symmetric matrix $i(e + w)/(e - w)$.

Now the positive real axis is the image of the half-circle in $\mathbf{S}^1$, where $\operatorname{Im} z \leqslant 0$, under the homographic transformation

$$\mathbf{S}^1 \ni (\exp i\theta) \mapsto i\frac{1 + \exp i\theta}{1 - \exp i\theta} \in \mathbf{R}. \tag{4.13}$$

We orient it positively: it forms a chain χ in $\mathbf{S}^1$. Since (4.13) transforms the eigenvalues of w into those of $i(e + w)/(e - w)$,

$$\operatorname{Inert}(\lambda, X, X^*) = \operatorname{KI}[\chi, \operatorname{sp}(w)].$$

Let τ be a chain in $\mathbf{S}^1$ with boundary

$$\partial\tau = \operatorname{sp}(w) - \operatorname{sp}(ie).$$

Then

$$\begin{aligned} \operatorname{Inert}(\lambda, X, X^*) &= \operatorname{KI}[\chi, \partial\tau] = -\operatorname{KI}[\tau, \partial\chi] \\ &= \operatorname{KI}[\tau, (-1)] - \operatorname{KI}[\tau, (1)]. \end{aligned}$$

Let σ and σ^* be 1-chains in $\mathbf{S}^1$ such that $\sigma - \tau$ and $\sigma^* - \tau$ are defined as follows:

$$\partial(\sigma - \tau) = \operatorname{sp}(ie) - \operatorname{sp}(e),$$

$\sigma - \tau$ belongs to the arc $\theta \in [0, \pi/2]$ in $\mathbf{S}^1 = \{(\exp i\theta)\}$;

$$\partial(\sigma^* - \tau) = \operatorname{sp}(ie) - \operatorname{sp}(-e),$$

$\sigma^* - \tau$ belongs to the arc $\theta \in [\pi/2, \pi]$ in $\mathbf{S}^1$.

Since

$$\mathrm{KI}[\sigma - \tau, (-1)] = \mathrm{KI}[\sigma^* - \tau, (1)] = 0,$$

the preceding expression for Inert(λ, X, X^*) is equivalent to (4.12). Now σ and σ^* satisfy conditions (4.10) and (4.11) and are determined by them up to the addition of 1-cycles γ and γ^* in $\mathbf{S}^1$ such that $\gamma - \gamma^*$ belongs to the half-circle in $\mathbf{S}^1$, where $\operatorname{Im} z > 0$. Since γ and γ^* are then homologous,

$$\mathrm{KI}[\gamma, (-1)] = \mathrm{KI}[\gamma^*, (1)].$$

Thus (4.12) holds for any pair (σ, σ^*) satisfying (4.10) and (4.11).

5. The Maslov Index m on $\Lambda^2_\infty(l)$

We are going to supplement Arnold's definition [1], but first we shall use the following preliminary definition.

Definition 5.1 *of the index M on $\Lambda^4_\infty(l)$.* Let (v_∞, v'_∞) be an element of $\Lambda_\infty(l) \times \Lambda_\infty(l) = \Lambda^2_\infty(l)$, that is, a pair of elements of $\Lambda_\infty(l)$. Let v and $w(v)$ be the natural images of v_∞ in $\Lambda(l)$ and $W(l)$ [cf. (2.3)], and let v' and $w' = w(v')$ be the images of v'_∞. By lemma 2.1, 3°), the condition that v and v' be transverse can be expressed as

$$(1) \notin \operatorname{sp}(ww'^{-1});$$

$\operatorname{sp}(ww'^{-1})$, which we denote $\operatorname{sp}(v, v')$, is a 0-chain in $\mathbf{S}^1$ (see Lefschetz [9]). The mapping

$$\Gamma \ni (v_\infty, v'_\infty) \mapsto \operatorname{sp}(v, v')$$

maps the arc Γ onto a 1-chain in $\mathbf{S}^1$ denoted $\operatorname{sp}(\Gamma)$. If

$$\partial\Gamma = (\lambda_\infty, \lambda'_\infty) - (\mu_\infty, \mu'_\infty),$$

then

$$\partial \operatorname{sp}(\Gamma) = \operatorname{sp}(\lambda, \lambda') - \operatorname{sp}(\mu, \mu').$$

Suppose

$$\lambda \text{ and } \lambda' \text{ are transverse, } \mu \text{ and } \mu' \text{ are transverse;} \tag{5.1}$$

then by lemma 2.1, 3°),

$$(1) \notin |\partial \operatorname{sp}(\Gamma)|.$$

Thus KI[sp(Γ), (1)] is defined. It is an integer depending only on the homotopy class of Γ, that is, on $\partial\Gamma$. We denote it

$$M[\lambda_\infty, \lambda'_\infty; \mu_\infty, \mu'_\infty] = \mathrm{KI}[\mathrm{sp}(\Gamma), (1)] \in \mathbf{Z}. \tag{5.2}$$

Remark. By Remark 1, M is independent of the choice of the hermitian structure on $Z(l)$ used in its definition.

Properties of M. M is defined under hypothesis (5.1). M is *locally constant* on its domain of definition. M has the obvious *additive property*

$$M[\lambda_\infty, \lambda'_\infty; \mu_\infty, \mu'_\infty] + M[\mu_\infty, \mu'_\infty; \nu_\infty, \nu'_\infty] = M[\lambda_\infty, \lambda'_\infty; \nu_\infty, \nu'_\infty], \tag{5.3}$$

which implies

$$M[\lambda_\infty, \lambda'_\infty; \lambda_\infty, \lambda'_\infty] = 0. \tag{5.4}$$

M has the *invariance property*

$$M[S\lambda_\infty, S\lambda'_\infty; \mu_\infty, \mu'_\infty] = M[\lambda_\infty, \lambda'_\infty; \mu_\infty, \mu'_\infty], \quad \forall S \in \mathrm{Sp}_\infty(l). \tag{5.5}$$

If β is the generator of $\pi_1[\Lambda(l)]$ (cf. definition 3), then

$$M[\beta^r\lambda_\infty, \beta^{r'}\lambda'_\infty; \mu_\infty, \mu'_\infty] = M[\lambda_\infty, \lambda'_\infty; \mu_\infty, \mu'_\infty] + r - r' \qquad \forall r, r' \in \mathbf{Z}. \tag{5.6}$$

Proof of (5.5). Assuming hypothesis (5.1), $s\lambda$ and $s\lambda'$ are transverse, and the mapping

$$\mathrm{Sp}_\infty(l) \ni S \mapsto M[S\lambda_\infty, S\lambda'_\infty; \mu_\infty, \mu'_\infty] \in \mathbf{Z}$$

is defined and locally constant. Therefore it is constant since $\mathrm{Sp}_\infty(l)$ is connected.

Proof of (5.6). Choose the arc Γ to be differentiable and such that

$$\partial\Gamma = (\beta^r\lambda_\infty, \beta^{r'}\lambda'_\infty; \lambda_\infty, \lambda'_\infty).$$

Clearly one can define l continuous functions

$$\theta_j : \Gamma \ni (\nu_\infty, \nu'_\infty) \mapsto \theta_j(\nu_\infty, \nu'_\infty) \in \mathbf{R}, \qquad j = 1, \ldots, l,$$

such that

$$\mathrm{sp}(\nu, \nu') = \sum_j (\exp i\theta_j);\ \theta_j(\beta^r\lambda_\infty, \beta^r\lambda'_\infty) = \theta_j(\lambda_\infty, \lambda'_\infty) \bmod 2\pi.$$

Definition (5.2) of M gives

$$M[\beta^r\lambda_\infty, \beta^{r'}\lambda'_\infty; \lambda_\infty, \lambda'_\infty] = \sum_j \frac{1}{2\pi}\int_\Gamma d\theta_j = \frac{1}{2\pi}\int_\Gamma d\Big(\sum_j \theta_j\Big)$$
$$= \frac{1}{2\pi i}\int_\Gamma \frac{d\det(ww'^{-1})}{\det(ww'^{-1})}$$
$$= \frac{1}{2\pi i}\int_\Gamma \frac{d(\det w)}{\det w} - \frac{1}{2\pi i}\int_\Gamma \frac{d(\det w')}{\det w'} = r - r'$$

by (3.5) and definition 3. Hence (5.6) follows by the additive property (5.3) of M.

In order to recover assumption (4.11) of lemma 4.2, we need the following definitions.

Definition 5.2 X_∞ is the point of $\Lambda_\infty(l)$ that projects onto X in $\Lambda(l)$ and that may be joined to X^*_∞ by an arc γ of $\Lambda_\infty(l)$ whose spectrum $\mathrm{sp}(\gamma)$ belongs to the half-circle in $\mathbf{S}^1$, where $\mathrm{Im}(z) \geqslant 0$.

Definition 5.3. The *Maslov index* is the function m given by

$$m(\lambda_\infty, \lambda'_\infty) = M(\lambda_\infty, \lambda'_\infty; X^*_\infty, X_\infty).$$

This allows us to formulate lemma 4.2 as follows.

LEMMA 5.1. For any triple $\lambda_\infty, \lambda'_\infty, \lambda''_\infty$ of elements of $\Lambda_\infty(l)$,

$$\mathrm{Inert}(\lambda, \lambda', \lambda'') = m(\lambda_\infty, \lambda'_\infty) - m(\lambda_\infty, \lambda''_\infty) + m(\lambda'_\infty, \lambda''_\infty). \tag{5.7}$$

Proof. By (5.6), the right-hand side of (5.7) depends only on $(\lambda, \lambda', \lambda'')$. By Remark 4.2 and the invariance property (5.5) it suffices to establish the following special case of (5.7):

$$\mathrm{Inert}(\lambda, X, X^*) = m(\lambda_\infty, X_\infty) - m(\lambda_\infty, X^*_\infty) + m(X_\infty, X^*_\infty). \tag{5.8}$$

Definition 5.3 of m and the additive property (5.3) of M give

$$m(\lambda_\infty, X_\infty) = M[\lambda_\infty, X_\infty; X^*_\infty, X_\infty],$$
$$m(\lambda_\infty, X^*_\infty) - m(X_\infty, X^*_\infty) = M(\lambda_\infty, X^*_\infty; X_\infty, X^*_\infty].$$

Definition 5.1 of these two values of M makes use of two arcs Γ and Γ^* of $\Lambda^2_\infty(l)$ such that

$$\partial\Gamma = (\lambda_\infty, \mathrm{X}_\infty) - (X^*_\infty, X_\infty), \qquad \partial\Gamma^* = (\lambda_\infty, X^*_\infty) - (X_\infty, X^*_\infty).$$

We choose these arcs as cartesian products,

$$\Gamma = \gamma \times X_\infty, \qquad \Gamma^* = \gamma^* \times X^*_\infty,$$

where γ and γ^* are then arcs of $\Lambda_\infty(l)$ such that

$$\partial\gamma = \lambda_\infty - X^*_\infty, \qquad \partial\gamma^* = \lambda_\infty - X_\infty.$$

We have

$$\partial(\gamma - \gamma^*) = X_\infty - X^*_\infty.$$

Definition 5.2 of X_∞ allows us to choose γ and γ^* such that $\mathrm{sp}(\gamma - \gamma^*)$ belongs to the half-circle in $\mathbf{S}^1$, where $\mathrm{Im}(z) \geqslant 0$.

Denote

$$\sigma = \mathrm{sp}(\gamma), \qquad \sigma^* = \mathrm{sp}(\gamma^*).$$

Since the homeomorphism $w(\cdot)$ defined by (2.3) has the values

$$w(X) = -e, \qquad w(X^*) = e,$$

definition 5.1 of M gives

$$M[\lambda_\infty, X_\infty; X^*_\infty, X_\infty] = \mathrm{KI}[\sigma, (-1)],$$

$$M[\lambda_\infty, X^*_\infty; X_\infty, X^*_\infty] = \mathrm{KI}[\sigma^*, (1)]$$

The formula (5.8) to be proved is thus identical to formula (4.12), which holds because $\sigma - \sigma^* = \mathrm{sp}(\gamma - \gamma^*)$ satisfies condition (4.11).

LEMMA 5.2. We have

$$m(\lambda_\infty, \lambda'_\infty) + m(\lambda'_\infty, \lambda_\infty) = l. \tag{5.9}$$

Proof. Substitute the expression (5.7) for Inert into (4.4).

The following lemma supplements lemma 5.1.

LEMMA 5.3. Every function

$$n: (\lambda_q, \lambda'_q) \mapsto n(\lambda_q, \lambda'_q)$$

- defined for $\lambda_q, \lambda'_q \in \Lambda_q(l)$, λ and λ' transverse,
- with values in an abelian group,
- locally constant on its domain of definition,
- such that, for pairwise transverse $\lambda, \lambda', \lambda''$,

$$n(\lambda_q, \lambda'_q) - n(\lambda_q, \lambda''_q) + n(\lambda'_q, \lambda''_q) = 0, \tag{5.10}$$

is identically zero.

Proof. By (5.10), n is constant in a neighborhood of any pair (λ_q, λ'_q), transverse or not. Therefore n is constant on $\Lambda_q^2(l)$, which is connected. By (5.10), its value is 0.

We have proved the following theorem.

THEOREM 5. 1°) *The Maslov index (definitions 5.1–5.3) is the only function*

$$m: (\lambda_\infty, \lambda'_\infty) \mapsto m(\lambda_\infty, \lambda'_\infty) \in \mathbf{Z}$$

defined on pairs of transverse elements of $\Lambda_\infty(l)$ that is locally constant on its domain and makes the following decomposition of the index of inertia possible:

$$\mathrm{Inert}(\lambda, \lambda', \lambda'') = m(\lambda_\infty, \lambda'_\infty) - m(\lambda_\infty, \lambda''_\infty) + m(\lambda'_\infty, \lambda''_\infty). \tag{5.7}$$

2°) *Taking into account definition 5.2, this function has the following properties:*

$$m(\lambda_\infty, \lambda'_\infty) + m(\lambda'_\infty, \lambda_\infty) = l; \tag{5.9}$$

$$m(S\lambda_\infty, S\lambda'_\infty) = m(\lambda_\infty, \lambda'_\infty) \qquad \forall S \in \mathrm{Sp}_\infty(l); \tag{5.11}$$

$$m(\beta^r\lambda_\infty, \beta^{r'}\lambda'_\infty) = m(\lambda_\infty, \lambda'_\infty) + r - r' \qquad \forall r, r' \in \mathbf{Z}; \tag{5.12}$$

$$m(X^*_\infty, X_\infty) = 0, \qquad m(X_\infty, X^*_\infty) = l. \tag{5.13}$$

Proof of 1°). Lemmas 5.1 and 5.3.

Proof of (5.9). Lemma 5.2.

Proof of (5.11) *and* (5.12). Definition 5.3 of m, (5.5), and (5.6).

Proof of (5.13). Definition 5.3 of m and (5.4); (5.9).

Remark 5.1. Formula (5.7) clearly implies the following: If $\lambda, \lambda', \lambda'', \lambda'''$ are four pairwise transverse elements of $\Lambda(l)$, then

$$\mathrm{Inert}(\lambda, \lambda', \lambda'') - \mathrm{Inert}(\lambda, \lambda', \lambda''') + \mathrm{Inert}(\lambda, \lambda'', \lambda''') - \mathrm{Inert}(\lambda', \lambda'', \lambda''') = 0.$$

Remark 5.2. We call any transverse pair $(\mu_\infty, \mu'_\infty) \in \Lambda^2_\infty(l)$ such that

$$M[\lambda_\infty, \lambda'_\infty; \mu_\infty, \mu'_\infty] = m(\lambda_\infty, \lambda'_\infty) \qquad \forall \lambda_\infty, \lambda'_\infty \in \Lambda_\infty(l)$$

a *basis*. For example, (X^*_∞, X_∞) is a basis. By the additive property (5.3) of M, the condition that $(\mu_\infty, \mu'_\infty)$ be a basis can be expressed as

$$m(\mu_\infty, \mu'_\infty) = 0.$$

6. The Jump of the Maslov Index $m(\lambda_\infty, \lambda'_\infty)$ at a Point (λ, λ'), Where $\dim \lambda \cap \lambda' = 1$

Maslov [11] defined his index, up to an additive constant, by the expression of its jump across the *hypersurface* $\Sigma_{\Lambda^2_\infty}$ in $\Lambda^2_\infty(l)$, which is the set of pairs $(\lambda_\infty, \lambda'_\infty)$ of nontransverse elements of $\Lambda_\infty(l)$. Theorem 6 will make the expression for this jump explicit.

First of all let us establish some properties of $U(l)$. By γ we denote a differentiable arc of $U(l)$:

$$\gamma: [-1, 1] \ni \theta \mapsto u(\theta) \in U(l), \qquad u_\theta = \frac{du}{d\theta} \neq 0.$$

LEMMA 6.1. Let $\exp(i\psi(\theta))$ be a simple eigenvalue of $u(\theta)$ and let $z(\theta)$ be a corresponding eigenvector. Then

$$\| z(\theta) \|^2 \psi_\theta = \left(\frac{1}{i} u^{-1}(\theta) u_\theta z(\theta) \,|\, z(\theta) \right). \tag{6.1}$$

Remark 6.1. Recall that if U is a Lie group with generic element u, infinitesimal transformations X_k, and Maurer-Cartan forms ω_k, then $u^{-1}\,du$ (hence, in particular, $u^{-1}u_\theta$) may be written

$$u^{-1}\,du = \sum_k \omega_k(u, du) X_k;$$

see E. Cartan [4].

Remark 6.2. Since $U(l)$ is the unitary group, $(1/i)u^{-1}(\theta)u_\theta$ is an arbitrary $l \times l$ self-adjoint matrix characterizing the vector u_θ tangent to $U(l)$ at $u(\theta)$.

Proof. Since $\exp(i\psi(\theta))$ is a simple eigenvalue, $\psi(\theta)$ is a differentiable function of θ, and the vector $z(\theta)$, which is defined up to a scalar factor, may be chosen to be a differentiable function of θ. Then differentiating the relation

$$u(\theta)z(\theta) = z(\theta)\exp(i\psi(\theta))$$

and multiplying on the left by $(1/i)u^{-1}(\theta)$, we obtain

$$\frac{1}{i}u^{-1}u_\theta z + \frac{1}{i}z_\theta = \frac{1}{i}u^{-1}z_\theta \exp(i\psi) + z(\theta)\psi_\theta.$$

Taking the scalar product with z eliminates z_θ and gives (6.1) because

$$(u^{-1}z_\theta \exp(i\psi) \,|\, z) = (z_\theta \,|\, \exp(-i\psi)\, uz) = (z_\theta \,|\, z)$$

since u is unitary.

Denote by Σ_U *the set of* $u \in U(l)$ *such that* $(\mathbf{1}) \in \mathrm{sp}(u)$; recall that $(\mathbf{1})$ denotes the point of $\mathbf{S}^1 \subset \mathbf{C}$ with coordinate 1.

The following lemma has the sole purpose of interpreting the assumption of lemma 6.3 geometrically.

LEMMA 6.2. If $u(o) \in \Sigma_U$, (1) is a simple value of $\mathrm{sp}\, u(o)$, and $z(o)$ is a corresponding eigenvector, then the condition that u_θ define a direction tangent to Σ_U at $u(o)$ takes the form

$$\left(\frac{1}{i}u^{-1}(o)u_\theta(o)z(o) \,\middle|\, z(o)\right) = 0. \tag{6.2}$$

Proof. $u(o)$ is a regular point of Σ_U. By lemma 6.1, every direction tangent to Σ_U at $u(o)$ satisfies (6.2). The hyperplane of directions satisfying (6.2) thus contains the tangent plane to Σ_U at $u(o)$; but Σ_U is a hypersurface in $U(l)$.

LEMMA 6.3. If the arc γ in $U(l)$ intersects Σ_U only at the point $u(o)$, if the eigenvalue 1 of $u(o)$ is simple, and if the corresponding eigenvector $z(o)$ satisfies

$$\left(\frac{1}{i}u^{-1}(o)u_\theta z(o) \,\middle|\, z(o)\right) \neq 0$$

(i.e., if γ is not tangent to Σ_U), then

$$\mathrm{KI}[\mathrm{sp}\,\gamma, (\mathbf{1})] = \mathrm{sign}\left(\frac{1}{i}u^{-1}(o)u_\theta z(o) \,\middle|\, z(o)\right). \tag{6.3}$$

Proof. Let $\exp(i\psi(\theta))$ be the eigenvalue of $u(\theta)$ near 1. Evidently

$$\mathrm{KI}[\mathrm{sp}\,\gamma, (\mathbf{1})] = \mathrm{sign}\,\psi_\theta(o) \quad \text{if } \psi_\theta(o) \neq 0,$$

so (6.3) follows from (6.1).

Notation. Σ_{Λ^2} denotes the set of nontransverse (λ, λ'):

$$\Sigma_{\Lambda^2} \subset \Lambda^2(l).$$

Γ denotes a differentiable arc in $\Lambda^2(l)$:

$$[-1, 1] \ni \theta \mapsto (\lambda(\theta), \lambda'(\theta)) \in \Lambda^2(l).$$

$w(\theta)$ and $w'(\theta)$ denote the natural images of $\lambda(\theta)$ and $\lambda'(\theta) \in \Lambda(l)$ in $W(l)$: see (2.3).

LEMMA 6.4. If the arc Γ in $\Lambda^2(l)$ intersects Σ_{Λ^2} only at the point $(\lambda(o), \lambda'(o))$ and if

$$\dim \lambda(o) \cap \lambda'(o) = 1, \qquad z(o) \in \lambda(o) \cap \lambda'(o),$$

$$\left(\frac{1}{i} w_\theta w^{-1}(o) z(o) \,\middle|\, z(o)\right) \neq \left(\frac{1}{i} w'_\theta w'^{-1}(o) z(o) \,\middle|\, z(o)\right),$$

then

$$\mathrm{KI}[\mathrm{sp}\,\Gamma, (1)] = \mathrm{sign}\left\{\left(\frac{1}{i} w_\theta w^{-1}(o) z(o) \,\middle|\, z(o)\right) - \left(\frac{1}{i} w'_\theta w'^{-1}(o) z(o) \,\middle|\, z(o)\right)\right\}. \tag{6.4}$$

Proof. The image of the arc $\Gamma \in \Lambda^2(l)$ in $U(l)$ is the arc

$$\gamma : [-1, 1] \ni \theta \mapsto u(\theta) = w(\theta) w'^{-1}(\theta).$$

By definition,

$$\mathrm{KI}[\mathrm{sp}\,\Gamma, (1)] = \mathrm{KI}[\mathrm{sp}\,\gamma, (1)].$$

By lemma 2.1,3°), γ intersects Σ_U only at the point $\theta = o$, and 1 is a simple value of sp $u(o)$. By lemma 2.1,2°),

$$z(o) + w(o)\overline{z(o)} = o, \qquad z(o) + w'(o)\overline{z(o)} = o,$$

so

$$z(o) = u(o) z(o).$$

Now

$$u^{-1} u_\theta = u^{-1} w_\theta w^{-1} u - w'_\theta w'^{-1};$$

thus, because u is unitary,

$$(u^{-1}(o)u_\theta z(o)\,|\,z(o)) = (w_\theta w^{-1}(o)z(o)\,|\,z(o)) - (w'_\theta w'^{-1}(o)z(o)\,|\,z(o)).$$

Hence, by lemma 6.3,

$$\mathrm{KI}[\mathrm{sp}\,\gamma, (1)] = \mathrm{sign}\left\{\left(\frac{1}{i} w_\theta w^{-1}(o)z(o)\,\middle|\,z(o)\right) - \left(\frac{1}{i} w'_\theta w'^{-1}(o)z(o)\,\middle|\,z(o)\right)\right\},$$

and (6.4) follows.

LEMMA 6.5. Let

$$\theta \mapsto \lambda(\theta), \qquad \theta \mapsto z(\theta)$$

be differentiable mappings of $[-1, 1]$ into $\Lambda(l)$ and $Z(l)$ such that

$$z(\theta) \in \lambda(\theta).$$

Then

$$\left(\frac{1}{2i} w_\theta w^{-1} z\,\middle|\,z\right) = [z_\theta, z(\theta)]. \tag{6.5}$$

Proof. By lemma 2.1,2°),

$$z + w\bar{z} = o,$$

so

$$z_\theta + w_\theta \bar{z} + w\bar{z}_\theta = o.$$

Hence

$$(z_\theta\,|\,z) - (w_\theta w^{-1} z\,|\,z) + (w\bar{z}_\theta\,|\,z) = o,$$

where

$$(w\bar{z}_\theta\,|\,z) = (\bar{z}_\theta\,|\,w^{-1}z) = -(\bar{z}_\theta\,|\,\bar{z}) = -\overline{(z_\theta\,|\,z)}.$$

From (1.1), (6.5) follows.

The left-hand side of (6.4) can be expressed in terms of the Maslov index by (5.2) and definition 5.3; the right-hand side of (6.4) can be expressed in terms of the symplectic form $[\cdot, \cdot]$ by lemma 6.5. Hence the following theorem, which will allow us to establish theorem 3.2 of §3.

THEOREM 6. *Let*

$[-1, 1] \ni \theta \mapsto (\lambda_\infty, \lambda'_\infty) \in \Lambda^2_\infty(l)$,

$[-1, 1] \ni \theta \mapsto (z, z') \in Z^2(l)$

be two differentiable mappings such that

$z \in \lambda, \qquad z' \in \lambda'$ *for each* θ;

$z = z' \neq 0$ *for* $\theta = 0$; $\qquad \lambda$ *and* λ' *are transverse for* $\theta \neq 0$;

the function

$[-1, 1] \ni \theta \mapsto [z, z'] \in \mathbf{R}$

vanishes only at $\theta = 0$ *and vanishes only to first order. Then there exists a constant* $c \in \mathbf{Z}$ *such that*

$$m(\lambda_\infty, \lambda'_\infty) = \begin{cases} c & \text{for } [z, z'] < 0 \\ 1 + c & \text{for } [z, z'] > 0. \end{cases}$$

7. The Maslov Index on $\mathrm{Sp}_\infty(l)$; the Mixed Inertia

Let

$S, S', S'' \in \mathrm{Sp}_\infty(l) \backslash \Sigma_{\mathrm{Sp}_\infty}$ be such that

$$S S' S'' = E \text{ (the identity element)}; \tag{7.1}$$

let s, s', s'' be their projections onto $\mathrm{Sp}(l)$. Formula (4.7) of theorem 4.1, which relates the definitions of the inertia on $\Lambda(l)$ and $\mathrm{Sp}(l)$, and formula (5.7) of theorem 5, which relates the inertia on $\Lambda(l)$ to the Maslov index, yield, by the invariance property of the Maslov index (5.11),

$$\mathrm{Inert}(s, s', s'') = m(SX^*_\infty, X^*_\infty) - m(S'^{-1}X^*_\infty, X^*_\infty) + m(S''X^*_\infty, X^*_\infty).$$

We rewrite this formula as (7.3), by virtue of the following definition:

Definition 7.1 *of the Maslov index on* $\mathrm{Sp}_\infty(l)$. If $S \in \mathrm{Sp}_\infty(l) \backslash \Sigma_{\mathrm{Sp}_\infty}$ we define

$$m(S) = m(SX^*_\infty, X^*_\infty). \tag{7.2}$$

This definition makes sense by Remark 4.1: $S \notin \Sigma_{\mathrm{Sp}_\infty}$ is equivalent to the condition that SX^* and X^* be transverse.

Let us supplement this formula with a lemma analogous to lemma 5.3:

LEMMA 7. Every function

$n: S \mapsto n(S)$

- defined for $S \in \mathrm{Sp}_q(l) \backslash \Sigma_{\mathrm{Sp}_q}$,
- with values in an abelian group,
- locally constant on its domain of definition,
- such that

$n(S) - n(S'^{-1}) + n(S'') = 0$ when $SS'S'' = E$,

is identically zero.

This lemma, combined with theorem 5 and (3.12) gives the following theorem.

THEOREM 7.1. 1°) *The Maslov index defined by* (7.2) *is the only function*

$m: \mathrm{Sp}_\infty(l) \backslash \Sigma_{\mathrm{Sp}_\infty} \to \mathbf{Z}$

that is locally constant on its domain and that makes the following decomposition of the index of inertia possible: under hypothesis (7.1),

$$\mathrm{Inert}(s, s', s'') = m(S) - m(S'^{-1}) + m(S''). \tag{7.3}$$

2°) *This function has the following properties:*

$$m(S) + m(S^{-1}) = l, \tag{7.4}$$

$$m(\alpha^r S) = m(S) + 2r, \tag{7.5}$$

where α *is the generator of* $\pi_1[\mathrm{Sp}(l)]$ (*definition* 3).

Definition 7.2 *of the mixed inertia.* Let

$$s \in \mathrm{Sp}(l) \backslash \Sigma_{\mathrm{Sp}};\ \lambda, \lambda' \in \Lambda(l), \text{ transverse to } X^* \text{ and such that } \lambda = s\lambda'. \tag{7.6}$$

The *mixed inertia* is defined by the formula

$$\mathrm{Inert}(s, \lambda, \lambda') = \mathrm{Inert}(sX^*, X^*, \lambda) = \mathrm{Inert}(X^*, s^{-1}X^*, \lambda'), \tag{7.7}$$

the last two terms being equal because of the invariance (4.5) of the inertia.

The following theorem is evident.

THEOREM 7.2. *Under assumption* (7.6),

$$\mathrm{Inert}(s^{-1}, \lambda', \lambda) = l - \mathrm{Inert}(s, \lambda, \lambda'), \tag{7.8}$$

$$\mathrm{Inert}(s, \lambda, \lambda') = m(S) - m(\lambda_\infty, X_\infty^*) + m(\lambda'_\infty, X_\infty^*) \tag{7.9}$$

if $\lambda_\infty = S\lambda'_\infty$ *and if* s *is the projection of* $S \in \mathrm{Sp}_\infty(l)$.

Now we express the mixed inertia in terms of the inertia of a quadratic form, as we did for $\mathrm{Inert}(s, s', s'')$ (§1,definition 2.4) and $\mathrm{Inert}(\lambda, \lambda', \lambda'')$ (§2,4). This result will be used in section 10.

THEOREM 7.3. *Under assumption* (7.6), *we have* $s = s_A$(§1,1) *and, by* (2.5), *the equation of* λ' *is* $p' = \varphi'_{x'}(x')$, *where* φ' *is a quadratic form on* X. *Then*

$$\mathrm{Inert}(s, \lambda, \lambda') = \mathrm{Inert}(A(o, \cdot) + \varphi'(\cdot)). \tag{7.10}$$

Proof. By (7.7) and the definition of $\mathrm{Inert}(\lambda, \lambda', \lambda'')$ (section 4),

$$\mathrm{Inert}(s, \lambda, \lambda') = \mathrm{Inert}(X^*, s^{-1}X^*, \lambda')$$

is the inertia of the quadratic form

$$x' \mapsto [z, z']$$

for $z = (x, p)$, $z' = (x', p')$,

$$(x, p) \in X^*, \qquad s(x', p') \in X^*, \qquad (x + x', p + p') \in \lambda'. \tag{7.11}$$

Relations (7.11) may be rewritten

$$x = o, \qquad p' = -A_{x'}(o, x'), \qquad p + p' = \varphi'_{x'}(x + x'),$$

whence

$$x = o, \qquad p = A_{x'}(o, x') + \varphi'_{x'}(x').$$

This implies

$$[z, z'] = \langle p, x' \rangle = 2A(o, x') + 2\varphi'(x'),$$

and therefore (7.10).

8. Maslov Indices on $\Lambda_q(l)$ and $\mathrm{Sp}_q(l)$

Let $\lambda_q, \lambda'_q \in \Lambda_q(l)$ and $S \in \mathrm{Sp}_q(l)$ be the projections of $\lambda_\infty, \lambda'_\infty \in \Lambda_\infty(l)$ and $S_\infty \in \mathrm{Sp}_\infty(l)$. By (5.12) and (7.5) (where α and β are the generators of $\pi_1[\mathrm{Sp}(l)]$ and $\pi_1[\Lambda(l)]$), the relations

$$m(\lambda_q, \lambda'_q) = m(\lambda_\infty, \lambda'_\infty) \bmod q, \qquad m(S) = m(S_\infty) \bmod 2q, \tag{8.1}$$

define functions m. They are again called *Maslov indices*. From theorems 5, 7.1, and 7.2 and lemmas 5.3 and 7, they evidently have the following properties.

THEOREM 8. 1°) *The Maslov index defined by* $(8.1)_1$ *is the only function*

$$m:(\lambda_q, \lambda'_q) \mapsto m(\lambda_q, \lambda'_q) \in \mathbf{Z}_q$$

defined on transverse pairs of elements of $\Lambda_q(l)$ *that is locally constant on its domain and that makes the following decomposition of the index of inertia possible:*

$$\operatorname{Inert}(\lambda_q, \lambda'_q, \lambda''_q) = m(\lambda_q, \lambda'_q) - m(\lambda_q, \lambda''_q) + m(\lambda'_q, \lambda''_q) \bmod q. \tag{8.2}$$

2°) *The Maslov index defined by* $(8.1)_2$ *is the only function*

$$m: \mathrm{Sp}_q(l)\backslash\Sigma_{\mathrm{Sp}_q} \mapsto \mathbf{Z}_{2q}$$

that is locally constant on its domain and that makes the following decomposition of the index of inertia possible:

$$\operatorname{Inert}(s, s', s'') = m(S) - m(S'^{-1}) + m(S'') \bmod 2q. \tag{8.3}$$

3°) *These functions have the following properties:*

$$m(\lambda_q, \lambda'_q) + m(\lambda'_q, \lambda_q) = l \bmod q, \qquad m(S) + m(S^{-1}) = l \bmod 2q, \tag{8.4}$$

$$\operatorname{Inert}(s, \lambda, \lambda') = m(S) - m(\lambda_{2q}, X^*_{2q}) + m(\lambda'_{2q}, X^*_{2q}) \bmod 2q \tag{8.5}$$

if $S \in \mathrm{Sp}_q$, $\lambda_{2q} = S\lambda'_{2q}$. *We recall that* $\mathrm{Sp}_q(l)$ *acts on* $\Lambda_{2q}(l)$.

By §1,theorem 2,3°), assuming $q = 2$, theorem 8,2°) identifies:

- the Maslov index that formula (2.15) of §1 defines on $\mathrm{Sp}_2(l)$ mod 4,
- the Maslov index that (7.2) and (8.1) define on $\mathrm{Sp}_q(l)$ mod $2q$.

Definition 8. Let $A \in \mathbf{A}$ (§1,definition 1.2). Let us give a definition of $S_A \in \mathrm{Sp}_q(l)\backslash\Sigma_{\mathrm{Sp}_q}$ that coincides with that of §1 for $q = 2$.

Recall that $m(s_A)$ is defined mod 2. By formula (2.15) of §1, $m(s_A) = m(A)$ mod 2. Then by (7.5), there exists a unique element of $\mathrm{Sp}_q\backslash\Sigma_{\mathrm{Sp}_q}$, *denoted* S_A, such that

$$\text{its projection onto } \mathrm{Sp}(l) \text{ is } s_A; \qquad m(S_A) = m(A) \bmod 2q. \tag{8.6}$$

The mapping

$\mathbf{A} \ni A \mapsto S_A \in \mathrm{Sp}_q \backslash \Sigma_{\mathrm{Sp}_q}$

is clearly surjective and, if $q = \infty$, bijective. If $q \neq \infty$, the condition

$S_A = S_{A'}$

is equivalent to the following:

$$A(x, x') = A'(x, x') \qquad \forall x, x' \in X, \qquad m(A) = m(A') \bmod 2q.$$

9. Lagrangian Manifolds

An *isotropic manifold* in $Z(l)$ is a manifold on which

$$d\langle p, dx\rangle = 0; \text{ i.e., } \sum_j dp_j \wedge dx^j = 0; \text{ i.e., } d[z, dz] = 0. \tag{9.1}$$

Its tangent plane is isotropic, and hence it has dimension $\leqslant l$.

A *lagrangian manifold* V is an isotropic manifold such that

$\dim \mathrm{V} = l.$

Let $\check{V}$ be the universal covering space of V (see Steenrod [17], 14.7). Evidently (9.1) means that there exist functions

$\varphi : \check{V} \to \mathbf{R}, \qquad \psi : \check{V} \to \mathbf{R},$

defined up to additive constants, such that

$$d\varphi = \langle p, dx\rangle, \qquad \psi(x, p) = \varphi(x, p) - \tfrac{1}{2}\langle p, x\rangle, \qquad d\psi = \tfrac{1}{2}[z, dz]; \tag{9.2}$$

φ is the *phase* of V and ψ its *lagrangian phase.*

The l-plane $\lambda(z)$ tangent to V at $z = (x, p) \in V$ is obviously lagrangian. The apparent contour Σ_V of V is the set of $z \in V$ such that $\lambda(z)$ is not transverse to X^*. On $V \backslash \Sigma_V$ and on its tangent l-planes, x may serve as a local coordinate. By (9.2), the equations of V and its tangent l-planes are then

$$V : p = \varphi_x(x), \qquad \lambda(z) : dp_j = \sum_{k=1}^{l} \varphi_{x^j x^k}\, dx^k. \tag{9.3}$$

Each element of Sp(l),

$s : Z(l) \ni z' \mapsto sz' = z \in Z(l),$

maps every lagrangian manifold V' in $Z(l)$ into a lagrangian manifold V in $Z(l)$, with lagrangian phase

$$\psi(z) = \psi'(z') + \text{const.} \quad \text{for } z = sz'.$$

If $z = (x, p)$ and $z' = (x', p')$, the phases φ and φ' of V and V' are then related by

$$\varphi(x) - \tfrac{1}{2}\langle p, x\rangle = \varphi'(x') - \tfrac{1}{2}\langle p', x'\rangle + \text{const.} \tag{9.4}$$

If $s = s_A \notin \Sigma_{\mathrm{Sp}(l)}$, then $p = A_x$, $p' = -A_{x'}$ by §1,(1.11), and so

$$\begin{aligned} &\varphi(x) = \varphi'(x') + A(x, x') + \text{const,} \\ &\quad p = \varphi_x = A_x, \qquad p' = \varphi'_{x'} = -A_{x'}. \end{aligned} \tag{9.5}$$

LEMMA 9. Let $s_A \in \mathrm{Sp}(l)\backslash\Sigma_{\mathrm{Sp}}$; let $z' = (x', p') \in V'\backslash\Sigma_{V'}$ be such that

$$s_A z' = z = (x, p) \in V\backslash\Sigma_V, \quad \text{where } V = s_A V'.$$

Evidently these relations define a diffeomorphism $x' \mapsto x$ of the local coordinates of V and V'. Its jacobian takes the form

$$\frac{d^l x}{d^l x'} = \frac{\mathrm{Hess}_{x'}[A(x, x') + \varphi'(x')]}{\Delta^2(A)}. \tag{9.6}$$

(In the calculation of $\mathrm{Hess}_{x'}$, x and x' are viewed as independent.)

Proof. By (9.5), the diffeomorphism $x' \mapsto x$ satisfies

$$\varphi'_{x'}(x') + A_{x'}(x, x') = 0.$$

Then (9.6) follows, because, from §1,1,

$$\Delta^2(A) = \det_{jk}(-A_{x^j x'^k}).$$

10. q-Orientation ($q = 1, 2, 3, \ldots, \infty$)

The notions defined in this section enable us to supply a complement to formula (9.6). This is theorem 10, which will be crucial in the sequel.

By a q-orientation of $\lambda \in \Lambda(l)$ we mean a choice of a $\lambda_{2q} \in \Lambda_{2q}(l)$ with natural projection λ.

A q-oriented lagrangian manifold, denoted V_q, is a lagrangian manifold V together with a continuous mapping

$$V \ni z \mapsto \lambda_{2q}(z) \in \Lambda_{2q}(l)$$

that, when composed with the natural mapping

$\Lambda_{2q}(l) \to \Lambda(l)$,

gives the mapping

$V \ni z \mapsto \lambda(z) \in \Lambda(l)$, where $\lambda(z)$ is the tangent l-plane to V at z.

Each element of $Sp_q(l)$ maps a q-oriented lagrangian manifold V_q into another.

A q-orientation of V is characterized by the values taken by the locally constant function

$V\backslash\Sigma_V \ni z \mapsto m(X^*_{2q}, \lambda_{2q}) \in \mathbf{Z}_{2q}$.

By (5.12), a change in this q-orientation is equivalent to the addition of a constant $\in \mathbf{Z}_{2q}$ to the function m.

The statement of theorem 10 will be simplified by the following definition.

Definition 10. The *argument of* $d^l x$ at a point of V_q, where the tangent l-plane is $\lambda_{2q} \in \Lambda_{2q}(l)$, is defined by the formula

$$\arg d^l x = \pi m(X^*_{2q}, \lambda_{2q}) \mod 2q\pi. \tag{10.1}$$

For example, by (5.13),

$$\arg d^l x = 0 \text{ on } X_{2q}. \tag{10.2}$$

Let

$S_A \in Sp_q(l)\backslash\Sigma_{Sp_q}$ (definition 8),

and let V'_q be a q-oriented lagrangian manifold in $Z(l)$; denote

$V_q = S_A V'_q$.

Let λ'_{2q} and λ_{2q} be the tangent planes to V'_q and V_q at z' and $z = s_A z'$. In formula (8.5), taking $S = S_A$ and $s = s_A$, we have, by (8.6) and §1, definition 1.2,

$$m(S) = m(A) = \frac{2}{\pi} \arg(\Delta(A)).$$

By theorem 7.3 and equation (9.3) for λ' we obtain that $\mathrm{Inert}(s, \lambda, \lambda')$ is the inertia of the symmetric matrix

$$\left(\frac{\partial^2}{\partial x'^j\,\partial x'^k}[A(x, x') + \varphi'(x')]\right).$$

By §1,definition 2.3 of arg Hess, this is

$$\frac{1}{\pi}\arg \operatorname{Hess}_{x'}[A(x, x') + \varphi'(x')].$$

Thus, by (8.4), formula (8.5) may be written

$$m(X^*_{2q}, \lambda_{2q}) - m(X^*_{2q}, \lambda'_{2q}) = \frac{1}{\pi}\arg \operatorname{Hess}_{x'}[A + \varphi'] - \frac{1}{\pi}\arg \Delta^2(A) \mod 2q,$$

whence, by definition 10, we have the following theorem.

THEOREM 10. *If V_q and V'_q are two q-oriented lagrangian manifolds such that*

$$V_q = S_A V'_q, \text{ where } S_A \in \mathrm{Sp}_q(l)\backslash\Sigma_{\mathrm{Sp}_q},$$

then not only are the two terms of (9.6) *equal, but so are their arguments* mod $2q\pi$.

It is the special case $q = 2$ of this theorem that will be used (see §3, corollary 3).

§3. Symplectic Spaces

0. Introduction

Symplectic geometry makes it possible to state the preceding results in the following form, which will be used in chapter II.

$Z(l)$ has been provided with a symplectic structure, and moreover with a particular frame consisting of a pair (X, X^*) of transverse lagrangian l-planes. It is important to state conclusions that are independent of this choice of a particular frame.

1. Symplectic Space Z

A *symplectic space* Z consists of $\mathbf{R}^{2l}$ together with a symplectic form, that is, a form

$$[\cdot,\cdot]: \mathbf{R}^{2l} \times \mathbf{R}^{2l} \ni (z, z') \mapsto [z, z'] \in \mathbf{R}$$

that is bilinear, alternating, real valued, and nondegenerate. We then have

$$[z, z'] = -[z', z];$$

$[z, z'] = 0$ for a given z and all $z' \in \mathbf{R}^{2l}$ only if $z = 0$.

$Z(l)$ is provided with the symplectic structure defined in §1,1; the isomorphism of the symplectic structures of Z and $Z(l)$ is obvious and has the following consequences.

The subspaces of Z on which $[\cdot,\cdot]$ vanishes identically are called *isotropic*; their dimension is $\leqslant l$.

The isotropic subspaces of dimension l are called *lagrangian subspaces.* The collection of lagrangian subspaces $\Lambda(Z)$ is called the *lagrangian grassmannian* of Z; $\Lambda(Z)$ is homeomorphic to $\Lambda(l)$ and therefore has a unique covering space $\Lambda_q(Z)$ of order $q \in \{1, 2, \ldots, \infty\}$.

The projection of $\lambda_q \in \Lambda_q(Z)$ onto $\Lambda(Z)$ is denoted λ. The following definition (see §2,4) makes sense on Z.

Definition 1.1. Let $\lambda, \lambda', \lambda'' \in \Lambda(Z)$ be pairwise transverse.
$\operatorname{Inert}(\lambda, \lambda', \lambda'')$ is the *index of inertia* of the nondegenerate quadratic form

$$z \mapsto [z, z'] = [z', z''] = [z'', z], \tag{1.1}$$

where

$$z \in \lambda, \qquad z' \in \lambda', \qquad z'' \in \lambda'', \qquad z + z' + z'' = 0.$$

Clearly its values belong to $\{0, 1, \ldots, l\}$, and

$$\operatorname{Inert}(\lambda, \lambda', \lambda'') = \operatorname{Inert}(\lambda', \lambda'', \lambda) = l - \operatorname{Inert}(\lambda, \lambda'', \lambda'). \tag{1.2}$$

Let $\Sigma_{\Lambda_q^2}$ denote the set in $\Lambda_q^2(Z)$ consisting of pairs of nontransverse elements of $\Lambda_q(Z)$. By theorem 8 and (8.1), we have the following theorem.

THEOREM 1. *For each $q \in \{1, 2, \ldots, \infty\}$, there is a unique function*

$$m: \Lambda_q^2(Z) \backslash \Sigma_{\Lambda_q^2} \to \mathbf{Z}_q,$$

called the Maslov index, that is locally constant on its domain and that satisfies

$$\operatorname{Inert}(\lambda, \lambda', \lambda'') = m(\lambda_q, \lambda'_q) - m(\lambda_q, \lambda''_q) + m(\lambda'_q, \lambda''_q) \bmod q. \tag{1.3}$$

It has the following properties:

$$m(\lambda_q, \lambda'_q) + m(\lambda'_q, \lambda_q) = l \bmod q, \tag{1.4}$$

$$m(\lambda_q, \lambda'_q) = m(\lambda_\infty, \lambda'_\infty) \bmod q. \tag{1.5}$$

Theorem 6 of §2 applies to the jump of this Maslov index across $\Sigma_{\Lambda_q^2}$.

Definition 1.2. A *lagrangian manifold* V in Z is a manifold of dimension l on which

$$d[z, dz] = 0; \tag{1.6}$$

its tangent l-plane is lagrangian. In other words, on the universal covering space $\check{V}$ there exists a function

$$\psi : \check{V} \to \mathbf{R} \text{ such that } d\psi = \tfrac{1}{2}[z, dz], \tag{1.7}$$

called the *lagrangian phase*, which is defined up to an additive constant.

2. The Frames of Z

A *frame*[3] of Z is an isomorphism

$$R : Z \to Z(l) \tag{2.1}$$

respecting $[\cdot, \cdot]$. If R and R' are two such frames, then RR'^{-1} is called the change of frame:

$$RR'^{-1} \in \mathrm{Sp}(l). \tag{2.2}$$

Clearly, if $\lambda, \lambda', \lambda'' \in \Lambda(Z)$, then $R\lambda, R\lambda', R\lambda'' \in \Lambda(l)$ and

$$\mathrm{Inert}(\lambda, \lambda', \lambda'') = \mathrm{Inert}(R\lambda, R\lambda', R\lambda''). \tag{2.3}$$

If V is a lagrangian manifold in Z, then RV is a lagrangian manifold in $Z(l)$ with phase φ_R given by

$$\varphi_R(z) = \psi(z) + \tfrac{1}{2}\langle p, x\rangle, \text{ where } Rz = (x, p) \in X \oplus X^* = Z(l). \tag{2.4}$$

The *apparent contour of V relative to the frame* R is the set Σ_R of points in V at which the tangent l-plane is not transverse to $R^{-1}X^*$.

On $V \backslash \Sigma_R$, x may serve as a local coordinate.

By lemma 9 of §2 we have the following theorem.

[3] The use of the letter R to denote a frame comes from the initial of the French word *repère*. [Translator's note]

THEOREM 2. *Let x and x′ be the local coordinates defined on* $V\backslash(\Sigma_R \cup \Sigma_{R'})$ *by two frames R and R′ such that* $RR'^{-1} \in \mathrm{Sp}(l)\backslash\Sigma_{\mathrm{Sp}}$. *Then there exists* A *(§1,2) such that* $s_A = RR'^{-1}$; *let* $\varphi'(x') = \varphi_{R'}(z)$ *be the phase of R′V. Then the local diffeomorphism* $X \ni x' \mapsto x \in X$ *has the jacobian*

$$\frac{d^l x}{d^l x'} = \frac{\mathrm{Hess}_{x'}[A(x, x') + \varphi'(x')]}{\Delta^2(A)} \tag{2.5}$$

(In the calculation of $\mathrm{Hess}_{x'}$, x and x' are viewed as independent.)

Remark 2. The frames we have just defined do not allow us to fix the q-orientations if $q > 1$.

3. The q-Frames of Z

The q-frames of Z do allow this: each q-frame ($q \in \{1, 2, \ldots, \infty\}$) consists of

i. an isomorphism $j_R: Z \to Z(l)$ respecting $[\cdot, \cdot]$;
ii. a homeomorphism $h_R: \Lambda_{2q}(Z) \to \Lambda_{2q}(l)$ whose natural image is the homeomorphism $\Lambda(Z) \to \Lambda(l)$ induced by j_R.

If R and R' are two such frames, then the change of frame RR'^{-1} consists of

i. $s = j_R j_{R'}^{-1} \in \mathrm{Sp}(l)$,
ii. $H = h_R h_{R'}^{-1}$, a homeomorphism of $\Lambda_{2q}(l)$ whose natural image is the homeomorphism of $\Lambda_2(l)$ induced by s (cf. §2, example 3.1).

To define H knowing s it suffices to give $HX_{2q}^* \in \Lambda_{2q}(l)$ with the projection $sX_2^* \in \Lambda_2(l)$.

There are q elements of $\mathrm{Sp}_q(l)$ with image $s \in \mathrm{Sp}(l)$. By §2,(3.12), they map X_{2q}^* into the q elements of $\Lambda_{2q}(l)$ with image $sX_2^* \in \Lambda_2(l)$.

The *unique element* $S \in \mathrm{Sp}_q(l)$ with image $s \in \mathrm{Sp}(l)$ and *such that* $SX_{2q}^* = HX_{2q}^*$ thus induces the homeomorphism H of $\Lambda_{2q}(l)$. It *characterizes* RR'^{-1}, which we denote S:

$$RR'^{-1} \in \mathrm{Sp}_q(l). \tag{3.1}$$

R denotes either j_R or h_R; we write

$$R: Z \ni z \mapsto Rz = (x, p) \in X \oplus X^* = Z(l),$$

$$R: \Lambda_{2q}(Z) \ni \lambda_{2q} \mapsto R\lambda_{2q} \in \Lambda_{2q}(l).$$

Evidently

$$m(\lambda_{2q}, \lambda'_{2q}) = m(R\lambda_{2q}, R\lambda'_{2q}) \bmod 2q \qquad \forall \lambda_{2q}, \lambda'_{2q} \in \Lambda_{2q}(Z). \tag{3.2}$$

R maps a q-oriented lagrangian manifold V in Z into another in $Z(l)$. If $\lambda_{2q} \in \Lambda_{2q}(Z)$ is the tangent plane to V at z, we define

$$m_R(z) = m(R^{-1}X^*_{2q}, \lambda_{2q}) \in \mathbf{Z}_{2q}. \tag{3.3}$$

If x is the local coordinate of $V \backslash \Sigma_R$ (q-oriented) defined by the q-frame R, we define

$$\arg d^l x = \pi m_R(z) \bmod 2q\pi. \tag{3.4}$$

Example 3. On $R^{-1}X_{2q}$, $\arg d^l x = 0 \bmod 2q\pi$ by §2,(10.2).

Theorem 10 of §2 evidently allows us to supplement theorem 2 as follows.

THEOREM 3.1. *In the statement of theorem* 2, *suppose that* $V = V_q$ *is* q*-oriented and that* R *and* R' *are* q*-frames. Then*

$$RR'^{-1} = S_A \in \mathrm{Sp}_q(l) \tag{3.5}$$

and the arguments of the two sides of (2.5) *are equal* mod $2q\pi$.

Let us recall that §1 gave the definitions 1.2 and 2.3 of $\arg \Delta(A)$ and arg Hess. By §2,(8.6),

$$m(A) = m(S_A) \bmod 2q.$$

The following special case of this theorem will be used in part (iii) of the proof of theorems 4.1–4.3 of II,§1.

COROLLARY 3. *If* $q = 2$, *the half-measure* $[d^l x]^{1/2}$ *on the lagrangian 2-oriented manifold* V *is defined by*

$$\arg[d^l x]^{1/2} = \frac{\pi}{2} m_R(z) \bmod 2\pi. \tag{3.6}$$

We have

$$[d^l x]^{1/2} = \frac{1}{\Delta(A)} \{\mathrm{Hess}_{x'}[A(x, x') + \varphi'(x')]\}^{1/2} [d^l x']^{1/2}. \tag{3.7}$$

In a neighborhood of a point of Σ_R, where $\dim \lambda \cap R^{-1}X^* = 1$, $m_R(z)$ has the following expression, which follows from §2, theorem 6.

THEOREM 3.2. *Let $\lambda(z)$ be the tangent l-plane to the lagrangian manifold V at z; $z \in \Sigma_R$ means*

$$\dim(\lambda \cap R^{-1}X^*) > 0.$$

We stay in a neighborhood of a point of Σ_R at which

$$\dim(\lambda \cap R^{-1}X^*) = 1;$$

then for $z \in \Sigma_R$, the projection of $R\lambda(z)$ onto X parallel to X^* is a hyperplane:

$$\sum_{j=1}^{l} c_j dx^j = 0. \tag{3.8}$$

1°) *There exists a regular measure ϖ on V such that for $z \in \Sigma_R$, for all j and all k*

$$dx^1 \wedge \cdots \wedge dx^{j-1} \wedge dp_K \wedge dx^{j+1} \wedge \cdots \wedge dx^l = c_j c_K \varpi. \tag{3.9}$$

2°) *If $V = V_q$ is q-oriented, there exists a constant $c \in \mathbf{Z}_{2q}$ such that*

$$\begin{aligned} m_R(z) &= c \mod 2q \text{ for } d^lx/\varpi < 0, \\ m_R(z) &= 1 + c \mod 2q \text{ for } d^lx/\varpi > 0, \end{aligned} \tag{3.10}$$

provided that d^lx/ϖ vanishes to the first order where it vanishes (namely on Σ_R).

Remark. A regular measure ϖ on V is a differential form of maximal degree l on V with everywhere nonvanishing coefficient.

Proof of 1°). The c_j entering into (3.8) are not all zero. Say $c_1 \neq 0$; then, on V at z,

$$dx^2 \wedge \cdots \wedge dx^l \neq 0; \exists k \text{ such that } dp_k \wedge dx^2 \wedge \cdots \wedge dx^l \neq 0. \tag{3.11}$$

Since $\Sigma_j dp_j \wedge dx^j = 0$ on V, we have, for all k,

$$\begin{aligned} &dp_1 \wedge dx^1 \wedge dx^2 \wedge \cdots \widehat{dx_k} \cdots \wedge dx^l \\ &+ dp_k \wedge dx^k \wedge dx^2 \wedge \cdots \widehat{dx^k} \cdots \wedge dx^l = 0, \end{aligned} \tag{3.12}$$

where the cap suppresses the term it covers. By (3.8)

$$dx^1 = -\frac{c_k}{c_1} dx^k \mod (dx^2, \ldots, \widehat{dx^k}, \ldots, dx^l).$$

Denoting

$$\varpi = \frac{1}{c_1^2} dp_1 \wedge dx^2 \wedge \cdots \wedge dx^l,$$

we may write (3.12) as

$$c_1 c_k \varpi = dp_k \wedge dx^2 \wedge \cdots \wedge dx^l, \tag{3.13}$$

which is (3.9) for $j = 1$.

By (3.11), (3.13) implies

$\varpi \neq 0$ on V at z.

Using the expression

$$dx^j = -\frac{c_1}{c_j} dx^1 \mod (dx^2, \ldots, \widehat{dx^j}, \ldots, dx^l)$$

in the right-hand side of (3.13), we obtain (3.9) for $j > 1$ and $c_j \neq 0$. If $j > 1$ and $c_j = 0$, then the two sides of (3.9) are obviously zero.

Proof of 2°). Suppose $c_1 \neq 0$ at a point of Σ_R. By (3.9), for $z \in V$ near this point, we can choose the coordinates $(p_1, x^2, \ldots, x^l)$ of Rz in $Z(l)$ as coordinates on V. Then on V, x^1 and p are functions of $(p_1, x^2, \ldots, x^l)$. The equation of Σ_R is

$$\Sigma_R : \frac{\partial x^1(p_1, x^2, \ldots, x^l)}{\partial p_1} = 0.$$

Let $\zeta'(z)$ be the vector in $\lambda(z)$ having the coordinates

$$(dp_1 = 1, dx^2 = \cdots = dx^l = 0).$$

Then in $Z(l)$, $R\zeta'(z)$ has the coordinates

$$\left(dx^1 = \frac{\partial x^1(p_1, x^2, \ldots, x^l)}{\partial p_1}, dx^2 = 0, \ldots, dx^l = 0, \right.$$
$$\left. dp_j = \frac{\partial p_j(p_1, x^2, \ldots, x^l)}{\partial p_1}\right).$$

Let $\zeta(z)$ be the vector on $R^{-1}X^*$ such that

$$R\zeta(z) = \left(dx = 0, dp = \frac{\partial p(p_1, x^2, \ldots, x^l)}{\partial p_1}\right);$$

$\zeta(z)$ is a function of $z \in V$. Then

$$\zeta(z) = \zeta'(z) \textit{ for } z \in \Sigma_R;$$

$$[\zeta(z), \zeta'(z)] = [R\zeta(z), R\zeta'(z)] = \frac{\partial x^1}{\partial p_1} = \frac{1}{c_1^2}\frac{d^l x}{\varpi} \tag{3.14}$$

because, in view of (3.9), the obvious relation

$$dx^1 \wedge \cdots \wedge dx^l = \frac{\partial x^1(p_1, x_2, \ldots, x^l)}{\partial p_1} dp_1 \wedge dx^2 \wedge \cdots \wedge dx^l \text{ on } V$$

means

$$\frac{\partial x^1}{\partial p_1} = \frac{d^l x}{c_1^2 \varpi}.$$

Now by §2, theorem 6, for $z \in V \backslash \Sigma_R$,

$$m_R(z) = m(R^{-1}X_\infty^*, \lambda_\infty(z)) = \begin{cases} c & \text{for } [\zeta(z), \zeta'(z)] < 0 \\ 1 + c & \text{for } [\zeta(z), \zeta'(z)] > 0, \end{cases}$$

c a constant $\in \mathbf{Z}$. Considering (3.14), (3.10) follows.

4. q-Symplectic Geometries

If R is a q-frame, clearly the group

$$R^{-1}\,\mathrm{Sp}_q(l)R = \mathrm{Sp}_q(Z)$$

acts on Z and its lagrangian manifolds while preserving their q-orientations; this group is independent of R by (3.1).

F. Klein defined a geometry by specifying a manifold and a Lie group acting on that manifold.

Each of the groups $\mathrm{Sp}_q(Z)$, where $q \in \{1, 2, \ldots, \infty\}$, thus defines a geometry on Z which it is convenient to call the *q-symplectic geometry.*

Conclusion

This chapter in §1 *defined a unitary representation of the group* $\mathrm{Sp}_2(l)$ *in the Hilbert space* $\mathscr{H}(X)$, *where* $\dim X = l$. *As a result of* §2, *in* §3 *the study of this group was subsituted by that of the isomorphic group* $\mathrm{Sp}_2(Z)$ *that defines the* 2-*symplectic geometry of* Z, *where* $\dim Z = 2l$. *Earlier* §2 *and §3 detailed the properties of the index of inertia and the Maslov index, already introduced in §1. The interest of these various properties is to make possible the definition and study of a new structure, lagrangian analysis, which is the subject of the next chapter.*

II Lagrangian Functions; Lagrangian Differential Operators

Introduction

Summary. In Chapter I we studied only differential operators with polynomial coefficients and functions defined on all of $\mathbf{R}^l = X$.

The aim of chapter II is to extend those results. In §2, we consider a symplectic space Z of dimension $2l$ provided with a 2-symplectic geometry. We define

- *lagrangian functions* and *lagrangian distributions*, on which $\mathrm{Sp}_2(Z)$ acts locally;
- *lagrangian operators*, more general than differential operators, which are transformed by $\mathrm{Sp}_2(Z)$ like differential operators with polynomial coefficients (see I,§1,theorem 3.1).

Each lagrangian function U is defined on a lagrangian manifold V in Z; in each 2-frame R, U has an expression U_R, a function with formal values defined on $V \backslash \Sigma_R$. In each frame R, a lagrangian operator has an expression

$$a_R = \overset{+}{a}_R\left(\nu, x, \frac{1}{\nu}\frac{\partial}{\partial x}\right)$$

that is a formal differential operator of order $\leqslant \infty$, acting *locally* on the U_R.

A change of frame

$S = RR'^{-1} \in \mathrm{Sp}_2(l)$ (see I,§3,3)

is a local operator that transforms

$U_{R'}$ into $U_R = SU_{R'}$, $\qquad a_{R'}$ into $a_R = Sa_{R'}S^{-1}$,

and hence transforms

$a_{R'}U_{R'}$ into $a_R U_R = S(a_{R'}U_{R'})$.

It follows that *lagrangian operators act on lagrangian functions and on lagrangian distributions.*

All the expressions a_R of a single lagrangian operator a are readily obtained from a single formal function a^0 defined on Z.

In geometry, a change of frame leaves invariant (or changes linearly) the value of a scalar (or vector) function at a point; but *an essential characteristic of lagrangian analysis is the following*: at each point z of V, *the group* $\mathrm{Sp}_2(l)$ *of changes of frame acts on the germs of expressions* U_R *of a lagrangian function* U and not on the values at z of these expressions. $U_R = SU_{R'}$ may have singularities on Σ_R, but has none on $\Sigma_{R'} \backslash \Sigma_{R'} \cap \Sigma_R$, whereas the singularities of $U_{R'}$ are on $\Sigma_{R'}$. Thus *the singularities of the expressions* U_R *of* U are not singularities of U; they *may be described as apparent singularities.* Their nature can be made explicit by making use of the *Maslov index.*

Historical account. V. P. Maslov [10] elucidated this essential characteristic of lagrangian analysis, even though he only studied the projections of lagrangian functions onto X without defining either lagrangian operators or lagrangian functions explicitly. He only used a subgroup of $\mathrm{Sp}_2(l)$, depending on a choice of coordinates of X: the subgroup generated by Fourier transforms in a single coordinate.

§1. Formal Analysis

0. Summary

In section 1 we define and study *asymptotic equivalence classes* of functions.

In section 2 we define formal functions; formal functions defined on X with compact support are asymptotic equivalence classes. Then we may integrate them (section 3) and transform them by $\mathrm{Sp}_2(l)$ and by differential operators, whose definition can be generalized (sections 4 and 6). We deduce (sections 4 and 6) that $\mathrm{Sp}_2(l)$ and formal differential operators act locally on formal functions defined on lagrangian manifolds V or their covering spaces $\check{V}$; these functions may be compactly supported or not. In section 5 we study their scalar product.

Formal functions on $\check{V}$, which are no longer asymptotic equivalence classes, enable us to define lagrangian functions in §2.

1. The Algebra $\mathscr{C}(X)$ of Asymptotic Equivalence Classes

The algebra $\mathscr{B}(X)$ Let $\mathbf{I}$ be the purely imaginary half-line

$\mathbf{I} = i[1, \infty[\in \mathbf{C}.$

$\mathscr{B}(X)$ denotes the algebra of mappings

$f : \mathbf{I} \times X \ni (\nu, x) \mapsto f(\nu, x) \in \mathbf{C}$

all of whose x-derivatives are continuous in (v, x) and satisfy

$$\operatorname*{Sup}_{(v,x)\in \mathbf{I}\times X}\left|x^q\left(\frac{1}{v}\frac{\partial}{\partial x}\right)^r f(v,x)\right| < \infty \qquad \forall q, r \in \mathbf{N}^l; \tag{1.1}$$

$Sp_2(l)$ and the differential operators with polynomial coefficients in $(1/v, x)$,

$$a = a^+\left(v, x, \frac{1}{v}\frac{\partial}{\partial x}\right) = a^-\left(v, \frac{1}{v}\frac{\partial}{\partial x}, x\right),$$

act on $\mathscr{B}(X)$.

If $l = \dim X = 0$, $\mathscr{B}(X)$ is denoted $\mathscr{B}$: thus $\mathscr{B}$ is the algebra of bounded continuous mappings $\mathbf{I} \to \mathbf{C}$.

The algebra $\mathscr{C}(X)$ Let $N \in \mathbf{R}_+$. Let $\mathscr{B}_N(X)$ be the set of $f \in \mathscr{B}(X)$ such that the mapping

$$(v, x) \mapsto v^N f(v, x)$$

still belongs to $\mathscr{B}(X)$. Clearly $\mathscr{B}_N(X)$ is an ideal in $\mathscr{B}(X)$ that shrinks as N grows; hence $\mathscr{B}_\infty(X) = \bigcap_{N\in\mathbf{R}_+}\mathscr{B}_N(X)$ is an ideal in $\mathscr{B}(X)$. We define the algebra $\mathscr{C}(X)$ as follows:

$$\mathscr{C}(X) = \mathscr{B}(X)/\mathscr{B}_\infty(X). \tag{1.2}$$

$\mathscr{C}(X)$ is an algebra over $\mathbf{C}$; its elements are called *asymptotic equivalence classes*. The condition that the elements f and f' of $\mathscr{B}(X)$ belong to the same class is

$$f - f' \in \mathscr{B}_N(X) \qquad \forall N \in \mathbf{R}_+.$$

Clearly,

$$\mathscr{C}_N(X) = \mathscr{B}_N(X)/\mathscr{B}_\infty(X) \tag{1.3}$$

is an ideal in $\mathscr{C}(X)$ that shrinks as N grows:

$$\bigcap_{N\in\mathbf{R}_+}\mathscr{C}_N(X) = \{0\}. \tag{1.4}$$

Let f and $f' \in \mathscr{B}(X)$ with classes $\tilde{f}$ and $\tilde{f}' \in \mathscr{C}(X)$. To express the equivalent relations

$$f - f' \in \mathscr{B}_N(X), \qquad \tilde{f} - \tilde{f}' \in \mathscr{C}_N(X), \text{ where } N \leqslant \infty, \tag{1.5}$$

we shall write one of the following:

$$f = f' \mod \frac{1}{\nu^N}, \qquad \tilde{f} = \tilde{f}' \mod \frac{1}{\nu^N}, \qquad f' \in \tilde{f} \mod \frac{1}{\nu^N}. \tag{1.6}$$

If $l = \dim X = 0$, $\mathscr{C}(X)$ is denoted $\mathscr{C}$. Clearly *$\mathscr{C}(X)$ is a subalgebra of the algebra of functions $X \to \mathscr{C}$.*

Thus an element $\tilde{f}$ of $\mathscr{C}(X)$ has a restriction to any subset of X and a *support*

$$\operatorname{Supp}(\tilde{f}) \subset X. \tag{1.7}$$

Remark 1. Let F be an entire holomorphic function

$F: \mathbf{C}^j \to \mathbf{C}$ such that $F(0) = 0$;

for every $f_j \in \tilde{f}_j$, the $F(f_1, \ldots, f_J)$ are in the same class, denoted $F(\tilde{f}_1, \ldots, \tilde{f}_J)$. Thus $\forall \tilde{f}_j \in \mathscr{C}(X)(j = 1, \ldots, J)$, $F(\tilde{f}_1, \ldots, \tilde{f}_J) \in \mathscr{C}(X)$.

Let F be a Hölder function

$F: \mathbf{C}^J \to \mathbf{C}$ such that $F(0) = 0$.

We similarly define

$$F(\tilde{f}_1, \ldots, \tilde{f}_j) \in \mathscr{C} \qquad \forall f_j \in \mathscr{C}, \quad j = 1, \ldots, J.$$

For example, if $\tilde{f} \in \mathscr{C}$, then $|\tilde{f}|$, $\operatorname{Re} \tilde{f}$, and $\operatorname{Im} \tilde{f}$ are defined. $\tilde{f} \in \mathscr{C}$ is real if $\operatorname{Im} \tilde{f} = 0$; then $\tilde{f}_+$ (the positive part) and $\tilde{f}_-$ are defined. $\tilde{f}_- = 0$ is written $0 \leqslant \tilde{f}$ and defines a *partial ordering* on $\operatorname{Re}\mathscr{C}$, the set of real elements of $\mathscr{C}$: let $\tilde{f}$ and $\tilde{g} \in \operatorname{Re}\mathscr{C}$; $\tilde{f} \geqslant 0$ and $\tilde{g} \geqslant 0$ imply $\tilde{f} + \tilde{g} \geqslant 0$ and $\tilde{f} \cdot \tilde{g} \geqslant 0$; $\tilde{f}^2 \geqslant 0$; $\tilde{f} \leqslant 0$ (that is, $-\tilde{f} \geqslant 0$) excludes $\tilde{f} > 0$ (that is, $0 \leqslant \tilde{f} \neq 0$). However, $\tilde{f}$ may satisfy neither $\tilde{f} \leqslant 0$ nor $\tilde{f} > 0$.

Let F be a Hölder (and increasing) function

$F: \mathbf{R} \to \mathbf{R}$ such that $F(0) = 0$;

$\forall \tilde{f} \in \operatorname{Re}\mathscr{C}$ we have $F(\tilde{f}) \in \operatorname{Re}\mathscr{C}$ (the order relation on $\operatorname{Re}\mathscr{C}$ being preserved).

In particular we have the Schwarz inequality

$$\left|\sum_j \tilde{f}_j \tilde{g}_j\right| \leqslant \left[\sum_j |\tilde{f}_j|^2\right]^{1/2} \cdot \left[\sum_j |\tilde{g}_j|^2\right]^{1/2} \qquad \forall \tilde{f}_j, \tilde{g}_j \in \mathscr{C}.$$

The differential operators with polynomial coefficients in $(1/v, x)$ given by

$$a = a^+\left(v, x, \frac{1}{v}\frac{\partial}{\partial x}\right) = a^-\left(v, \frac{1}{v}\frac{\partial}{\partial x}, x\right)$$

are clearly endomorphisms of $\mathscr{B}(X)$, $\mathscr{B}_N(X)$, $\mathscr{C}(X)$, and $\mathscr{C}_N(X)$. They act locally: the restriction of $a\tilde{f}$ to an open set Ω in X only depends on the restriction of $\tilde{f}$ to this open set:

$$\operatorname{Supp}(a\tilde{f}) \subset \operatorname{Supp}(\tilde{f}).$$

As in §1,3, the differential operator

$$a = a^+\left(v, x, \frac{1}{v}\frac{\partial}{\partial x}\right) = a^-\left(v, \frac{1}{v}\frac{\partial}{\partial x}, x\right)$$

is *associated* to the polynomial a^0 in $(1/v, x, p)$ given by

$$\begin{aligned} a^0(v, x, p) &= \left[\exp -\frac{1}{2v}\left\langle\frac{\partial}{\partial x}, \frac{\partial}{\partial p}\right\rangle\right]a^+(v, x, p) \\ &= \left[\exp \frac{1}{2v}\left\langle\frac{\partial}{\partial x}, \frac{\partial}{\partial p}\right\rangle\right]a^-(v, p, x). \end{aligned} \tag{1.8}$$

Integration is defined on $\mathscr{C}(X)$: if $f \in \mathscr{B}(X)$, the function

$$v \mapsto \int_X f(v, x)\, d^l x$$

belongs to $\mathscr{B}$. Its asymptotic equivalence class, which depends only on the class $\tilde{f}$ of f, is denoted

$$\tilde{\int_X} \tilde{f}(v, x)\, d^l x \in \mathscr{C};$$

$\tilde{\int}$ is called the *asymptotic integral.*

The *scalar product* $(\cdot|\cdot)$ is a sesquilinear form on $\mathscr{C}(X) \times \mathscr{C}(X)$ with values in $\mathscr{C}$ defined by

$$(\tilde{f}, \tilde{g}) = \tilde{\int_X} f(v, x)\overline{g(v, x)}\, d^l x, \quad \text{where } f \in \tilde{f} \in \mathscr{C}(X), \quad g \in \tilde{g} \in \mathscr{C}(X); \tag{1.9}$$

$\overline{g(v, x)}$ is the complex conjugate of $g(v, x)$; $\bar{v} = -v$. Thus

$$(\tilde{g}, \tilde{f}) = \overline{(\tilde{f}, \tilde{g})}.$$

The *seminorm* $\| \cdot \|$ is the function $\mathscr{C}(X) \to \mathscr{C}$ with values $\geqslant 0$ defined by

$$\| \tilde{f} \|^2 = (\tilde{f}, \tilde{f});$$

clearly

$$\| \tilde{f} + \tilde{g} \| \leqslant \| \tilde{f} \| + \| \tilde{g} \|, \qquad |(\tilde{f}, \tilde{g})| \leqslant \| \tilde{f} \| \cdot \| \tilde{g} \| \text{ in } \mathscr{C}.$$

LEMMA 1.1. $\mathrm{Sp}_2(l)$ forms a group of automorphisms of $\mathscr{C}(X)$ that are unitary, that is, preserve seminorms and scalar products.

Proof. Let $S_A \in \mathrm{Sp}_2(l) \backslash \Sigma_{\mathrm{Sp}_2}$. Its definition in I,§1 by the integral (1.10) shows that S_A transforms the asymptotic equivalence class $\tilde{f}'$ into the class $S_A \tilde{f}'$ defined by the asymptotic integral

$$(S_A \tilde{f}')(x) = \left(\frac{|v|}{2\pi i} \right)^{l/2} \Delta(A) \tilde{\int}_X e^{vA(x,x')} \tilde{f}'(x') \, d^l x'.$$

This formula makes sense because multiplication by e^{vA}, where vA is purely imaginary, clearly preserves asymptotic equivalence. The S_A act on $\mathscr{C}(X)$, composition being the same as when they act on $\mathscr{S}(X)$. Now they generate the group $\mathrm{Sp}_2(l)$; thus this group acts on $\mathscr{C}(X)$.

Since $\mathrm{Sp}_2(l)$ is unitary on $\mathscr{S}(X)$, it is unitary on $\mathscr{C}(X)$.

Theorems 3.1 and 3.2 of I,§1 and their corollaries give the following.

LEMMA 1.2. The transform SaS^{-1} of the differential operator a by $S \in \mathrm{Sp}_2(l)$ is the differential operator associated to $a^0 \circ s^{-1}$; s denotes the image of S in $\mathrm{Sp}(l)$; s acts on the space $Z(l)$ of vectors (x, p).

LEMMA 1.3. A necessary and sufficient condition for the differential operators a and b to be adjoint is that

$$b^0(v, x, p) = \overline{a^0(v, x, p)}.$$

In particular, a is self-adjoint when $a^0(v, x, p)$ has real values for

$$v \in \mathbf{I}, \qquad (x, p) \in Z(l).$$

S transforms adjoint operators into adjoint operators.

2. Formal Numbers; Formal Functions

Theorem 2.2 connects the following definitions with the previous ones.

A formal number is a formal series

$$u = u(v) = \sum_{j\in J} \sum_{r\in \mathbf{N}} \frac{\alpha_{jr}}{v^r} e^{v\varphi_j}, \tag{2.1}$$

where J is a finite set, $\alpha_{jr} \in \mathbf{C}$, $\varphi_j \in \mathbf{R}$, and $\varphi_j \neq \varphi_k$ if $j \neq k$.

The expression (2.1) for a formal number is *unique*, by definition.

The set of formal numbers is a *commutative algebra* $\mathscr{F}$ for which the addition and multiplication rules are obvious.

We give $\mathscr{F}$ the topology defined by the following neighborhood system $\mathscr{N}(N, \varepsilon)$ of the origin:

$$N \in \mathbf{N}, \quad \varepsilon \in \dot{\mathbf{R}}_+, \quad u \in \mathscr{N}(N, \varepsilon) \text{ means } \sum_{j\in J} |\alpha_{jr}| < \varepsilon \qquad \forall r \leqslant N.$$

A formal function on X is a mapping $u: X \to \mathscr{F}$ that can be put into the form

$$u = u(v) = \sum_{j\in J} \sum_{r\in \mathbf{N}} \frac{\alpha_{jr}}{v^r} e^{v\varphi_j} \tag{2.2}$$

where J is a finite set, $\alpha_{jr}: X \to \mathbf{C}$, $\varphi_j: X \to \mathbf{R}$, and α_{jr} and φ_j are infinitely differentiable.

The mapping of u may be put in the form (2.2) in many ways at a point where two of the φ_j are equal in a neighborhood of that point. The value of $u(v)$ at x is denoted

$$u(v, x) = \sum_{j\in J} \sum_{r\in \mathbf{N}} \frac{\alpha_{jr}(x)}{v^r} e^{v\varphi_j(x)}. \tag{2.3}$$

Moreover, $u(v)$ has a support

$$\operatorname{Supp} u(v) \subset \overline{\bigcup_{j,r} \operatorname{Supp} \alpha_{j,r}}. \tag{2.4}$$

The set $\mathscr{F}(X)$ of formal functions on X is *an algebra* over $\mathscr{F}$, which is *commutative*. The set of formal functions on X with *compact support* is a subalgebra $\mathscr{F}_0(X)$ of $\mathscr{F}(X)$.

The φ_j are called *phases*, the $\alpha_j = \Sigma_{r\in\mathbf{N}}(\alpha_{jr}/v^r)$ *amplitudes*.

The set $\mathscr{F}^0$ [respectively, $\mathscr{F}^0(X)$] of elements of $\mathscr{F}$ [respectively, $\mathscr{F}(X)$] with *vanishing phase* forms a subalgebra of $\mathscr{F}$ [respectively, $\mathscr{F}(X)$].

Let Z denote a symplectic space with a 2-symplectic geometry, V a lagrangian manifold in Z (I,§3,1 and 4), and $\check{V}$ the universal covering of V.

A formal function $\check{U}_R$ on $\check{V}$ consists of

i. a 2-frame R of Z
ii. a mapping

$$\check{U}_R = \check{U}_R(\nu) = \sum_{r \in \mathbf{N}} \frac{\alpha_r}{\nu^r} e^{\nu\varphi_R}: \check{V} \to \mathscr{F}, \tag{2.5}$$

where

φ_R is the phase of RV (I,§3,2),
$\alpha_r: \check{V} \to \mathbf{C}$ is infinitely differentiable.

Clearly,

$$\operatorname{Supp} \check{U}_R = \overline{\bigcup_r \operatorname{Supp} \alpha_r} \subset \check{V}. \tag{2.6}$$

Given R and V, the set of these formal functions $\check{U}_R$ is a vector space over $\mathscr{F}^0$, denoted $\mathscr{F}(\check{V}, R)$; the set of those of its functions with *compact* support is a subspace $\mathscr{F}_0(\check{V}, R)$.

We give $\mathscr{F}(\check{V}, R)$ the *topology* defined by the following neighborhood system $\mathscr{N}(K, p, r, \varepsilon)$ of the origin:

K is a compact set in V; $p, r \in \mathbf{N}$, $\varepsilon \in \dot{\mathbf{R}}_+$;
$\check{U}_R \in \mathscr{N}(K, p, r, \varepsilon)$ means that the derivatives of the α_s ($s \leqslant r$) of orders $\leqslant p$ have modulus $\leqslant \varepsilon$ on K.

Σ_R denotes the *apparent contour* of V for the frame R (I,§3,2). $\check{\Sigma}_R$ denotes the apparent contour of $\check{V}$, that is, the set of points of $\check{V}$ that project onto Σ_R. We shall deduce the properties of $\mathscr{F}(\check{V} \backslash \check{\Sigma}_R, R)$ from those of $\mathscr{F}_0(\check{V} \backslash \check{\Sigma}_R, R)$. The latter will be deduced from the properties of $\mathscr{F}_0(X)$ using the morphism resulting from the composition of the two morphisms that will be defined in theorems 2.1 and 2.2, respectively.

Notation. If $z \in Z$ and $Rz = (x, p)$, then we write $x = R_X z$. The composition of the natural projection $\check{V} \to V$ and the restriction of R_X to V is denoted $\check{R}_X: \check{V} \to X$.

THEOREM 2.1. *There exists a natural morphism*

$$\Pi_R:\mathscr{F}_0(\check{V}\backslash\check{\Sigma}_R, R) \to \mathscr{F}_0(X) \tag{2.7}$$

called the ***projection****. It is defined as follows. Let* $\check{U}_R \in \mathscr{F}_0(\check{V}\backslash\check{\Sigma}_R, R)$; $u = \Pi_R \check{U}_R$ *is given by*

$$u(v, x) = \sum_{\check{z}\in\check{R}_X^{-1}x} \check{U}_R(v, \check{z}). \tag{2.8}$$

Remark 2. Π_R is a monomorphism if any period of φ_R is nonvanishing.

Proof. Formula (2.8) makes sense because $\check{R}_X^{-1}x \cap \operatorname{Supp} \check{U}_R$ is a finite set, since $\operatorname{Supp} \check{U}_R$ is compact in $\check{V}\backslash\check{\Sigma}_R$ and $\check{R}_X$ is a local homeomorphism $\check{V}\backslash\check{\Sigma}_R \to X$.

THEOREM 2.2. *There exists a natural monomorphism of the algebra* $\mathscr{F}_0(X)$ *into the algebra* $\mathscr{C}(X)$ *that allows the convention*

$$\mathscr{F}_0(X) \subset \mathscr{C}(X). \tag{2.9}$$

It is defined as follows. Let

$$u = \sum_{j\in J} \sum_{r\in\mathbf{N}} \frac{\alpha_{jr}}{v^r} e^{v\varphi_j} \in \mathscr{F}_0(X); \tag{2.10}$$

put

$$u_N = \sum_{j\in J} \sum_{r=0}^{N-1} \frac{\alpha_{jr}}{v^r} e^{v\varphi_j}: \mathbf{I} \times X \to \mathbf{C}. \tag{2.11}$$

1°) *There exist functions* $f \in \mathscr{B}(X)$ *such that* $f - u_N = 0 \bmod 1/v^N \ \forall N$.
2°) *All these functions belong to the same asymptotic equivalence class* $\tilde{f} \in \mathscr{C}(X)$; *the mapping* $u \mapsto \tilde{f}$ *defines a natural morphism* $\mathscr{F}_0(X) \mapsto \mathscr{C}(X)$.
3°) *This morphism is a monomorphism.*

Proof of 1°). It suffices to consider the case in which (2.10) reduces to

$$u(v, x) = \sum_{r\in\mathbf{N}} \frac{\alpha_r(x)}{v^r} e^{v\varphi(x)}. \tag{2.12}$$

For all $g \in \mathscr{B}(X)$ vanishing outside a compact set of X we use the norm

$$|g|_r = \operatorname*{Sup}_{v,x,s} \left| \left(\frac{1}{v}\frac{\partial}{\partial x}\right)^s g \right|, \quad \text{where } v \in \mathbf{I}, \quad x \in X, \quad |s| \leqslant r. \tag{2.13}$$

We choose an increasing sequence of numbers

$0 \leqslant \mu_0 \leqslant \mu_1 \leqslant \cdots$

such that

$$2^r |\alpha_r e^{\nu\varphi}|_r \leqslant \mu_r, \qquad \lim_{r \to \infty} \mu_r = \infty, \tag{2.14}$$

and we choose a sequence of continuous functions $\varepsilon_0, \varepsilon_1, \ldots : \mathbf{I} \to \mathbf{R}^+$ such that

$$\begin{aligned} &0 \leqslant \varepsilon_r(\nu) \leqslant 1, \\ &\varepsilon_r(\nu) = 0 \quad \text{for } |\nu| \leqslant \mu_r, \\ &\varepsilon_r(\nu) = 1 \quad \text{for } 2\mu_r \leqslant |\nu|. \end{aligned} \tag{2.15}$$

Now we define

$$f(\nu, x) = \sum_{r=0}^{\infty} \varepsilon_r(\nu) \frac{\alpha_r(x)}{\nu^r} e^{\nu\varphi(x)}. \tag{2.16}$$

This series converges since, from $(2.14)_2$ and $(2.15)_2$, the number of non-zero terms is finite on any bounded interval in the domain of definition of ν. In accordance with (2.11), we define

$$u_N(\nu, x) = \sum_{r=0}^{N-1} \frac{\alpha_r(x)}{\nu^r} e^{\nu\varphi(x)}. \tag{2.17}$$

By $(2.15)_3$, for $2\mu_N \leqslant |\nu|$,

$$|f(\nu) - u_{N+1}(\nu)|_N \leqslant \sum_{r=N+1}^{\infty} \frac{\varepsilon_r(\nu)}{|\nu|^r} |\alpha_r e^{\nu\varphi}|_r.$$

Now $(2.14)_1$, $(2.15)_1$, and $(2.15)_2$ imply

$$\varepsilon_r(\nu) 2^r |\alpha_r e^{\nu\varphi}|_r \leqslant |\nu|.$$

Hence, for $2\mu_N \leqslant |\nu|$,

$$|f(\nu) - u_{N+1}(\nu)|_N \leqslant \sum_{r=N+1}^{\infty} \frac{1}{2^r |\nu|^{r-1}} = \frac{1}{2^N |\nu|^N} \frac{1}{2 - (1/|\nu|)} \leqslant \frac{1}{|\nu|^N}$$

because $1 \leqslant |\nu|$; thus

$$|f - u_N|_N = 0 \bmod \frac{1}{\nu^N}. \tag{2.18}$$

Let us prove a more general statement:

$$|f - u_N|_M = 0 \bmod \frac{1}{\nu^N} \qquad \forall M, N \in \mathbf{N}. \tag{2.19}$$

If $M \leqslant N$, (2.19) follows from (2.18) and the fact that $|\cdot|_M$ increases with M. If $N < M$, (2.19) follows from the relations

$$|f - u_N|_M \leqslant |f - u_M|_M + |u_M - u_N|_M,$$

$$|f - u_M|_M = 0 \bmod \frac{1}{\nu^M}, \text{ and hence mod} \frac{1}{\nu^N},$$

$$|u_M - u_N|_M = 0 \bmod \frac{1}{\nu^N} \text{ because } u_M - u_N = \sum_{r=N}^{M-1} \frac{\alpha_r}{\nu^r} e^{\nu\varphi}.$$

Since the supports of f and of the u_N belong to $\operatorname{Supp} u = \overline{\bigcup_r \operatorname{Supp} \alpha_r}$, which is compact, (2.19) implies

$$|x^q(f - u_N)|_M = 0 \bmod \frac{1}{\nu^N} \qquad \forall M, N, q \in \mathbf{N};$$

hence

$$f \in \mathscr{B}(X), \qquad f - u_N = 0 \bmod \frac{1}{\nu^N}.$$

Proof of 2°). If f and $f' \in \mathscr{B}(X)$ and satisfy

$$f - u_N = f' - u_N = 0 \bmod \frac{1}{\nu^N} \qquad \forall N,$$

then

$$f - f' = 0 \bmod \frac{1}{\nu^N} \qquad \forall N.$$

Proof of 3°). Assume

$$u \mapsto \tilde{f} = 0;$$

then

$$u_N(\nu, x) = 0 \bmod \frac{1}{\nu^N} \qquad \forall N \in \mathbf{N}, \quad \forall x \in X.$$

It has to be proved that

$$u(\nu, x) = 0 \qquad \forall x \in X.$$

It suffices to consider the following case: u is a formal *number*. Assume that in expression (2.1)

$$\alpha_{jr} = 0 \text{ for } r < N \qquad \forall j \in J;$$

then we have

$$u_{N+1}(\nu) = \sum_{j\in J} \frac{\alpha_{jN}}{\nu^N} e^{\nu\varphi_j} = 0 \mod \frac{1}{\nu^{N+1}},$$

whence

$$\lim_{\nu\to i\infty} \sum_{j\in J} \alpha_{jN} e^{\nu\varphi_j} = 0,$$

which implies

$$\alpha_{jN} = 0 \qquad \forall j.$$

Theorem 2.2 follows.

COROLLARY 2. *Asymptotic integration, elements of* $\mathrm{Sp}_2(l)$, *and differential operators with polynomial coefficients are morphisms*

$$\mathscr{F}_0(X) \to \mathscr{C}, \qquad \mathscr{F}_0(X) \to \mathscr{C}(X), \qquad \mathscr{F}_0(X) \to \mathscr{F}_0(X).$$

The end of this section describes the properties of these morphisms. We use the following theorem in §2, 3 to study the norm of lagrangian functions.

THEOREM 2.3. 1°) *Let*

$$u = \sum_{r\in \mathrm{N}} \frac{\alpha_r}{\nu^r} \in \mathscr{F}^0$$

be a formal number with vanishing phase. The condition that it be real, $u \in \mathrm{Re}\, \mathscr{F}^0$, *is*

$$\frac{\alpha_r}{\nu^r} \text{ is real, that is, } i^{-r}\alpha_r \text{ is real } \forall r.$$

The condition $u > 0$ *is*

$0 < \frac{\alpha_s}{v^s}$, *that is,* $i^{-s}\alpha_s > 0$,

where s *is the first* r *such that* $\alpha_r \neq 0$. *Thus every* $u \in \operatorname{Re}\mathscr{F}^0$ *satisfies either* $u > 0$ *or* $u \leqslant 0$. *Hence* $\geqslant$ *defines an order relation on* $\operatorname{Re}\mathscr{F}^0$.

2°) *The condition* $u^{1/2} \in \operatorname{Re}\mathscr{F}^0$ *is equivalent to the following:*

$u \in \mathscr{F}^0$; $\quad u > 0$; $\quad$ *the integer* s *(defined above) is even.*

Notation. $u^{1/2}$ may be chosen in two ways; we take $u^{1/2} \geqslant 0$.

Proof. Let $u = \Sigma_{r\in\mathbf{N}}(\alpha_r/v^r) \in \mathscr{F}^0$; let s be the first r such that $\alpha_r \neq 0$. If $\alpha_r v^{-r}$ is real valued for every r, the proof of theorem 2.2 constructs a real-valued function $f \in u$, so that $u \in \operatorname{Re}\mathscr{F}^0$. It follows that

$$\operatorname{Re} u = \sum_{r\in\mathbf{N}} \operatorname{Re}\frac{\alpha_r}{v^r}, \qquad \operatorname{Im} u = \sum_{r\in\mathbf{N}} \operatorname{Im}\frac{\alpha_r}{v^r}.$$

Thus $u \in \operatorname{Re}\mathscr{F}^0$ if and only if $\alpha_r v^{-r}$ is real for every r.

Let us study $u^{1/2}$. Since

$$u^2 = \frac{\alpha_s^2}{v^{2s}} \qquad \operatorname{mod}\frac{1}{v^{2s+1}},$$

the assumption $u^{1/2} \in \operatorname{Re}\mathscr{F}^0$ implies:

s is even, $\quad \alpha_s/i^s > 0, \quad u \in \operatorname{Re}\mathscr{F}^0$ (see remark 1).

For the converse, it evidently suffices to prove the following:

$u^{1/2} \in \operatorname{Re}\mathscr{F}^0$ when $\alpha_0 > 0$ and $u \in \operatorname{Re}\mathscr{F}^0$.

This is clear because the condition

$$\left(\sum_r \frac{\beta_r}{v^r}\right)^2 = \sum_r \frac{\alpha_r}{v^r}$$

is equivalent to the condition $\beta_0^2 = \alpha_0$ and to certain conditions giving $\beta_0\beta_r$ (for each $r = 1, 2, \ldots.$) as a real linear function of α_r and $\beta_t\beta_{r-t}$ $(0 < t < r)$.

Now $u^{1/2} \in \operatorname{Re}\mathscr{F}^0$ implies $u \geqslant 0$ (see remark 1). Hence, if $u \in \operatorname{Re}\mathscr{F}^0$, then $\alpha_s v^{-s} > 0$ implies $u > 0$. Thus $u \in \operatorname{Re}\mathscr{F}^0$ satisfies either $u > 0$ or $u \leqslant 0$.

3. Integration of Elements of $\mathscr{F}_0(X)$

The essential properties of formal functions will be deduced from the following theorem, which supplements some classical results (the *method of stationary phase*). Its proof requires the use of lemmas 3.4 and 3.5 in place of the lemma of *Marston Morse* [12], by which a function with a nondegenerate critical point is transformed locally into a quadratic form by a change of coordinates. We briefly recall the other lemmas needed for the proof of this theorem.

By corollary 2, asymptotic integration is a morphism

$$\tilde{\int_X} : \mathscr{F}_0(X) \to \mathscr{C}.$$

Clearly,

$$\tilde{\int_X} u(v, x)\, d^l x \in \mathscr{F}$$

when the phases of u are all constant. Theorem 3 shows that

$$v^{l/2} \tilde{\int_X} u(v, x)\, d^l x \in \mathscr{F}$$

in the important case considered in corollary 3.

THEOREM 3. *Let $u \in \mathscr{F}_0(X)$.*

1°) *The value of $\tilde{\int}_X u(v, x) d^l x$ depends only on the restriction of u to an arbitrarily small neighborhood of the set of critical points of the phases of u; this value is 0 if this set is empty.*

2°) *Let*

$$u(v) = \sum_{r \in \mathrm{N}} \frac{\alpha_r}{v^r} e^{v\varphi} \in \mathscr{F}_0(X), \tag{3.1}$$

where φ has a single nondegenerate critical point x_C on Supp u. *Let $\varphi_\mathrm{C} = \varphi(x_\mathrm{C})$ be the critical value of φ; $\arg i^{l/2} = \pi l/4$ and $\sqrt{\mathrm{Hess}_\mathrm{C}\, \varphi}$ be the value of $\sqrt{\mathrm{Hess}\, \varphi}$ at x_C* (I,§1,definition 2.3). *Then*

$$\tilde{\int_X} u(v, x)\, d^l x = \left(\frac{2\pi i}{|v|}\right)^{l/2} (\mathrm{Hess}_\mathrm{C}\, \varphi)^{-1/2} \sum_{r \in \mathrm{N}} \frac{1}{v^r} I_r(\varphi ; \alpha) e^{v\varphi_\mathrm{C}}, \tag{3.2}$$

where

$$I_0(\varphi;\alpha) = \alpha_0(x_C), \tag{3.3$_0$}$$

$$I_r(\varphi,\alpha) = \sum_{s=0}^{r}\sum_{j=0}^{2(r-s)} \frac{1}{(r-s+j)!j!}\left\{\Phi^{*r-s+j}\left(\frac{\partial}{\partial x}\right)[\Theta^j(x)\alpha_s(x)]\right\}_{x=x_C}; \tag{3.3$_r$}$$

Φ, Φ^*, and Θ have the following definitions.

Notation. The second-order Taylor expansion of φ at x_C is written

$$\varphi(x) = \varphi_C + \Phi(x - x_C) + \Theta(x), \tag{3.4}$$

where Θ vanishes to third order at x_C and Φ is a quadratic form $X \to \mathbf{R}$; Φ^* is its "dual" form, that is,

$$\Phi^*(p) \text{ is the critical value of the function } x \mapsto \Phi(x) + \langle p, x\rangle. \tag{3.5}$$

In other words, if M is the value of the matrix $(\varphi_{x^jx^k})$ at x_C, then

$$\Phi(x) = \tfrac{1}{2}\langle Mx, x\rangle, \qquad \Phi^*(p) = -\tfrac{1}{2}\langle p, M^{-1}p\rangle,$$
$$\text{where } M = {}^tM : X \to X^*. \tag{3.6}$$

The following is evident.

COROLLARY 3. *Let $u \in \mathscr{F}_0(X)$. If all the critical points of all the phases of u are nondegenerate on* Supp u, *then*

$$\nu^{l/2}\tilde{\int_X} u(\nu, x)\,d^lx \in \mathscr{F};$$

the phases of $\nu^{l/2}\tilde{\int}_X u\,d^lx$ are the critical values of the phases of u on Supp u.

It clearly suffices to prove theorem 3 when u is replaced by a function f given by

$$f(\nu, x) = \alpha(x)e^{\nu\varphi(x)}, \tag{3.7}$$

where $\alpha \in \mathscr{D}(X)$ (that is, α is infinitely differentiable and has compact support). The theorem then results from the following lemmas, the first of which proves part 1 of the theorem.

LEMMA 3.1. We denote by x_C the critical points of φ belonging to Supp f. If they are all nondegenerate, then

$$\int_X \alpha(x)e^{\nu\varphi(x)}d^lx \mod \frac{1}{\nu^N}$$

is a linear function of the values of the derivatives of α of order $<2N$ at the points x_C.

Proof. This amounts to proving the following:

If α vanishes to order $2N$ at the points x_C, then $\int_X f\,d^l x = 0 \mod \frac{1}{\nu^N}$.

$$(3.8)_N$$

Now $(3.8)_0$ is obvious. Suppose $(3.8)_N$ holds. If α vanishes to order $2(N+1)$ at the points x_C, then there exist $\beta_j \in \mathscr{D}$ such that

$$\alpha = \sum_j \beta_j \varphi_{x^j}, \qquad \beta_j \text{ vanishes to order } 2N+1 \text{ at the points } x_C.$$

Hence by $(3.8)_N$,

$$\int_X \alpha e^{\nu\varphi}\, d^l x = -\int_X \frac{1}{\nu} \sum_j \frac{\partial \beta_j}{\partial x^j} e^{\nu\varphi}\, d^l x = 0 \mod \frac{1}{\nu^{N+1}},$$

which proves $(3.8)_{N+1}$.

The next lemma is proved similarly.

LEMMA 3.2. Let $\alpha \in \mathscr{S}(X)$ (Schwartz space; see I,§1,1). Let Φ be a nondegenerate quadratic form $X \to \mathbf{R}$. Then

$$\int_X \alpha(x) e^{\nu\Phi(x)}\, d^l x \mod \frac{1}{\nu^N}$$

is a linear function of the values of the derivatives of α of order $<2N$ *at* 0.

More explicit is the following.

LEMMA 3.3. Let $\alpha \in \mathscr{S}(X)$. Let Φ be a nondegenerate quadratic form $X \to \mathbf{R}$. Then

$$\int_X \alpha(x) e^{\nu\Phi(x)}\, d^l x$$

$$= \left(\frac{2\pi i}{|\nu|}\right)^{l/2} [\operatorname{Hess} \Phi]^{-1/2} \sum_{r \in \mathbf{N}} \frac{1}{r!\,\nu^r} \left[\Phi^{*r}\left(\frac{\partial}{\partial x}\right)\alpha(x)\right]_{x=0} \mod \frac{1}{\nu^\infty}.$$

Proof. We know (see I,§1, proof of lemma 2.2) that

$$\int_{-\infty}^{\infty} e^{-\mu x^2/2}\, dx = \sqrt{2\pi}\, \mu^{-1/2} \text{ for } |\arg \mu| < \pi/2.$$

Hence, by differentiating with respect to μ,

$$\int_{-\infty}^{\infty} x^{2r} e^{-\mu x^2/2} dx = \sqrt{2\pi}\,\frac{\mu^{-1/2}}{(2\mu)^r}\frac{(2r)!}{r!}$$
$$= \sqrt{2\pi}\,\mu^{-1/2}\left\{\left[\exp\frac{1}{2\mu}\frac{d^2}{dx^2}\right]x^{2r}\right\}_{x=0}.$$

Moreover,

$$\int_{-\infty}^{\infty} x^{2r+1} e^{-\mu x^2/2}\, dx = 0 \qquad \forall r \in \mathbf{N}.$$

Thus for any $l \times l$ diagonal complex matrix M_{c} with nonzero eigenvalues whose arguments $\in \left]-\pi/2, \pi/2\right[$, and for any polynomial $P: X \to \mathbf{C}$,

$$\int_X P(x) e^{-\Phi_{\mathrm{c}}(x)} d^l x = (2\pi)^{l/2}[\operatorname{Hess}\Phi_{\mathrm{c}}]^{-1/2}[e^{\Psi_{\mathrm{c}}(\partial/\partial x)}P(x)]_{x=0}, \tag{3.9}$$

where

$$\Phi_{\mathrm{c}}(x) = \tfrac{1}{2}\langle M_{\mathrm{c}}x, x\rangle, \qquad \Psi_{\mathrm{c}}(p) = \tfrac{1}{2}\langle p, M_{\mathrm{c}}^{-1}p\rangle;$$

$\arg\operatorname{Hess}\Phi_{\mathrm{c}}$ is the sum of the arguments of the eigenvalues of Φ_{c}. More explicitly, we write

$$\Phi_{\mathrm{c}} = \Phi_+ - \nu\Phi,$$

where $\nu \in \boldsymbol{i}[1, \infty[$ and Φ_+ and Φ are two quadratic forms $X \to \mathbf{R}$, independent of ν; Φ_+ is positive definite; we assume Φ is nondegenerate.

The assumption that the matrix M_{c} is diagonal is superfluous because a change of coordinates reduces two such forms to

$$\Phi_+(x) = \sum_j x_j^2, \qquad \Phi(x) = \sum_j c_j x_j^2, \quad \text{where } c_j \neq 0.$$

Let $\varepsilon \in \mathscr{D}(X)$ be such that $\varepsilon(x) = 1$ for x near 0. Lemma 3.2 and (3.9) give

$$\left(\frac{-\nu}{2\pi}\right)^{l/2}\int_X \varepsilon(x)P(x)e^{\nu\Phi(x)-\Phi_+(x)}d^l x$$
$$= \left[\operatorname{Hess}\left(\Phi - \frac{1}{\nu}\Phi_+\right)\right]^{-1/2}[e^{\Psi_{\mathrm{c}}(\partial/\partial x)}P(x)]_{x=0} \mod \frac{1}{\nu^\infty}; \tag{3.10}$$

$\arg\operatorname{Hess}(\Phi - (1/\nu)\Phi_+)$ is the sum of the arguments of the eigenvalues of $\Phi - (1/\nu)\Phi_+$, which belong to $]0, \pi[$. The argument of $-\nu = |\nu|/i$ is $-\pi/2$. The right-hand side is a function of $1/\nu$, which is holomorphic for $1/\nu = 0$. Hence, mod $1/\nu^\infty$, it is an element of $\mathscr{F}$ whose limit, when Φ_+ vanishes, is

$$[\operatorname{Hess}\Phi]^{-1/2}[e^{(1/\nu)\Phi^*(\partial/\partial x)}P(x)]_{x=0} \mod \frac{1}{\nu^\infty}$$

from the Taylor series of this function; arg Hess Φ is given by I,§1, definition 2.3. Thus, mod $1/\nu^\infty$, the left-hand side of (3.10) is an element of $\mathscr{F}$; by lemma 3.2, its limit, when Φ_+ vanishes, is

$$\left(\frac{|\nu|}{2\pi i}\right)^{l/2}\int_X \varepsilon(x)P(x)e^{\nu\Phi(x)}d^lx.$$

Thus

$$\int_X \varepsilon(x)P(x)e^{\nu\Phi(x)}d^lx$$
$$= \left(\frac{2\pi i}{|\nu|}\right)^{l/2}[\operatorname{Hess}\Phi]^{-1/2}[e^{(1/\nu)\Phi^*(\partial/\partial x)}P(x)]_{x=0} \mod \frac{1}{\nu^\infty}; \tag{3.11}$$

lemma 3.3 follows by approximating α at the origin by its partial Taylor series P and applying lemma 3.2.

Lemma 3.3 proves part 2 of theorem 3 in the particular case where the phase φ is a quadratic form Φ. The general case is deduced from this special case by applying the following lemma to the remainder of a partial Taylor series. The proof of the lemma is analogous to that of lemma 3.1.

LEMMA 3.4. Let

$$\theta \in [0, 1], \qquad x \in X.$$

Let

$$\alpha : (\theta, x) \mapsto \alpha(\theta, x) \in \mathbf{C}$$

be infinitely differentiable with compact support, vanishing to order $2N$ when x vanishes. Let

$$\varphi : (\theta, x) \mapsto \varphi(\theta, x) \in \mathbf{R}$$

be infinitely differentiable and such that on Supp α, for every θ,

$\varphi : x \mapsto \varphi(\theta, x)$ has only one critical point, which occurs at the origin and is nondegenerate. Then

$$\int_0^1 d\theta \int_X \alpha(\theta, x) e^{\nu\varphi(\theta,x)} d^l x = 0 \bmod \frac{1}{\nu^N}. \tag{3.12}$$

From this lemma we deduce the following.

LEMMA 3.5. Let $\varphi : X \to \mathbf{R}$ be infinitely differentiable with only one critical point, which occurs at the origin and is nondegenerate, and critical value zero:

$$\varphi(0) = 0, \qquad \varphi_x(0) = 0, \qquad \operatorname{Hess} \varphi(0) \neq 0.$$

Let

$$\varphi(x) = \Phi(x) + \Theta(x)$$

be its second-order Taylor expansion at the origin: Φ is a quadratic form; Θ vanishes to third order at the origin.

Let $\alpha : X \to \mathbf{C}$ be infinitely differentiable, with support that is compact. Then

$$\int_X \alpha(x) e^{\nu\varphi(x)} d^l x = \sum_{j=0}^{\infty} \frac{\nu^j}{j!} \int_X \Theta^j(x) \alpha(x) e^{\nu\Phi(x)} d^l x \bmod \frac{1}{\nu^\infty}. \tag{3.13}$$

This series converges in $\mathscr{F}$ since, by lemma 3.3,

$$\nu^j \int_X \Theta^j(x) \alpha(x) e^{\nu\Phi(x)} d^l x = 0 \bmod \nu^{-(l/2)-[(j+1)/2]}$$

where $[\cdots]$ denotes the integer part of $\cdots$, that is, the greatest integer $\leqslant \cdots$.

Proof. By (3.12) and Taylor's formula,

$$e^{\nu\varphi} = \sum_{j=1}^{2N-1} \frac{\nu^j \Theta^j}{j!} e^{\nu\Phi} + \nu^{2N} \int_0^1 \frac{(1-\theta)^{2N-1}}{(2N-1)!} \Theta^{2N} e^{\nu(\Phi+\theta\Theta)} d\theta. \tag{3.14}$$

Lemma 3.4 gives

$$\nu^{2N} \int_X d^l x \int_0^1 (1-\theta)^{2N-1} \Theta^{2N}(x) \alpha(x) e^{\nu(\Phi+\theta\Theta)} d\theta = 0 \bmod \frac{1}{\nu^N} \tag{3.15}$$

since Θ^{2N} vanishes to order $6N$ at the origin. When N tends to infinity, (3.14) and (3.15) imply (3.13).

Proof of theorem 3,2°). We may replace u by a function f given by (3.7); φ has a single critical point that is nondegenerate. We may assume that this critical point is the origin and the critical value is zero. Then by (3.13) and lemma 3.3,

$$\int_X \alpha(x)e^{\nu\varphi(x)}\,d^l x$$
$$= \left(\frac{2\pi i}{|\nu|}\right)^{l/2} (\text{Hess}\,\Phi)^{-1/2} \sum_{r,j} \frac{1}{r!j!\nu^{r-j}} \left\{\Phi^{*r}\left(\frac{\partial}{\partial x}\right)[\Theta^j(x)\alpha(x)]\right\}_{x=0} \mod \frac{1}{\nu^\infty}, \tag{3.16}$$

where $r, j \in \mathbf{N}$. Denote the exponent of ν by $q = r - j$. Since $\Theta(x) = 0 \mod |x|^3$, the condition $\{\cdots\}_{x=0} \neq 0$ implies $3j \leqslant 2r$, that is $j \leqslant 2q$. Let

$$I_q(\varphi;\alpha) = \sum_{j=0}^{2q} \frac{1}{(q+j)!j!} \left\{\Phi^{*q+j}\left(\frac{\partial}{\partial x}\right)[\Theta^j(x)\alpha(x)]\right\}_{x=0};$$

in particular,

$$I_0(\varphi;\alpha) = \alpha(0).$$

Now (3.16) can be written

$$\int_X \alpha(x)e^{\nu\varphi(x)}\,d^l x = \left(\frac{2\pi i}{|\nu|}\right)^{l/2} (\text{Hess}\,\Phi)^{-1/2} \sum_{q\in\mathbf{N}} \frac{1}{\nu^q} I_q(\varphi;\alpha),$$

whence part 2 of the theorem follows.

4. Transformation of Formal Functions by Elements of $\mathrm{Sp}_2(l)$

Each $S \in \mathrm{Sp}_2(l)$ induces an automorphism

$$S : \mathscr{C}(X) \to \mathscr{C}(X)$$

(see section 1) whose restriction to

$\mathscr{F}_0(X) \subset \mathscr{C}(X)$ (theorem 2.2)

is a *monomorphism:*

$$S : \mathscr{F}_0(X) \to \mathscr{C}(X). \tag{4.1}$$

Theorems 4.1, 4.2, and 4.3 develop the properties of this monomorphism.

THEOREM 4.1. *Let Z be a 2-symplectic space, V a lagrangian manifold in Z, R' a 2-frame of Z, and*

$S \in \mathrm{Sp}_2(l)$;

$R = SR'$ *is then a 2-frame of* Z (I,§3,3).

1°) *There exists an isomorphism, induced by* S *and denoted* S,

$$S: \mathscr{F}_0(\check{V}\backslash\check{\Sigma}_R \cup \check{\Sigma}_{R'}, R') \to \mathscr{F}_0(\check{V}\backslash\check{\Sigma}_R \cup \check{\Sigma}_{R'}, R), \tag{4.2}$$

characterized by the two following properties:

i. *It is local.* That is, if

$$\check{U}_{R'} \in \mathscr{F}_0(\check{V}\backslash\check{\Sigma}_R \cup \check{\Sigma}_{R'}, R'), \qquad \check{U}_R = S\check{U}_{R'} \in \mathscr{F}_0(\check{V}\backslash\check{\Sigma}_R \cup \check{\Sigma}_{R'}, R),$$

then $\check{U}_R(\nu, \check{z})$, the value of $\check{U}_R$ at $\check{z} \in \check{V}$, is a linear function of the values of $\check{U}_{R'}$ and its derivatives at the same point $\check{z}$.

ii. *The following diagram is commutative:*

$$\begin{array}{ccc} \mathscr{F}_0(\check{V}\backslash\check{\Sigma}_R \cup \check{\Sigma}_{R'}, R') & \xrightarrow{S} & \mathscr{F}_0(\check{V}\backslash\check{\Sigma}_R \cup \check{\Sigma}_{R'}, R) \quad (4.2) \\ \Pi_{R'}\downarrow & & \downarrow\Pi_R \\ \mathscr{F}_0(X) & \xrightarrow{\quad S \quad} & \mathscr{C}(X) \quad (4.1) \end{array}$$

It thus defines a morphism

$$S \circ \Pi_{R'} = \Pi_R \circ S: \mathscr{F}_0(\check{V}\backslash\check{\Sigma}_R \cup \check{\Sigma}_{R'}, R') \to \mathscr{F}_0(X). \tag{4.3}$$

2°) *The composition law of the morphisms* (4.1) *and* (4.2) *is that of* $\mathrm{Sp}_2(l)$.

Remark 4. The restriction of (4.1),

$$S: \Pi_{R'}\mathscr{F}_0(\check{V}\backslash\check{\Sigma}_R \cup \check{\Sigma}_{R'}, R') \to \Pi_R\mathscr{F}_0(\check{V}\backslash\check{\Sigma}_R \cup \check{\Sigma}_{R'}, R) \subset \mathscr{F}_0(X), \tag{4.4}$$

is thus an isomorphism. While (4.2) is local, (4.4) is not local, but only pointwise: If

$$u' \in \Pi_{R'}\mathscr{F}_0(\check{V}\backslash\check{\Sigma}_R \cup \check{\Sigma}_{R'}, R'), \qquad u = Su',$$

then the value $u(\nu, x)$ at $x \in X$ is a linear function of the values of u' and its derivatives at the points of the finite set $R'_X R_X^{-1} x \cap \mathrm{Supp}\, u'$.

The local character of the morphism (4.2) is made explicit as follows.

THEOREM 4.2. *Let*

$$\check{U}_{R'} = \sum_{r \in \mathbf{N}} \frac{1}{\nu^r} \alpha_r e^{\nu\varphi_{R'}} \in \mathscr{F}_0(\check{V}\backslash\check{\Sigma}_R \cup \check{\Sigma}_{R'}, R'). \tag{4.5}$$

Then

$$\check{U}_R = S\check{U}_{R'} \in \mathscr{F}_0(\check{V}\backslash\check{\Sigma}_R \cup \check{\Sigma}_{R'}, R), \qquad \textit{where } R = SR',$$

is given at $\check{z}$ *by*

$$\check{U}_R(v, \check{z}) = \sum_{r\in\mathbf{N}} \left(\frac{d^l x'}{d^l x}\right)^{3r+1/2} \frac{1}{v^r} J_{R,r}(\check{z}; \alpha) e^{v\varphi_R(\check{z})}, \tag{4.6}$$

where

$$x = \check{R}_X \check{z}, \qquad x' = \check{R}'_X \check{z}, \tag{4.7}$$

$[d^l x]^{1/2}$ *is defined in* I,§3,*corollary* 3.

$J_{R,r}$ *is a linear function of the derivatives of the* α_s $(s \leqslant r)$ *of order* $\leqslant 2(r - s)$ *on* $\check{V}$. *Its coefficients are infinitely differentiable functions of* $\check{z} \in \check{V}\backslash\check{\Sigma}_{R'}$; *these functions depend on* V, R', *and* S.

$$J_{R,0}(\check{z}, \alpha) = \alpha_0(\check{z}) \tag{4.8}$$

Formula (4.6) retains a meaning for all $\check{U}_{R'} \in \mathscr{F}(\check{V}\backslash\check{\Sigma}_{R'}, R')$ and $\check{z} \in \check{V}\backslash\check{\Sigma}_R \cup \check{\Sigma}_{R'}$; therefore the following theorem results from the preceding one.

THEOREM 4.3. *Formula* (4.6) *defines an isomorphism induced by* S *and denoted* S:

$$S : \mathscr{F}(\check{V}\backslash\check{\Sigma}_R \cup \check{\Sigma}_{R'}, R') \to \mathscr{F}(\check{V}\backslash\check{\Sigma}_R \cup \check{\Sigma}_{R'}, R). \tag{4.9}$$

The composition law of these isomorphisms is that of the group $\mathrm{Sp}_2(l)$.

Proof of the preceding theorems. Let $\check{U}_{R'} \in \mathscr{F}_0(\check{V}\backslash\check{\Sigma}_R \cup \check{\Sigma}_{R'}, R')$ be such that the restriction

$$\check{R}'_X : \mathrm{Supp}\, \check{U}_{R'} \to X \text{ of } \check{R}'_X : \check{V} \to X$$

is a diffeomorphism, with inverse denoted $\check{R}'^{-1}_X$.

The case $S \notin \Sigma_{\mathrm{Sp}_2}$. Then $S = S_A$ (I,§1,1). Let $\check{U}_{R'}$ be expressed by (4.5);

$$u' = \Pi_{R'} \check{U}_{R'}$$

is given at x' by

$$u'(v, x') = \sum_{r\in\mathbf{N}} \frac{1}{v^r} \alpha'_r(x') e^{v\varphi'(x')}, \tag{4.10}$$

where

$$\alpha'_r(x') = \alpha_r(\check{R}'^{-1}_X x'), \qquad \varphi'(x') = \varphi_{R'}(\check{R}'^{-1}_X x') \text{ for } x' \in \operatorname{Supp} u',$$

$$\alpha'_r(x') = 0 \text{ for } x' \notin \operatorname{Supp} u' = \check{R}'_X \operatorname{Supp} \check{U}_{R'}.$$

Since $u' \in \mathscr{F}_0(X) \subset \mathscr{C}(X)$, we have $u = Su' \in \mathscr{C}(X)$. By I,§1,(1.10), we have the following expression for u at x:

$$u(v, x) = \left(\frac{|v|}{2\pi i}\right)^{l/2} \Delta(A) \tilde{\int_X} \sum_{r \in \mathbf{N}} \frac{1}{v^r} \alpha'_r(x') e^{v[A(x,x') + \varphi'(x')]} d^l x'. \tag{4.11}$$

We apply theorem 3 to this asymptotic integral.

i. For each $x \in X$ let us find *the critical points of the phase*

$$X \ni x' \mapsto A(x, x') + \varphi'(x') \in \mathbf{R} \tag{4.12}$$

in Supp u'; that is, the $x' \in \operatorname{Supp} u'$ such that

$$A_{x'} + \varphi'_{x'} = 0. \tag{4.13}$$

Let us write

$$\check{z} = \check{R}'^{-1}_X x', \qquad \check{R}'\check{z} = (x', p'); \text{ that is, } p' = \varphi'_{x'} \in X^*. \tag{4.14}$$

Relation (4.13) becomes

$$p' + A_{x'} = 0.$$

By I,§1,(1.11), this means that there exists $p \in X^*$ such that

$$(x, p) = s_A(x', p').$$

Thus (4.13) is equivalent to

$$x = R_X R'^{-1}(x', p') = \check{R}_X \check{R}'^{-1}(x', p').$$

By (4.14) this says that (4.13) is equivalent to

$$x = \check{R}_X \check{z}.$$

Therefore, on Supp u' the *critical points of the phase* (4.12) *are the points*

$$x' = \check{R}'_X \check{z}, \text{ where } \check{z} \in \check{R}_X^{-1} x \cap \operatorname{Supp} \check{U}_{R'}. \tag{4.15}$$

This set is finite since Supp $\check{U}_{R'}$ is a compact set in $\check{V}\backslash\check{\Sigma}_R$ and $\check{R}_X$ is a local diffeomorphism in a neighborhood of this compact set.

ii. Let us find *the critical values of the phase* (4.12). At a critical point x' defined by (4.15), let

$$\check{R}\check{z} = (x, p), \qquad \check{R}'\check{z} = (x', p');$$

from the preceding formulas,

$$p = A_X, \qquad p' = -A_{X'}.$$

Hence, since A is a quadratic form in (x, x'),

$$A(x, x') = \tfrac{1}{2}\langle p, x\rangle - \tfrac{1}{2}\langle p', x'\rangle,$$

that is, in view of I,§3,(2.4),

$$A(x, x') = \varphi_R(\check{z}) - \varphi_{R'}(\check{z}), \text{ where } \varphi_{R'}(\check{z}) = \varphi'(x').$$

Therefore *at the critical point* x' defined by (4.15), *the value of the phase* (4.12) *is* $\varphi_R(\check{z})$.

iii. The *square root of the Hessian of this phase* (4.12) is given by I,§3, (3.7):

$$\{\mathrm{Hess}_{x'}[A(x, x') + \varphi'(x')]\}^{1/2} = \Delta(A)\left[\frac{d^l x}{d^l x'}\right]^{1/2}. \tag{4.16}$$

While calculating $\mathrm{Hess}_{x'}$, x and x' are viewed as independent. Afterwards we give x and x' the values $x = \check{R}_X\check{z}$ and $x' = \check{R}'_X\check{z}$ so that x and x' are local coordinates of the same point $\check{z} \in \check{V}\backslash\check{\Sigma}_R \cup \check{\Sigma}_{R'}$. Taking these values for x and x' in the right-hand side, it follows from (4.16) that

$$\mathrm{Hess}_{x'}[A(x, x') + \varphi'(x')] \neq 0.$$

Consequently theorem 3 can be applied to the calculation of (4.11).

iv. *Completion of the proof.* This application of theorem 3 to (4.11) gives

$$u(v, x) = \sum_{\check{z}\in\check{R}_X^{-1}x} \check{U}_R(v, \check{z}), \tag{4.17}$$

where:

- $\check{U}_R$ has the same support as $\check{U}_{R'}$;
- defining $x = \check{R}_X\check{z}$ and $x' = \check{R}'_X\check{z}$, the value of $\check{U}_R$ at $\check{z}$ is

$$\check{U}_R(v, \check{z}) = \Delta(A)[\mathrm{Hess}_{x'}(A + \varphi')]^{-1/2} \sum_{r\in\mathbf{N}} \frac{1}{v^r} I_r(A + \varphi'; \alpha')\, e^{v\varphi_R(\check{z})}. \tag{4.18}$$

Now on one hand, (4.17) means

$$u = \Pi_R\check{U}_R,$$

but on the other hand, by (4.16), (4.18) is equivalent to (4.6). Indeed, by $(3.3)_r$, $I_r(A + \varphi'; \alpha')$ is the value of a linear function of the derivatives of the $\alpha'_s(s \leqslant r)$ of order $\leqslant 2(r - s)$; the coefficients of this function are rational functions of the derivatives of φ'; the common denominator of these rational functions is $[\mathrm{Hess}_{x'}(A + \varphi')]^{3r}$, whose value is given by (4.16).

Thus theorems 4.1 and 4.2 hold for $S_A \in \mathrm{Sp}_2(l)\backslash\Sigma_{\mathrm{Sp}_2}$.

The case where $S \in \Sigma_{\mathrm{Sp}_2}$. By I,§1,lemma 2.5, given $S \in \Sigma_{\mathrm{Sp}_2}$ there exist S_A and $S_{A'} \in \mathrm{Sp}_2(l)\backslash\Sigma_{\mathrm{Sp}_2}$ such that

$$S = S_{A'}S_A.$$

Since the morphisms (4.1) and (4.2) are composed like the elements of $\mathrm{Sp}_2(l)$ that induce them, the compositions of the morphisms induced by S_A and $S_{A'}$ depend only on $S = S_{A'}S_A$; these are then morphisms induced by S. Given $\check{U}_{R'}$ such that $\mathrm{Supp}\,\check{U}_{R'} \subset \check{V}\backslash\check{\Sigma}_{R'} \cup \check{\Sigma}_{SR'}$, we can choose S_A sufficiently close to the identity so that $\check{\Sigma}_{S_A R'} \cap \mathrm{Supp}\,\check{U}_{R'} = \varnothing$, whence follow theorems 4.1 and 4.2, which imply theorem 4.3.

5. Norm and Scalar Product of Formal Functions with Compact Support

Let V and V' be two lagrangian manifolds in Z, $\check{V}$ and $\check{V}'$ their universal covering spaces, R a 2-frame of Z, φ_R and φ'_R the phases of RV and RV', ψ and ψ' the lagrangian phases of V and V'. Let

$$\check{U}_R = \sum_{r\in\mathbf{N}} \frac{1}{\nu^r}\alpha_{R,r}e^{\nu\varphi_R} \in \mathscr{F}_0(\check{V}\backslash\check{\Sigma}_R, R),$$
$$\check{U}'_R = \sum_{s\in\mathbf{N}} \frac{1}{\nu^s}\alpha'_{R,s}e^{\nu\varphi'_R} \in \mathscr{F}_0(\check{V}'\backslash\check{\Sigma}'_R, R), \tag{5.1}$$

with projections

$$u = \Pi_R\check{U}_R, \qquad u' = \Pi_R\check{U}'_R.$$

Definition 5.1. Under assumption (5.1) the *scalar product* of $\check{U}_R$ and $\check{U}'_R$ is the asymptotic class

$$(\check{U}_R \,\check{|}\, \check{U}'_R) = (u \,|\, u') \in \mathscr{C}, \tag{5.2}$$

where $(u \,|\, u')$ is defined by the asymptotic integral (1.9). Thus

$(\check{U}_R \,\check{|}\, \check{U}'_R)$ and $(\check{U}'_R \,\check{|}\, \check{U}_R)$ are complex conjugates and the Schwarz inequality applies:

$$|(\check{U}_R \check{|} \check{U}'_R)| \leqslant (\check{U}_R \check{|} \check{U}_R)^{1/2} (\check{U}'_R \check{|} U'_R)^{1/2}. \tag{5.3}$$

The seminorm of $\check{U}_R$ is $(\check{U}_R \check{|} \check{U}_R)^{1/2} = \|u\| \geqslant 0$; this seminorm satisfies the triangle inequality.

THEOREM 5.1. 1°) *The value of* $(\check{U}_R \check{|} \check{U}'_R)$ *depends only on the behavior of* $\check{U}_R$ *and* $\check{U}'_R$ *at the pairs of points of* $\check{V}$ *and* $\check{V}'$ *projecting onto the same point of* $V \cap V'$. *This value is* 0 *if* $V \cap V' = \varnothing$.

2°) *If* $V = V'$, *which implies* $\varphi_R = \varphi'_R$, *then*

$$(\check{U}_R \check{|} \check{U}'_R) = \sum_{r,s} \frac{(-1)^s}{\nu^{r+s}} \int_{z \in V} \sum_{\check{z},\check{z}'} \alpha_{R,r}(\check{z}) \overline{\alpha'_{R,s}(\check{z}')} e^{\nu[\psi(\check{z}) - \psi(\check{z}')} d^l x \in \mathscr{F}, \tag{5.4}$$

where:

- $x = R_X z$, $d^l x > 0$;
- $\check{z}$ *and* $\check{z}'$ *are the points of* $\operatorname{Supp} \check{U}_R$ *and* $\operatorname{Supp} \check{U}'_R$ *projecting onto* z.

In (5.4), $\psi(\check{z}) - \psi(\check{z}')$ *is one of the periods of* ψ:

$$c_\gamma = \frac{1}{2} \int_\gamma [z, dz], \text{ where } \gamma \text{ is a cycle in } V.$$

Thus the phases of $(\check{U}_R \check{|} \check{U}'_R)$ *are these periods* c_γ *of* ψ.

3°) *If* V *and* V' *are transverse, then*

$$\nu^{l/2} (\check{U}_R \check{|} \check{U}'_R) \in \mathscr{F}. \tag{5.5}$$

If V *and* V' *are transverse and intersect at a single point* z *that is the projection of a single point* $\check{z}$ *of* $\operatorname{Supp} \check{U}_R$ *and a single point* $\check{z}'$ *of* $\operatorname{Supp} \check{U}'_R$, *then*

$$\left(\frac{|\nu|}{2\pi i}\right)^{l/2} (\check{U}_R \check{|} \check{U}'_R) = \sum_{r \in \mathbb{N}} \frac{1}{\nu^r} [\operatorname{Hess}(\varphi - \varphi')^{-3r - 1/2} P_r(\alpha_R, \alpha'_R) e^{\nu[\psi(\check{z}) - \psi'(\check{z}')]}, \tag{5.6}$$

where

- φ *and* φ' *are defined near* $R_X z$ *by* $\varphi(x) = \varphi_R(\check{R}_X^{-1} x \cap \check{V})$, $\varphi'(x) = \varphi'_R(\check{R}_X^{-1} x \cap \check{V}')$;
- $\operatorname{Hess}(\varphi - \varphi') = \{\operatorname{Hess}_x[\varphi(x) - \varphi'(x)]\}_{x = R_X z}$;

- P_r *is a sesquilinear function of the values of the derivatives of* $\alpha_{R,s}$ *and* $\alpha'_{R,s}$ $(s \leqslant r)$ *of order* $\leqslant 2(r-s)$ *at* $\check{z}$ *and* $\check{z}'$;
- *the coefficients of this function depend on the behavior of* V *and* V' *at* z;
- $P_0(\alpha_R, \alpha'_R) = \alpha_{R,0}(\check{z})\overline{\alpha'_{R,0}(\check{z}')}$. (5.7)

4°) [*Invariance under* $\mathrm{Sp}_2(l)$]. *Let* $S \in \mathrm{Sp}_2(l)$. *Under the assumption*

$$\check{U}_R \in \mathscr{F}_0(\check{V} \backslash \check{\Sigma}_R \cup \check{\Sigma}_{SR}, R), \qquad \check{U}'_R \in \mathscr{F}_0(\check{V}' \backslash \check{\Sigma}'_R \cup \check{\Sigma}'_{SR}, R),$$

which is stronger than (5.1), *we have*

$$(\check{U}_R \,\check{|}\, \check{U}'_R) = (S\check{U}_R \,\check{|}\, S\check{U}'_R). \tag{5.8}$$

Proof of 1°), 2°), *and* 3°). u and u' can be written in the form

$$u(v, x) = \sum_{j \in J} v_j(v, x) e^{v\varphi_j(x)}, \text{ where } v_j(v, x) = \sum_{r \in \mathbf{N}} \frac{1}{v^r} v_{j,r}(x);$$

$$u'(v, x) = \sum_{k \in K} v'_k(v, x) e^{v\varphi'_k(x)}, \text{ where } v'_k(v, x) = \sum_{r \in \mathbf{N}} \frac{1}{v^r} v'_{k,r}(x);$$

J and K are finite sets. By part 1 of theorem 3,

$$(u \,|\, u') = \tilde{\int_X} u(v, x)\overline{u'(v, x)}\, d^l x$$

depends only on the behavior of the pairs $v_j e^{v\varphi_j}$, $v'_k e^{v\varphi'_k}$ at the critical points of $\varphi_j - \varphi'_k$, that is, at the points

$x \in X$ such that $\varphi_{j,x} = \varphi'_{k,x}$.

In other words, $(\check{U}_R \,\check{|}\, \check{U}'_R)$ depends only on the behavior of $\check{U}_R$ and $\check{U}'_R$ at pairs of points $(\check{z}, \check{z}') \in \check{V} \times \check{V}'$ such that

$$\check{R}\check{z} = \check{R}\check{z}' = (x, \varphi_{j,x}) = (x, \varphi'_{k,x}).$$

These pairs are the pairs of points of $\check{V} \times \check{V}'$ projecting onto a single point z of $V \cap V'$. 1°) and 2°) follow, as does 3°), by (3.2) and (3.3).

Proof of 4°). Lemma 1.1 and part 1 of theorem 4.1.

The invariance of $(\check{U}_R \,\check{|}\, \check{U}'_R)$ under $\mathrm{Sp}_2(l)$ stated in part 4 of theorem 5.1 raises the following problem: to give invariant expressions of the right-hand sides of (5.4) and (5.6). Theorems 5.2 and 5.3 will give such expressions mod$(1/v)$.

Notation. Let η and η' be regular positive measures on V and V'; we define

$\arg\eta = \arg\eta' = 0.$

We assume V and V' are both given 2-orientations: If $\check{z} \in \check{V}$, $x = \check{R}_X\check{z}$, then

$$\frac{d^l(\check{R}_X\check{z})}{\eta} = \frac{d^l x}{\eta} \text{ is a function } \check{V} \to \mathbf{R}$$

vanishing on $\check{\Sigma}_R$ whose argument is given by I, §3, corollary 3:

$$\arg\frac{d^l(\check{R}_X\check{z})}{\eta} = \pi m_R(\check{z}) \mod 4\pi.$$

We define $\beta_0 : \check{V} \to \mathbf{C}$ by

$$\beta_0(\check{z}) = \left[\frac{d^l(\check{R}_X\check{z})}{\eta}\right]^{1/2} \alpha_{R,0}(\check{z}), \text{ where } \check{z} \in \check{V}. \tag{5.9}$$

Definition 5.2. From formulas (4.6) and (4.8) of theorem 4.2, β_0 is *invariant* when we replace R by SR without changing V; β_0 is called the *lagrangian amplitude* of $\check{U}_R$. This amplitude depends on the choice of the measure η and the 2-orientation of $\check{V}$.

Substituting the expression (5.9) for $\alpha_{R,0}$ and $\alpha'_{R,0}$ into (5.4), we obtain the following.

THEOREM 5.2. *Suppose $V = V'$, which implies $\psi = \psi'$; let β_0 and β'_0 denote the lagrangian amplitudes of $\check{U}_R$ and $\check{U}'_R$. Then*

$$(\check{U}_R \,|\, \check{U}'_R) = \int_{z\in V} \sum_{\check{z},\check{z}'} \beta_0(\check{z})\overline{\beta'_0(\check{z}')e^{\nu[\psi(\check{z})-\psi(\check{z}')]}}\eta \mod \frac{1}{\nu}, \tag{5.10}$$

where $\check{z}$ and $\check{z}'$ are the points of Supp $\check{U}_R$ *and* Supp $\check{U}'_R$ *projecting onto z.*

Notation. (Continued) Let ψ and ψ' be the lagrangian phases of V and V' [I, §3, (1.7)]. We make the same assumptions as in part 3 of theorem 5.1. Let λ_4 and $\lambda'_4 \in \Lambda_4(Z)$ be tangent to the 2-oriented manifolds $\check{V}$ and $\check{V}'$ at $\check{z}$ and $\check{z}'$, respectively; let λ and λ' be their natural images in $\Lambda(Z)$. Let η_0 (and η'_0) be the measure on λ (and λ') that is translation invariant and equal to η (and η') at $\check{z}$ (and $\check{z}'$). V and V' are assumed transverse (as in part 3 of theorem 5.1); thus

$$Z = \lambda \oplus \lambda'; \qquad \eta_0 \wedge \eta'_0 \text{ is a measure on } Z. \tag{5.11}$$

By (5.1) and part 1 of theorem 5.1, the two sides of (5.6) are zero unless we assume

$$\check{z} \in \check{V} \backslash \check{\Sigma}_R, \qquad \check{z}' \in \check{V}' \backslash \check{\Sigma}'_R, \tag{5.12}$$

that is,

$R\lambda$ and $R\lambda'$ are transverse to X^*.

Substituting the expression (5.9) for $\alpha_{R,0}$ and $\alpha'_{R,0}$ into (5.6) we obtain, under assumption (5.12),

$$\left(\frac{|v|}{2\pi i}\right)^{l/2} (\check{U}_R \check{|} \check{U}'_R)$$
$$= [\operatorname{Hess}(\varphi - \varphi')]^{-1/2} \left[\frac{\eta}{d^l(\check{R}_X \check{z})}\right]^{1/2} \overline{\left[\frac{\eta'}{d^l(\check{R}_X \check{z}')}\right]^{1/2}} \beta_0(\check{z}) \overline{\beta'_0(\check{z}')} e^{v[\psi(\check{z}) - \psi'(\check{z}')]}$$
$$\operatorname{mod} \frac{1}{v}. \tag{5.13}$$

Lemma 5 transforms this formula into the following.

THEOREM 5.3. *If V and V' are each given a 2-orientation, are transverse, and intersect only at a point z that is the projection of a single point $\check{z}$ of* Supp $\check{U}_R$ *and of a single point $\check{z}'$ of* Supp $\check{U}'_R$, *then*

$$\left(\frac{|v|}{2\pi i}\right)^{l/2} (\check{U}_R \check{|} \check{U}'_R)$$
$$= \left|\frac{\eta_0 \wedge \eta'_0}{d^{2l} z}\right|^{1/2} \beta_0(\check{z}) \overline{\beta'_0(\check{z}')} e^{v[\psi(\check{z}) - \psi'(\check{z}')] - (\pi i/2) m(\lambda'_4, \lambda_4)} \operatorname{mod} \frac{1}{v}, \tag{5.14}$$

where m is the Maslov index, defined mod 4 (I,§3,3) *and $\eta_0 \wedge \eta'_0$ and $d^{2l}z$ are the measures on Z defined, respectively, by* (5.11) *and* (5.15).

LEMMA 5. 1°) The symplectic structure of Z defines a measure on Z that is invariant under translations and under $\operatorname{Sp}(Z)$:

$$d^{2l} z = \frac{1}{l!} (dx \wedge dp)^l = (-1)^{l(l-1)/2} d^l x \wedge d^l p, \tag{5.15}$$

where

$$Rz = (x, p), \qquad dx \wedge dp = \sum_{j=1}^{l} dx^j \wedge dp_j;$$

$\{x^j\}$ and $\{p_j\}$ are dual coordinates on X and X^*.

2°) Under assumption (5.12), with the notation (5.9),

$$\left| \text{Hess}(\varphi - \varphi') \frac{d^l(\check{R}_X \check{z})}{\eta} \frac{d^l(\check{R}_X \check{z}')}{\eta'} \right| = \left| \frac{d^{2l} z}{\eta_0 \wedge \eta'_0} \right|, \tag{5.16}$$

$$\arg\left[\text{Hess}(\varphi - \varphi') \frac{d^l(\check{R}_X \check{z})}{\eta} \frac{d^l(\check{R}_X \check{z}')}{\eta'} \right] = \pi m(\lambda'_4, \lambda_4) \mod 4\pi. \tag{5.17}$$

Proof of 1°). The value of $dp \wedge dx$ on a pair of vectors z, z' is $[z, z']$.

Proof of (5.16). Since $R\lambda$ and $R\lambda'$ are lagrangian and transverse to X^* by (5.12), their equations have the form

$$R\lambda : p = \Phi_x(x), \qquad R\lambda' : p = \Phi'_x(x), \tag{5.18}$$

where Φ and Φ' are two quadratic forms $X \to \mathbf{R}$. Since λ and λ' are transverse,

$$\text{Hess}(\Phi - \Phi') \neq 0.$$

By (5.11), each $z \in Z$ decomposes uniquely as the sum of a vector in λ and a vector in λ':

$$Rz = (x + x', \Phi_x(x) + \Phi_{x'}(x')) \in Z(l),$$

where x and $x' \in X$ are unique. It follows by (5.15) that

$$\begin{aligned} d^{2l} z &= (-1)^{l(l-1)/2} d^l[x + x'] \wedge d^l[\Phi_x(x) + \Phi'_{x'}(x')] \\ &= (-1)^{l(l-1)/2} d^l[x + x'] \wedge d^l[\Phi_x(x) - \Phi'_x(x)] \\ &= (-1)^{l(l+1)/2} \text{Hess}(\Phi - \Phi') d^l x \wedge d^l x' \\ &= (-1)^{l(l+1)/2} \text{Hess}(\Phi - \Phi') \frac{d^l(R_X \zeta)}{\eta_0} \frac{d^l(R_X \zeta')}{\eta'_0} \eta_0 \wedge \eta'_0, \end{aligned}$$

where $\zeta \in \lambda$, $\zeta' \in \lambda'$; hence (5.16) follows.

Proof of (5.17). By definition [I,§3,corollary 3, (3.2) and (3.3)]

$$\arg \frac{d^l(\check{R}_X \check{z})}{\eta} = \pi m_R(\check{z}) = \pi m(X_4^*, R\lambda_4) \mod 4\pi;$$

thus

$$\pi m(\lambda_4', \lambda_4) - \arg\frac{d^l(\check{R}_X\check{z})}{\eta} + \arg\frac{d^l(\check{R}_X\check{z}')}{\eta'}$$
$$= \pi m(\lambda_4', \lambda_4) - \pi m(X_4^*, R\lambda_4) + m(X_4^*, R\lambda_4')$$
$$= \pi \operatorname{Inert}(X^*, R\lambda', R\lambda) \mod 4\pi$$

by I,§2,(8.2). Then, by the definition of arg Hess [I,§1,(2.1)], to prove (5.17) it suffices to show that

$$\operatorname{Inert}(\Phi - \Phi') = \operatorname{Inert}(X^*, R\lambda', R\lambda). \tag{5.19}$$

Proof of (5.19). Let us write down the equations of $R\lambda$, $R\lambda'$, and X^*:

$$R\lambda: p = \Phi_x(x), \qquad R\lambda': p' = \Phi'_{x'}(x'), \qquad X^*: x'' = 0.$$

The definition of Inert (I,§2,4) makes use of

$$z \in R\lambda, \qquad z' \in R\lambda', \qquad z'' \in X^* \text{ such that } z + z' + z'' = 0.$$

Define

$$z = (x, p), \qquad z' = (x', p'), \qquad z'' = (x'', p'');$$

hence

$$x + x' = 0, \qquad x'' = 0, \qquad p + p' + p'' = 0,$$

$$[z'', z] = \langle p'', x\rangle = -\langle p, x\rangle - \langle p', x\rangle$$
$$= -\langle \Phi_x(x), x\rangle + \langle \Phi'_x(x), x\rangle = 2\Phi'(x) - 2\Phi(x).$$

Consequently,

$$\operatorname{Inert}(R\lambda, R\lambda', X^*) = \operatorname{Inert}(\Phi' - \Phi).$$

(5.19) follows by I,§2,(4.4) and I,§1,definition 2.3, of the inertia of a form.

6. Formal Differential Operators

Definition 6.1. Let Ω be an open set in Z. Let $\mathscr{F}^0(\Omega)$ denote the set of formal functions with vanishing phase

$$a^0 = \sum_{r \in \mathbf{N}} \frac{1}{\nu^r} \alpha_r \tag{6.1}$$

such that the

$$\alpha_r : \Omega \to \mathbf{C}$$

are infinitely differentiable. We give the *vector space* $\mathscr{F}^0(\Omega)$ the topology defined by the following system of neighborhoods $\mathscr{N}(K, p, r, \varepsilon)$ of the origin:

K is compact in Ω; $\quad p, r \in \mathbf{N}, \quad \varepsilon \in \dot{\mathbf{R}}_+$;

$a^0 \in \mathscr{N}(K, p, r, \varepsilon)$ means that on K the derivatives of α_s $(s \leqslant r)$ of order $\leqslant p$ have modulus $< \varepsilon$.

Let us say that a^0 is a *polynomial* if

$$(v, z) \mapsto a^0(v, z)$$

is a polynomial in $1/v$ and in the $2l$ coordinates of z. By the *Weierstrass* theorem the set of polynomials in $\mathscr{F}^0(Z)$ is dense in $\mathscr{F}^0(\Omega)$.

$\mathscr{F}^0(\Omega(l))$ is defined similarly by replacing Ω by a domain $\Omega(l)$ of $Z(l) = X \oplus X^*$.

Every 1-frame R of Z (hence, a fortiori, every 2-frame R) induces an isomorphism

$$\mathscr{F}^0(\Omega) \ni a \mapsto a_R^0 \in \mathscr{F}^0(\Omega(l)), \text{ where } \Omega(l) = R\Omega,$$

defined by the formula

$$a^0(v, z) = a_R^0(v, Rz) \qquad \forall v, z. \tag{6.2}$$

The formulas (3.4) I,§1,

$$\begin{aligned} a_R^+(v, x, p) &= \left[\exp \frac{1}{2v}\left\langle \frac{\partial}{\partial x}, \frac{\partial}{\partial p}\right\rangle\right] a_R^0(v, x, p), \\ a_R^-(v, p, x) &= \left[\exp -\frac{1}{2v}\left\langle \frac{\partial}{\partial x}, \frac{\partial}{\partial p}\right\rangle\right] a_R^0(v, x, p) \end{aligned} \tag{6.3}$$

define two automorphisms of $\mathscr{F}^0(Z(l))$ that are inverse to each other:

$$a_R^0 \mapsto a_R^+, \qquad a_R^0 \mapsto a_R^-.$$

Definition 6.2 of formal differential operators rests on the following lemma.

LEMMA 6.1. Let

$$u = \alpha e^{\nu\varphi} \in \mathscr{F}(X), \text{ where } \alpha = \sum_{r\in\mathbf{N}} \frac{1}{\nu^r}\alpha_r. \tag{6.4}$$

At a critical point x of the phase φ we have

$$\left(\frac{1}{\nu}\frac{\partial}{\partial x}\right)^h u(\nu, x) = 0 \bmod \frac{1}{\nu^{|h|/2}}.$$

Proof. It suffices to consider the case $\alpha = \alpha_0$. Then at a critical point of φ,

$$\left(\frac{1}{\nu}\frac{\partial}{\partial x}\right)^h u = \frac{1}{\nu^{|h|}} P_h e^{\nu\varphi} \ (h \text{ is a multi-index}),$$

where P_h is a polynomial in the derivatives of α of order $\in [0, |h|]$ and in the derivatives of $\nu\varphi$ of order $\in [2, h]$ (since $\varphi_x = 0$). Each monomial in P_h is a product of derivatives whose orders sum to $|h|$. Hence the power of ν in such a monomial is $\leqslant |h|/2$. The lemma follows.

Definition 6.2. Let

$$\Theta(y, x) = \varphi(x) - \varphi(y) - \langle \varphi_y(y), x - y\rangle \tag{6.5}$$

denote the remainder of the first-order Taylor series of the phase φ of (6.4) at the point y. To an element a^0 of $\mathscr{F}^0(\Omega)$ and to a frame R, let us associate the local endomorphism a_R of $\mathscr{F}(X)$ and $\mathscr{F}_0(X)$, called a *formal differential operator*, defined by the two equivalent formulas

$$\begin{aligned}(a_R u)(\nu, y) &= \sum_h \frac{1}{h!}\left\{\frac{\partial^{|h|} a_R^+}{\partial p^h}(\nu, x, p)\left(\frac{1}{\nu}\frac{\partial}{\partial x}\right)^h [\alpha(\nu, x)e^{\nu\Theta(y,x)}]\right\}_{\substack{x=y\\ p=\varphi_y(y)}} e^{\nu\varphi(y)} \\ &= \sum_h \frac{1}{h!}\left\{\left(\frac{1}{\nu}\frac{\partial}{\partial x}\right)^h \left[\frac{\partial^{|h|} a_R^-}{\partial p^h}(\nu, p, x)\alpha(\nu, x)e^{\nu\Theta(y,x)}\right]\right\}_{\substack{x=y\\ p=\varphi_y(y)}} e^{\nu\varphi(y)}.\end{aligned} \tag{6.6}$$

By lemma 6.1, each of these two formulas makes sense when $(y, \varphi_y) \in \Omega(l)$: we say that a_R is defined on $\Omega(l) = R\Omega$.

Proof of the equivalence of these two formulas. Leibniz's formula gives (compare I,§1,proof of lemma 3.1)

$$\sum_h \frac{1}{h!}\left(\frac{1}{\nu}\frac{\partial}{\partial x}\right)^h\left[\frac{\partial^{|h|} a_R^-(\nu, p, x)}{(\partial p)^h} v(\nu, x)\right]$$
$$= \sum_{j,k} \frac{1}{j!k!}\frac{\partial^{|2j+k|} a_R^-(\nu, p, x)}{(\partial p)^{j+k}(\nu\,\partial x)^j}\left(\frac{1}{\nu}\frac{\partial}{\partial x}\right)^k v(\nu, x)$$
$$= \sum_k \frac{1}{k!}\frac{\partial^{|k|} a_R^+(\nu, x, p)}{(\partial p)^k}\left(\frac{1}{\nu}\frac{\partial}{\partial x}\right)^k v(\nu, x)$$

because

$$\sum_j \frac{1}{j!}\frac{\partial^{|2j|} a_R^-(\nu, p, x)}{(\partial p)^j(\nu\,\partial x)^j} = \left[\exp\frac{1}{\nu}\left\langle\frac{\partial}{\partial x}, \frac{\partial}{\partial p}\right\rangle\right] a_R^-(\nu, p, x) = a_R^+(\nu, x, p).$$

Justification of definition 6.2. Let us show that it is equivalent to the definition in section 1 when a^0 is a polynomial, that is, when a_R is a classical differential operator with polynomial coefficients in $(1/\nu, x)$. In this case (6.6) means

$$(a_R u)(\nu, x) = \left\{a_R^+\left(\nu, x, p + \frac{1}{\nu}\frac{\partial}{\partial x}\right)\left[u(\nu, x)e^{-\nu\langle p, x-y\rangle}\right]\right\}_{\substack{y=x\\ p=\varphi_x}}.$$

Now

$$a_R^+\left(\nu, x, p + \frac{1}{\nu}\frac{\partial}{\partial x}\right)\left[u(\nu, x)e^{-\nu\langle p, x-y\rangle}\right] = e^{-\nu\langle p, x-y\rangle} a_R^+\left(\nu, x, \frac{1}{\nu}\frac{\partial}{\partial x}\right)u(\nu, x).$$

Then, taking $a^+(\nu, x, \partial v/\partial x) = a^-(\nu, \partial v/\partial x, x)$ in the sense of section 1, we have

$$(a_R u)(\nu, x) = a_R^+\left(\nu, x, \frac{1}{\nu}\frac{\partial}{\partial x}\right)u(\nu, x) = a_R^-\left(\nu, \frac{1}{\nu}\frac{\partial}{\partial x}, x\right)u(\nu, x).$$

Notation. Denote

$$(a_R u)(\nu, x) = a_R^+\left(\nu, x, \frac{1}{\nu}\frac{\partial}{\partial x}\right)u(\nu, x) = a_R^-\left(\nu, \frac{1}{\nu}\frac{\partial}{\partial x}, x\right)u(\nu, x).$$

Definition 6.3. Let $V \subset \Omega$, $a^0 \in \mathscr{F}^0(\Omega)$. Let a_R denote the unique *local endomorphism* of $\mathscr{F}_0(\check{V}\backslash\check{\Sigma}_R, R)$ such that

$$a_R \Pi_R = \Pi_R a_R. \tag{6.7}$$

a_R, being local, extends to an endomorphism of $\mathscr{F}(\check{V}\backslash\check{\Sigma}_R, R)$, which has the following properties.

THEOREM 6. 1°) *The formal differential operator* a_R *is local (in the sense of part 1 of theorem 4.1); hence*

$$\mathrm{Supp}(a_R U_R) \subset \mathrm{Supp}\, U_R \cap \mathrm{Supp}(a^0).$$

2°) *The mapping*

$$\mathscr{F}^0(\Omega) \times \mathscr{F}(\check{V}\backslash\check{\Sigma}_R, R) \ni (a_R, U_R) \mapsto a_R U_R \in \mathscr{F}(\check{V}\backslash\check{\Sigma}_R, R) \tag{6.8}$$

is continuous.

3°) *Each* $S \in \mathrm{Sp}_2(l)$ *transforms* a_R *into an operator*

$$a_{SR} = S a_R S^{-1}$$

defined as follows:

$$a_{SR}(S\check{U}_R) = S(a_R \check{U}_R) \text{ for all } \check{U}_R \in \mathscr{F}(\check{V}\backslash\check{\Sigma}_R \cup \check{\Sigma}_{SR}, R).$$

a_R *and its transform* a_{SR} *are associated to the same function* a^0, *which satisfies*

$$a^0(v, z) = a_R^0(v, Rz) = a_{SR}^0(v, SRz) \qquad \forall v, x. \tag{6.9}$$

4°) *The formal differential operators* a_R *and* b_R *are adjoint, that is,*

$$(a_R \check{U}_R, \check{U}'_R) = (\check{U}_R, b_R \check{U}'_R) \qquad \forall \check{U}_R, \check{U}'_R,$$

if and only if

$$b^0(v, z) = \overline{a^0(v, z)} \qquad \forall v, z.$$

5°) *If* a_R *and* b_R *are two differential operators associated to elements* a^0 *and* b^0 *of* $\mathscr{F}^0(Z)$, *then their composition* $c_R = a_R \circ b_R$ *is associated to the element* c^0 *of* $\mathscr{F}^0(Z)$ *defined by the formula*

$$c^0(v, z) = \left\{\left[\exp - \frac{1}{2v}\left\langle \frac{\partial}{\partial z}, \frac{\partial}{\partial z'} \right\rangle\right][a^0(v, z) b^0(v, z')]\right\}_{z=z'}; \tag{6.10}$$

there

$$\left[\frac{\partial}{\partial z}, \frac{\partial}{\partial z'}\right] = \left\langle \frac{\partial}{\partial x}, \frac{\partial}{\partial p'} \right\rangle - \left\langle \frac{\partial}{\partial x'}, \frac{\partial}{\partial p} \right\rangle, \tag{6.11}$$

where $Rz = (x, p)$, $Rz' = (x', p')$, *is clearly independent of* R.

Remark 6. $c^0 = a^0 \circ b^0 + (1/2v)(a^0, b^0) \bmod(1/v^2)$, $(\cdot, \cdot)$ being the Poisson bracket; see II,§3,(3.14).

Proof of 1°). The definition of a_R.

Proof of 2°). Use the definition of a_R and the definitions of the topologies of $\mathscr{F}^0(\Omega)$ (section 6) and of $\mathscr{F}(\check{V}\backslash\check{\Sigma}_R, R)$ (section 2).

Proof of 3°), 4°), *and* 5°). Apply I,§1 (theorem 3.1, theorem 3.2, and lemma 3.2) to the special case in which a^0 is a polynomial. This special case implies the general case, since (6.8) is continuous and the polynomials are everywhere dense in $\mathscr{F}^0(\Omega)$.

Corollaries 3.1 (adjoint of an operator) and 3.2 (self-adjoint operator) of I,§1 clearly apply to formal differential operators.

7. Formal Distributions

In sections 1 and 2, it is not possible to replace $\mathscr{S}(X)$ by $\mathscr{S}'(X)$: the proof of theorem 2.2 breaks down because $\mathscr{S}'(X)$, unlike $\mathscr{S}(X)$, does not have a countable fundamental system of neighborhoods of 0.

Definition 7.1. The *algebra* $\mathscr{D}(\check{V}\backslash\check{\Sigma}_R, R)$ is the set of functions

$$f: \check{V}\backslash\check{\Sigma}_R \to \mathbf{C}$$

that are infinitely differentiable and compactly supported. $x = \check{R}_X\check{z}$ is used as a local coordinate on $\check{V}\backslash\check{\Sigma}_R$.

Given a compact set $\check{K}$ of $\check{V}\backslash\check{\Sigma}_R$ and a positive number b_r for every l-index r, we let

$$B(\check{K}, \{b_r\}) = \left\{ f \in \mathscr{D}(\check{V}\backslash\check{\Sigma}_R, R) \,\middle|\, \operatorname{Supp} f \subset \check{K}, \left| \left(\frac{\partial}{\partial x}\right)^r f \right| < b_r \right\}.$$

The subsets of these $B(\check{K}, \{b_r\})$ are by definition the bounded sets of $\mathscr{D}(\check{V}\backslash\check{\Sigma}_R, R)$.

Definition 7.2. The *vector space* $\mathscr{D}'(\check{V}\backslash\check{\Sigma}_R, R)$ is the set of linear mappings

$$f': \mathscr{D}(\check{V}\backslash\check{\Sigma}_R, R) \to \mathbf{C}$$

that are bounded on each bounded set of $\mathscr{D}$.

These f' are distributions; they have supports and derivatives of all orders with respect to $x = \check{R}_X\check{z}$; these derivatives are distributions. We make the convention $\mathscr{D} \subset \mathscr{D}'$; $\mathscr{D}'$ is a vector space over the algebra $\mathscr{D}$.

The value of $f' \in \mathscr{D}'$ at $f \in \mathscr{D}$ is denoted

$$\int_{\check{V}} f'(\check{z}) f(\check{z}) d^l x, \text{ where } x = \check{R}_X \check{z}.$$

Then

$$\frac{1}{v}\frac{\partial}{\partial x}$$

is defined on $\mathscr{D}'$ by the requirement that it be self-adjoint:

$$\int_{\check{V}} \left[\frac{1}{v}\frac{\partial}{\partial x} f'(\check{z})\right] \overline{f(\check{z})}\, d^l x = \int_{\check{V}} f'(\check{z}) \overline{\left[\frac{1}{v}\frac{\partial}{\partial x} f(\check{z})\right]}\, d^l x.$$

Given a bounded set B in $\mathscr{D}(\check{V}\backslash\check{\Sigma}_R, R)$, we let

$$|f'|_B = \operatorname*{Sup}_{f \in B} \left| \int_{\check{V}} f'(\check{z}) f(\check{z})\, d^l x \right| < \infty.$$

The topology of $\mathscr{D}'(\check{V}\backslash\check{\Sigma}_R, R)$ is defined by the following fundamental system of neighborhoods of 0, each depending on a bounded set B of $\mathscr{D}$ and on a number $\varepsilon > 0$:

$$\mathscr{N}'(B, \varepsilon) = \{f' \,|\, |f'|_B \leqslant \varepsilon\}.$$

The f' with compact support form a subspace $\mathscr{D}'_0$ of $\mathscr{D}'$ that clearly is dense in $\mathscr{D}'$. It can be proved (as in L. Schwartz [13], chapter VI, §4, theorem IV, theorem XI and its comment) that $\mathscr{D}$ is dense in $\mathscr{D}'_0$, and thus

$\mathscr{D}(\check{V}\backslash\check{\Sigma}_R, R)$ is dense in $\mathscr{D}'(\check{V}\backslash\check{\Sigma}_R, R)$.

Definition 7.3. A *formal distribution* $\check{U}'_R$ on $\check{V}$ consists of

i. a 2-frame R of Z,
ii. a formal series

$$\check{U}'_R = \sum_{r \in \mathbf{N}} \frac{\alpha'_r}{v^r} e^{v\varphi_R},$$

where φ_R is the phase of RV and $\alpha'_r \in \mathscr{D}'(\check{V}\backslash\check{\Sigma}_R, R)$.

$$\operatorname{Supp} \check{U}'_R = \overline{\bigcup_r \operatorname{Supp} \alpha'_r} \subset \check{V}.$$

The set of formal distributions $\check{U}'_R$ on $\check{V}$ forms a vector space $\mathscr{F}'(\check{V}\backslash\check{\Sigma}_R, R)$. Its topology is defined by the following fundamental system of neighborhoods of 0, each depending on a bounded set B of $\mathscr{D}(\check{V}\backslash\check{\Sigma}_R, R)$ and on two numbers $\varepsilon > 0$, $N > 0$:

$$\mathscr{N}'(B, \varepsilon, N) = \{\check{U}'_R \,|\, |\alpha'_r|_B < \varepsilon \text{ for } r \leqslant N\}.$$

Clearly

$$\mathscr{F}_0(\check{V}\backslash\check{\Sigma}_R, R) \text{ is dense in } \mathscr{F}'(\check{V}\backslash\check{\Sigma}_R, R).$$

By (4.6), *the isomorphism* (4.9) *defined by* $S = RR'^{-1} \in Sp_2(l)$ extends to an isomorphism

$$S : \mathscr{F}'(\check{V}\backslash\check{\Sigma}_R \cup \check{\Sigma}_{R'}, R') \to \mathscr{F}'(V\backslash\check{\Sigma}_R \cup \check{\Sigma}_{R'}, R)$$

that is *continuous* and *local*. Similarly, by (6.6), the *formal differential operator* a_R extends to an endomorphism

$$a_R : \mathscr{F}'(\check{V}\backslash\check{\Sigma}_R, R) \to \mathscr{F}'(\check{V}\backslash\check{\Sigma}_R, R)$$

that is continuous and local and satisfies theorem 6.

The *scalar product* of $\check{U}_R$ and $\check{U}'_R$, a formal function and distribution *on the same* $\check{V}\backslash\check{\Sigma}_R$, is defined by formula (5.4) of theorem 5.1 when $\mathrm{Supp}\, \check{U}_R \cap \mathrm{Supp}\, \check{U}'_R$ is compact; it is *invariant* under $Sp_2(l)$ (part 4 of theorem 5.1) and satisfies theorem 5.2.

§2. Lagrangian Analysis

0. Summary

Synthesizing the properties of formal functions and formal differential operators from §1, we define and study *lagrangian operators*, *lagrangian functions* and *distributions*, and their *scalar products* in §2. These three notions make up the structure called *lagrangian analysis*.

In section 1, we define *lagrangian operators* and study their inverses.

In section 2, we define *lagrangian functions on the covering space* $\check{V}$ of a lagrangian manifold V; their scalar product $(\cdot \,\check{|}\, \cdot)$ is defined when their supports are compact.

This makes it possible to define *lagrangian functions on* V in section 3; their *scalar product* is defined when the intersection of their support is compact; section 3 requires *the datum of a number* $\nu_0 \in i]0, \infty[$; it uses the "restriction" of ν to the value ν_0 in the expressions U_R of a formal function U.

This process of defining lagrangian functions on V has a theoretical justification, which is the possibility of defining their scalar product $(\cdot|\cdot)$, and a pragmatic justification, namely, *the process of many "approximate calculations."* An example is quantum physics, where $2\pi i/h$ plays the role of our variable $\nu \in i[1, \infty[$, h being Planck's constant; here h is taken to be infinitely small without comparing the orders of magnitude of the numerical values taken by the terms coming into play when h is given its physical value; h is given this physical value *after* completing the calculations in which h is assumed to be infinitely small. Another historical example is celestial mechanics; *H. Poincaré* has elucidated how such a process is an approximate calculation capable of predicting celestial phenomena with extreme precision using divergent series in which ultimately only the first terms have to be retained. *V. P. Maslov* wants to obtain "asymptotics" by the same process, that is, by making approximate calculations. *In chapter III, we shall apply this process to cases in which it is certainly not an approximate calculation*, and we shall recover numerical results that are in agreement with experimental results. Therefore, *to make this process coherent is "a problem that poses itself and not a problem that one poses"* in the sense that H. Poincaré meant it.

1. Lagrangian Operators

Let Z be a symplectic space and Ω an open set in Z.

Definition 1.1. Let $a^0 \in \mathscr{F}^0(\Omega)$ (II,§1,definition 6.1). Let

$$a_R = a_R^+\left(\nu, x, \frac{1}{\nu}\frac{\partial}{\partial x}\right) = a_R^-\left(\nu, \frac{1}{\nu}\frac{\partial}{\partial x}, x\right)$$

be the formal differential operator associated to a^0 and to the 1-frame R (II,§1,definition 6.2). The *lagrangian operator associated to* a^0 is the collection $a = \{a_R\}$ of these formal differential operators a_R (a^0 given; R arbitrary); a_R is called the *expression of* a *in the frame* R.

Theorem 2.3 will justify this terminology.

We say that *a is defined on* Ω. We define

$$\operatorname{Supp}(a) = \operatorname{Supp}(a^0).$$

If

$$a(v, z) = c \in \mathbf{C} \qquad \forall v, z,$$

then a is denoted

$$a = c.$$

Formula (2.9) will justify the following definition.

Definition 1.2. Two lagrangian operators a and b associated to a^0 and $b^0 \in \mathscr{F}^0(\Omega)$ are *adjoint* when

$$b^0(v, z) = \overline{a^0(v, z)} \qquad \forall v, z;$$

equivalently, by II,§1,theorem 6,4°), a and b are adjoint when a_R and b_R are adjoint for every R.

Notation. The *adjoint of a* is denoted a^*.

Formula (1.2) and the resulting formula (2.2) will justify the following definition.

Definition 1.3. The *composition $a^0 \circ b^0$ of a^0 and $b^0 \in \mathscr{F}^0(\Omega)$* is given by

$$(a^0 \circ b^0)(v, z) = \left\{\left(\exp -\frac{1}{2v}\left[\frac{1}{\partial z}, \frac{1}{\partial z'}\right]\right)\Big/ a^0(v, z)\cdot b^0(v, z')\right\}_{z=z'},$$

$$\text{where } \left[\frac{\partial}{\partial z}, \frac{\partial}{\partial z'}\right] \text{ is defined by II,§1,(6.11).} \tag{1.1}$$

The *composition $a \circ b$ of the lagrangian operators a and b is the lagrangian operator associated to $a^0 \circ b^0$.*

By part 5 of II,§1,theorem 6,

$$(a \circ b)_R = a_R \circ b_R. \tag{1.2}$$

Since composition of the a_R is associative, composition of the a^0 and composition of the a are *associative*. It is easy to verify this directly by observing that (1.1) implies

$$(a^0 \circ b^0 \circ c^0)(v, z)$$

$$= \left\{\left(\exp -\frac{1}{2v}\left[\frac{\partial}{\partial z}, \frac{\partial}{\partial z'}\right] - \frac{1}{2v}\left[\frac{\partial}{\partial z}, \frac{\partial}{\partial z''}\right] - \frac{1}{2v}\left[\frac{\partial}{\partial z'}, \frac{\partial}{\partial z''}\right]\right)\right.$$

$$\left.\cdot a^0(v, z)\cdot b^0(v, z')\cdot c^0(v, z'')\right\}_{z=z'=z''}. \tag{1.3}$$

This formula easily extends to the composition of any number of elements of $\mathscr{F}^0(\Omega)$.

Clearly

$$\operatorname{Supp}(a \circ b) \subset \operatorname{Supp}(a) \cap \operatorname{Supp}(b), \tag{1.4}$$

$$(a^0 \circ b^0)(v, z) = a^0(v, z) \cdot b^0(v, z) \bmod \frac{1}{v}, \tag{1.5}$$

the right-hand side being the product in $\mathscr{F}^0$ of the values of a^0 and b^0:

$$(a^0 \circ a^0)(v, z) = \left\{\cosh \frac{1}{2v}\left[\frac{\partial}{\partial z}, \frac{\partial}{\partial z'}\right] a^0(v, z) \cdot a^0(v, z')\right\}_{z=z'}. \tag{1.6}$$

The set of lagrangian operators defined on Ω *is thus a noncommutative algebra* $\mathscr{F}^0(\Omega)$ over the algebra $\mathscr{F}^0$ of formal numbers with vanishing phase; $\mathscr{F}^0(\Omega)$ has *an identity element.*

THEOREM 1.1. (Inverse of a lagrangian operator) *The operator a has an inverse in the algebra of lagrangian operators defined on* Ω *if and only if*

$$a^0(v, z) \neq 0 \bmod \frac{1}{v} \qquad \forall z \in \Omega. \tag{1.7}$$

Recall that, by definition, $a^0(v, z)$ is a formal series:

$$a^0(v, z) = \sum_{r \in \mathbf{N}} \frac{1}{v^r} a_r(z). \tag{1.8}$$

Condition (1.7) means

$$a_0(z) \neq 0 \qquad \forall z \in \Omega. \tag{1.7a}$$

Proof. The equivalent conditions (1.7) and (1.7*a*) are necessary by (1.5).

Conversely, if these conditions hold, a right inverse a' of a is an operator associated to an element

$$a'(v, \cdot) = \sum_{r \in \mathbf{N}} \frac{1}{v^r} a'_r(\cdot)$$

of $\mathscr{F}^0(\Omega)$ satisfying

$$\left\{\left(\exp -\frac{1}{2v}\left[\frac{\partial}{\partial z}, \frac{\partial}{\partial z'}\right]\right) a^0(v, z) a'^0(v, z')\right\}_{z=z'} = 1, \tag{1.9}$$

that is,

$$a_0(z)a'_0(z) = 1, \tag{1.10$_0$}$$

$$a_0(z)a'_r(z) = c_r(z), \tag{1.10$_r$}$$

where $c_r(z)$ is a linear combination of the

$$\left[\left(\frac{\partial}{\partial z}\right)^s a_t(z)\right]\left[\left(\frac{\partial}{\partial z}\right)^{s'} a'_{t'}(z)\right]$$

such that $0 \leqslant |s| = |s'| = r - t - t', t' < r$. These conditions determine a'. Therefore a has a right inverse and, similarly, a left inverse; these are identical since the composition of elements of $\mathscr{F}^0(\Omega)$ is associative. The theorem follows.

The definition of scalar products (sections 2 and 3) will use the following.

Definition 1.4. A *partition of unity* is a collection $\{a_j\}$ of lagrangian operators defined on Z such that $\bigcup_j \operatorname{Supp} a_j = Z$ is a locally finite covering of Z and

$$\sum_j a_j = 1 \text{ (identity operator).} \tag{1.11}$$

This partition is said to be finer than an open covering $\bigcup_k \Omega_k = X$ of X when every $\operatorname{Supp} a_j$ belongs to at least one of the Ω_k.

THEOREM 1.2. (Partition of unity) 1°) *There exist partitions of unity finer than a given open covering of X.*

2°) *These can be chosen so that the operators a_j are self-adjoint.*

3°) *More precisely, these partitions of unity can be chosen such that each a_j has the form*

$$a_j = b_j^* \circ b_j, \qquad \operatorname{Supp} b_j = \operatorname{Supp} a_j,$$

where b_j is a lagrangian operator and b_j^* is its adjoint.

Proof of 1°) *and* 2°). It suffices to prove the theorem when the operators a_j are replaced by C^∞-functions $a_j^0 : Z \to \mathbf{R}$ that are independent of ν. It is thus a classical result.

Proof of 3°). Given a covering $\bigcup_k \Omega_k = X$, we choose C^∞-functions $b_j^0 : Z \to \mathbf{R}$ such that

$$\sum_j [b_j^0(z)]^2 = 1, \qquad \operatorname{Supp} b_j^0 \subset \Omega_k \text{ for some } k.$$

It follows by (1.6) that

$$\sum_j b_j^0 \circ b_j^0 = \sum_{r\in\mathbf{N}} \frac{1}{\nu^{2r}} \beta_r, \text{ where } \beta_r : Z \to \mathbf{R},\ \beta_0 = 1.$$

Let us try to find

$$c^0 = \sum_{r\in\mathbf{N}} \frac{1}{\nu^{2r}} \gamma_r, \text{ where } \gamma_r : Z \to \mathbf{R}, \qquad \gamma_0 = 1,$$

such that

$$c^0 \circ c^0 = \sum_j b_j^0 \circ b_j^0,$$

that is,

$$\left\{\cosh \frac{1}{2\nu}\left[\frac{\partial}{\partial z}, \frac{\partial}{\partial z'}\right] \sum_{t,t'} \frac{1}{\nu^{2(t+t')}} \gamma_t(z)\gamma_{t'}(z')\right\}_{z=z'} = \sum_r \frac{1}{\nu^{2r}} \beta_r.$$

This condition defines γ_r as a function of β_r and of the

$$\left[\left(\frac{\partial}{\partial z}\right)^s \gamma_t(z)\right]\left[\left(\frac{\partial}{\partial z}\right)^{s'} \gamma_{t'}(z)\right]$$

such that

$$0 \leqslant |s| - |s'| = 2(r - t - t'), \qquad 0 < t < r;$$

thus c^0 exists, is unique, and has an inverse by theorem 1.1. Now

$$b_j = b_j^*, \qquad c = c^*, \qquad \sum_j b_j \circ b_j = c \circ c;$$

therefore the

$$a_j = (b_j \circ c^{-1})^* \circ (b_j \circ c^{-1})$$

constitute a partition of unity finer than the given covering $\bigcup_k \Omega_k = X$.

2. Lagrangian Functions on $\check{V}$

Let Z be a symplectic space provided with a 2-symplectic geometry. Let V be a lagrangian manifold in Z and let $\check{V}$ be its universal covering space. Theorem 4.1 of §1 justifies the following definition.

Definition 2.1. A *lagrangian function* $\check{U}$ on $\check{V}$ is defined by the datum of a formal function $\check{U}_R$ on $\check{V}\backslash\check{\Sigma}_R$ for each 2-frame R, these formal functions satisfying

$$\check{U}_R = RR'^{-1}\check{U}_{R'} \text{ on } \check{V}\backslash\check{\Sigma}_R \cup \check{\Sigma}_{R'} \qquad \forall R, R'.$$

$\check{U}_R$ is called the *expression of* $\check{U}$ *in the frame* R.

The *support of* $\check{U}$, Supp $\check{U}$, is the subset of $\check{V}$ defined by the condition

$$\operatorname{Supp}\check{U}_R = (\check{V}\backslash\check{\Sigma}_R) \cap \operatorname{Supp}\check{U} \qquad \forall R.$$

This definition makes sense since $S = RR'^{-1}$ is local.

The set of lagrangian functions on $\check{V}$ and the subset of those with compact support are vector spaces over $\mathscr{F}$; they are denoted $\mathscr{F}(\check{V})$ and $\mathscr{F}_0(\check{V})$. Their dimension is infinite, as is proven by the following theorem.

THEOREM 2.1. (Existence) *Every* $\check{U}_{R'} \in \mathscr{F}_0(\check{V}\backslash\check{\Sigma}_{R'}, R')$ *is the expression in* R' *of a lagrangian function* $\check{U}$ *on* $\check{V}$ *with compact support.*

Proof. Let R be any 2-frame. Let

$$S = RR'^{-1} \in \mathrm{Sp}_2(l), \qquad \check{U}_R = S\check{U}_{R'};$$

$\check{U}_R$ is a formal function defined on $\check{V}\backslash\check{\Sigma}_R \cup \check{\Sigma}_{R'}$ which is identically zero in a neighborhood of $\check{\Sigma}_{R'}$. We extend its definition by making the following convention:

$$\check{U}_R = 0 \text{ on } \check{\Sigma}_{R'}\backslash\check{\Sigma}_R \cap \check{\Sigma}_{R'}.$$

Then $\check{U}_R$ becomes a formal function defined on $\check{V}\backslash\check{\Sigma}_R$; $\check{U} = \{\check{U}_R\}$ is clearly a lagrangian function and

$$\operatorname{Supp}\check{U} = \operatorname{Supp}\check{U}_{R'}.$$

II,§1,theorem 4.2 proves the following.

THEOREM 2.2. (Structure) *Let* ψ *be the lagrangian phase of* $\check{V}$, η *a positive regular measure on* V, *and* $x = \check{R}_X\check{z} \in X$, *where* $\check{z} \in \check{V}$. *We make the convention* $[\eta]^{1/2} > 0$. *We give* V *a 2-orientation; recall that* $[d^lx]^{1/2}$ *is defined by* I,§3,*corollary* 3 *using the Maslov index.*

We have

$$\check{U}_R(v, \check{z}) = \sum_{r\in\mathbf{N}} \left(\frac{\eta}{d^lx}\right)^{3r+1/2} \frac{1}{v^r}\beta_{Rr}(\check{z})e^{v\varphi_R(\check{z})} \tag{2.1}$$

on $\check{V}\backslash\check{\Sigma}_R$, *where the* β_{Rr} *are infinitely differentiable functions* $\check{V} \to \mathbf{C}$.

β_{R0} *is independent of* R, *is denoted* β_0, *and is called the* ***lagrangian amplitude.***

Remark 2.1. In (2.1), the exponent $3r + \frac{1}{2}$ cannot be decreased: see III,§3,theorem 4.2, where the polynomial f_r has degree $3r$ and the rational function g_r poles of order $3r$.

II,§1,theorem 6 justifies the following definition.

Definition 2.2. Let Ω be an open neighborhood of V. Let a be a lagrangian operator on Ω and $\check{U}$ a lagrangian function on $\check{V}$. Let R and R' be 2-frames and $S = RR'^{-1} \in \mathrm{Sp}_2(l)$. Then

$$a_R\check{U}_R = S(a_{R'}\check{U}_{R'});$$

thus the formal functions $a_R\check{U}_R$ form a lagrangian function, which we shall denote $a\check{U}$.

In other words

THEOREM 2.3. *The algebra of lagrangian operators* a, b *defined on* $\Omega \supset V$ *acts on the vector space* $\mathscr{F}(\check{V})$ [*and* $\mathscr{F}_0(\check{V})$] *of lagrangian functions* $\check{U}$ [*with compact support*] *defined on* $\check{V}$;

$$a(b\check{U}) = (a \circ b)\check{U}; \tag{2.2}$$

$$\mathrm{Supp}(a\check{U}) = \mathrm{Supp}\,\check{U} \cap \check{\Pi}^{-1}\,\mathrm{Supp}\,a, \tag{2.3}$$

$\check{\Pi}$ *being the projection of* $\check{V}$ *onto* $V \subset Z$.

Definition 2.3 *of the scalar product* $(\cdot\check{|}\cdot)$. Let $\check{U} \in \mathscr{F}_0(V)$ and $U' \in \mathscr{F}_0(\check{V}')$. Let us first assume that the following condition holds:

There exists a 2-frame R such that

$$\mathrm{Supp}\,\check{U} \cap \check{\Sigma}_R = \varnothing, \qquad \mathrm{Supp}\,\check{U}' \cap \check{\Sigma}'_R = \varnothing. \tag{2.4}$$

This condition means

$$\check{U}_R \in \mathscr{F}_0(\check{V}\backslash\check{\Sigma}_R, R), \qquad \check{U}'_R \in \mathscr{F}_0(\check{V}'\backslash\check{\Sigma}'_R, R).$$

We then define

$$(\check{U}\,\check{|}\,\check{U}') = (\check{U}_R\,\check{|}\,\check{U}'_R) \in \mathscr{C} \text{ for each } R \text{ satisfying (2.4).} \tag{2.5}$$

The right-hand side is an asymptotic class by II,§1,definition 5.1. It is independent of R by part 4 of II,§1,theorem 5.1.

Let $\check{U}$ and $\check{U}'$ no longer satisfy (2.4). Let

$$\sum_j a_j = 1, \qquad \sum_{j'} a'_{j'} = 1 \tag{2.6}$$

be two partitions of unity (definition 1.4) sufficiently fine (theorem 1.2) so that for every choice of (j, j') there correspond 2-frames $R_{j,j'}$ satisfying the condition

$$\operatorname{Supp} a_j \cap \Sigma_{R_{j,j}} = \varnothing, \qquad \operatorname{Supp} a'_{j'} \cap \Sigma_{R_{j,j'}} = \varnothing; \tag{2.7}$$

thus

$(a_j\check{U} \mid a'_{j'}\check{U}')$ is defined $\forall j, j'$.

We define

$$(\check{U} \mathbin{\check{|}} \check{U}') = \sum_{j,j'} (a_j\check{U} \mathbin{\check{|}} a'_{j'}\check{U}') \in \mathscr{C}. \tag{2.8}$$

The right-hand side is independent of the choice of partitions of unity (2.6); indeed, if

$$\sum_k b_k = 1, \qquad \sum_{k'} b'_{k'} = 1$$

is a second choice satisfying (2.7), then

$$\sum_{j,j'} (a_j\check{U} \mid a'_{j'}\check{U}')$$

$$= \sum_{j,j'} (a_j\check{U}_{R_{j,j'}} \mathbin{\check{|}} a'_{j'}\check{U}'_{R_{j,j'}}) = \sum_{j,j',k,k'} (a_j \circ b_k\check{U}_{R_{j,j'}} \mathbin{\check{|}} a'_{j'} \circ b'_{k'}\check{U}'_{R_{j,j'}})$$

$$= \sum_{k,k'} (b_k\check{U} \mathbin{\check{|}} b'_{k'}\check{U}'),$$

since $R_{j,j'}$ can be replaced by $R_{k,k'}$.

THEOREM 2.4. (Scalar product) 1°) *The scalar product* $(\cdot \mathbin{\check{|}} \cdot)$ *is a function of pairs* $\check{U}$ *and* $\check{U}'$ *of lagrangian functions on* V *and* V' *with compact support. It is sesquilinear over* $\mathscr{F}^0$ *and takes values in* $\mathscr{C}$ *(asymptotic equivalence classes).*

$(\check{U} \mathbin{\check{|}} \check{U}')$ *and* $(\check{U}' \mathbin{\check{|}} \check{U})$ *are complex conjugate and depend only on the* *behavior of $\check{U}$ and $\check{U}'$ at the pairs of points of $\check{V}$ and $\check{V}'$ projecting onto the same point of $V \cap V'$.*

If $V \cap V' = 0$, then $(\check{U} \,\check{|}\, \check{U}') = 0$.

$(a\check{U} \,\check{|}\, \check{U}') = (\check{U} \,\check{|}\, a^* \check{U}')$ *if the lagrangian operators a and a^* are adjoint.* (2.9)

2°) *If $V = V'$, which implies $\psi = \psi'$, then*

$$(\check{U} \,\check{|}\, \check{U}') \in \mathscr{F}, \tag{2.10}$$

the phases of $(\check{U} \,\check{|}\, \check{U}')$ being the periods of ψ (that is, the values of $\frac{1}{2}\int_\gamma [z, dz]$ on the cycles γ of V);

$$(\check{U} \,\check{|}\, \check{U}') = \int_{z \in V} \sum_{\check{z}, \check{z}'} \beta_0(\check{z}) \overline{\beta'_0(\check{z}')} e^{\nu[\psi(\check{z}) - \psi(\check{z}')]} \eta \mod \frac{1}{\nu}, \tag{2.11}$$

where the notation is that of theorem 2.2 and $\check{z}$ and $\check{z}'$ are the points of Supp $\check{U}$ *and* Supp $\check{U}'$ *projecting onto z in V. We have*

$$0 \leqslant (\check{U} \,\check{|}\, \check{U}) \in \mathscr{F}. \tag{2.12}$$

The seminorm of U,

$$0 \leqslant (\check{U} \,\check{|}\, \check{U})^{1/2} \in \mathscr{C},$$

satisfies the triangle and Schwarz inequalities:

$$|(\check{U} \,\check{|}\, \check{U}')| \leqslant (\check{U} \,\check{|}\, \check{U})^{1/2} \cdot (\check{U}' \,\check{|}\, \check{U}')^{1/2} \text{ in } \mathscr{C}.$$

3°) *If V and V' are transverse (where V and V' are given 2-orientations), then*

$$\nu^{l/2}(\check{U} \,\check{|}\, \check{U}') \in \mathscr{F}; \tag{2.13}$$

$$\left(\frac{|\nu|}{2\pi i}\right)^{l/2} (\check{U} \,\check{|}\, \check{U}')$$

$$= \sum_{z \in V \cap V'} \left| \frac{\eta_0 \wedge \eta'_0}{d^{2l} z} \right|^{1/2} \sum_{\check{z}, \check{z}'} \beta_0(\check{z}) \overline{\beta'_0(\check{z}')} e^{\nu[\psi(\check{z}) - \psi'(\check{z}')] - (\pi i/2) m(\lambda'_4, \lambda_4)} \mod \frac{1}{\nu}, \tag{2.14}$$

where

- $\eta_0 \wedge \eta'_0$ *and $d^{2l}z$ are the measures on Z defined at the point $z \in V \cap V'$ by* II,§1,(5.11) *and* II,§1,(5.15) *respectively,*
- $\check{z}$ *and $\check{z}'$ are the points of* Supp $\check{U}$ *and* Supp $\check{U}'$ *projecting onto z,*

• *m is the Maslov index defined* mod 4 (I,§3,theorem 1),
• λ_4 *and* λ'_4 *are the 2-oriented lagrangian planes tangent to* $\check{V}$ *and* $\check{V}'$ *at* $\check{z}$ *and* $\check{z}'$, *respectively.*

Proof of 1°). Use definition 2.3, part 1 of II,§1,theorem 5.1, part 4 of §1,theorem 6, and definition 1.2 of the adjoint of a lagrangian operator.

Proof of (2.10). Use §1,definition 2.3 and §1,(5.4).

Proof of (2.11). Use §1,theorem 5.2 and the fact that $a\check{U}$ mod$(1/\nu)$ is multiplication of $\check{U}$ by the function a_0^0.

Proof of (2.12). We use a partition of unity [part 3 of theorem 1.2]

$$\sum_j b_j^* \circ b_j = 1$$

sufficiently fine so that for every j there exists a frame R_j satisfying

$$\operatorname{Supp} b_j \cap \Sigma_{R_j} = \varnothing.$$

We have

$$(\check{U} \,\check{|}\, \check{U}) = \sum_j (b_j \check{U}_{R_j} \,\check{|}\, b_j \check{U}_{R_j}) \geqslant 0 \text{ (§1,definition 5.1).}$$

Proof of the Schwarz inequality. By §1,definition 5.1, we have the triangle and Schwarz inequalities in R [§1,(5.3)] and in $\mathscr{C}$ (§1,remark 1); hence,

$$\begin{aligned}
(\check{U} \,\check{|}\, \check{U}') &= \left| \sum_j (b_j \check{U}_{R_j} \,\check{|}\, b_j \check{U}'_{R_j}) \right| \\
&\leqslant \sum_j |(b_j \check{U}_{R_j} \,\check{|}\, b_j \check{U}'_{R_j})| \\
&\leqslant \sum_j (b_j \check{U}_{R_j} \,\check{|}\, b_j \check{U}_{R_j})^{1/2} \cdot (b_j \check{U}'_{R_j} \,\check{|}\, b_j \check{U}'_{R_j})^{1/2} \\
&\leqslant \left[\sum_j (b_j \check{U}_{R_j} \,\check{|}\, b_j \check{U}_{R_j}) \right]^{1/2} \cdot \left[\sum_j (b_j \check{U}'_{R_j} \,\check{|}\, b_j \check{U}'_{R_j}) \right]^{1/2} \\
&= (\check{U} \,\check{|}\, \check{U})^{1/2} \cdot (\check{U}' \,\check{|}\, \check{U}')^{1/2}.
\end{aligned}$$

The triangle inequality follows from the Schwarz inequality.

Proof of 3°). Use §1,theorem 5.3.

Remark 2.2. If $V = V'$, $(\check{U}\check{|}\check{U}')$ is defined by the right-hand side of §1, (5.4) under the assumption.

$$\operatorname{Supp}\check{U} \cap \check{\Sigma}_R = \varnothing, \qquad \operatorname{Supp}\check{U}' \cap \check{\Sigma}_R = \varnothing.$$

If this assumption is not satisfied, the right-hand side is a divergent integral in view of theorem 2.2 (structure).
Thus *definition* 2.3 of $(\cdot\check{|}\cdot)$ *gives a meaning to this divergent integral.*

3. Lagrangian Functions on V

We are interested in functions on V rather than functions on $\check{V}$, that is, multiforms on V. The definition of lagrangian functions on V requires the datum of a *number* v_0; it will be convenient (see, for instance, Maslov's quantization, II,§3,6) to choose $v_0 \in i]0, \infty[$.

The first homotopy group $\pi_1[V]$ acts on $\check{V}$: V is the quotient of $\check{V}$ by $\pi_1[V]$. Clearly

$\pi_1[V]$ leaves $\check{\Sigma}_R$ invariant $\forall R$;

$$\psi(\gamma\check{z}) = \psi(\check{z}) + c_\gamma, \qquad \varphi_R(\gamma\check{z}) = \varphi_R(\check{z}) + c_\gamma,$$

$$\text{where } c_\gamma = \tfrac{1}{2}\int_\gamma [z, dz] \quad \forall\gamma \in \pi_1[V];$$

that is,

$$\psi \circ \gamma^{-1} = \psi - c_\gamma, \qquad \varphi_R \circ \gamma^{-1} = \varphi_R - c_\gamma.$$

Definition 3.1 *of the group* $\pi_1[V]$ *of automorphisms of* $\mathscr{F}(\check{V})$. Let

$$\check{U}_R = \sum_{r\in\mathbf{N}} \frac{\alpha_{R,r}}{v^r} e^{v\varphi_R} \in \mathscr{F}(\check{V}\backslash\check{\Sigma}_R, R), \qquad \gamma \in \pi_1[V]. \tag{3.1}$$

Define the transform of $\check{U}_R$ by γ not as

$$\check{U}_R \circ \gamma^{-1} = e^{-vc_\gamma}\sum_{r\in\mathbf{N}} \frac{\alpha_{R,r} \circ \gamma^{-1}}{v^r} e^{v\varphi_R} \notin \mathscr{F}(\check{V}\backslash\check{\Sigma}_R, R)$$

but as

$$\gamma\check{U}_R = e^{(v-v_0)c_\gamma}\check{U}_R \circ \gamma^{-1} = e^{-v_0c_\gamma}\sum_{r\in\mathbf{N}} \frac{\alpha_{R,r} \circ \gamma^{-1}}{v^r} e^{v\varphi_R} \in \mathscr{F}(\check{V}\backslash\check{\Sigma}_R, R). \tag{3.2}$$

Clearly

$$\gamma'(\gamma \check{U}_R) = (\gamma'\gamma)\check{U}_R,$$

so $\pi_1[V]$ is a *group of automorphisms of* $\mathscr{F}(V\backslash\Sigma_R, R)$.

It clearly follows from definition (3.2) and the definition of the operator $S \in \mathrm{Sp}_2(l)$ in II,§1 (lemma 1.1, part 1 of theorem 4.1, and theorem 4.3) that S and γ commute: If

$$\check{U}_{R'} \in \mathscr{F}(\check{V}\backslash\check{\Sigma}_{R'} \cup \check{\Sigma}_R, R') \text{ and } R = SR',$$

then

$$\gamma(S\check{U}_{R'}) = S(\gamma\check{U}_{R'}). \tag{3.3}$$

If $\check{U}_R$ are the expressions of a lagrangian function $\check{U}$ on $\check{V}$, then $\gamma\check{U}_R$ are the expressions of a lagrangian function on $\check{V}$; denote it by $\gamma\check{U}$. Then $(\gamma\check{U})_R = \gamma\check{U}_R$ and $\pi_1[V]$ is a *group of automorphisms of* $\mathscr{F}(\check{V})$.

Clearly $\forall\gamma \in \pi_1[V]$, $\forall\check{U} \in \mathscr{F}(\check{V})$, $\forall a$ a lagrangian operator,

$$\mathrm{Supp}(\gamma\check{U}) = \gamma\,\mathrm{Supp}\,\check{U}; \tag{3.4}$$

γ and a commute, that is,

$$\gamma(a\check{U}) = a(\gamma\check{U}). \tag{3.5}$$

Definition 3.2. A *lagrangian function on* V is a lagrangian function U on $\check{V}$ having the following three properties, which are clearly pairwise equivalent and thus independent of the choice of the 2-frame R:

i. U is *invariant under* $\pi_1[V]$, that is,

$$\gamma U = U \qquad \forall\gamma \in \pi_1[V] \text{ (see definition 3.1);}$$

ii. $e^{-\nu_0 c_\gamma}\alpha_{R,r} \circ \gamma^{-1}$ is independent of $\gamma \in \pi_1[V]$ $\forall r \in \mathbf{N}$;
iii. the function

$$U_R^0 = \sum_{r\in\mathbf{N}} \frac{\alpha_{R,r}}{\nu^r} e^{\nu_0\varphi_R} = U_R e^{(\nu_0-\nu)\varphi_R}, \tag{3.6}$$

called the *restriction of* U_R *to* ν_0, takes the same value at all points of $\check{V}$ having the same projection onto V. Therefore we may write:

$$U_R^0 : V\backslash\Sigma_R \to \mathscr{F}^0.$$

The set of lagrangian functions on V[with compact support in V] is a Hence it is the set of points in $\check{V}$ projecting onto a closed subset of V, which will be called the *support of U in V* and denoted Supp U. Then

$$\operatorname{Supp} U_R^0 = \operatorname{Supp} U \backslash \Sigma_R \cap \operatorname{Supp} U, \qquad \forall R.$$

The set of lagrangian functions on V [with compact support in V] is a vector space over the algebra $\mathscr{F}^0$. It is denoted $\mathscr{F}(V)$ [$\mathscr{F}_0(V)$].

Clearly, by (3.5) the following theorem holds.

THEOREM 3.1. *The algebra of lagrangian operators a, b defined on* $\Omega \supset V$ *acts on the space* $\mathscr{F}(V)$ *of lagrangian functions on* V;

$$(a \circ b)U = a(bU), \qquad \operatorname{Supp}(aU) \subset \operatorname{Supp} a \cap \operatorname{Supp} U.$$

The scalar product of lagrangian functions on V will be defined using the following lemma and definitions.

LEMMA 3.1. There exists a natural epimorphism

$$\mathscr{F}_0(\check{V}) \ni \check{U} \mapsto U = \sum_{\gamma \in \pi_1[V]} \gamma \check{U} \in \mathscr{F}_0(V). \tag{3.7}$$

Proof. Definition (3.7) makes sense: it defines U by a finite sum, since for every $\check{z} \in V$, there are only finitely many γ such that $\gamma\check{z} \in \operatorname{Supp} \check{U}$.

Clearly U is invariant under $\pi_1[V]$ and has compact support.

Since γ commutes with partitions of unity (definition 1.4, theorem 1.2), to prove (3.7) is an epimorphism it suffices to prove the following: Every $U \in \mathscr{F}_0(V)$ with support in a simply connected subset ω of V is the image under (3.7) of an element $\check{U}$ of $\mathscr{F}_0(\check{V})$. This is proved as follows.

The connected components of the subset of $\check{V}$ projecting onto ω are the transforms of any one of these components $\check{\omega}$ by the elements γ of $\pi_1[V]$. Let $\check{U}$ be the lagrangian function on $\check{V}$ that vanishes outside $\check{\omega}$ and equals U in $\check{\omega}$; then $\gamma\check{U}$ vanishes outside $\gamma\check{\omega}$ and equals $U = \gamma U$ in $\gamma\check{\omega}$; hence

$$U = \sum_{\gamma \in \pi_1[V]} \gamma \check{U}.$$

Let us express the notion of restriction to ν_0 more explicitly.

Definition 3.3. Let $\mathscr{I}$ be the ideal of $\mathscr{F}$ generated by the elements

$$e^{\nu\varphi} - e^{\nu_0\varphi}, \qquad \varphi \in \mathbf{R}.$$

Each element of $\mathscr{I}$ may be written uniquely:

$$\sum_{j\in J} \alpha_j(e^{\nu\varphi_j} - e^{\nu_0\varphi_j}), \qquad J \text{ finite}, \quad \alpha_j \in \mathscr{F}^0, \quad \varphi_j \in \mathbf{R}.$$

$\mathscr{F}$ is the direct sum of $\mathscr{I}$ and $\mathscr{F}^0$. Let us call the projection of $\mathscr{F}$ onto $\mathscr{F}^0$ the *restriction* of $\mathscr{F}$ to ν_0:

$$\mathscr{F} \ni u = \sum_j \alpha_j e^{\nu\varphi_j} \mapsto u^0 = \sum_j \alpha_j e^{\nu_0\varphi_j} \in \mathscr{F}^0 = \mathscr{F}/\mathscr{I}.$$

In particular, if $U \in \mathscr{F}(V)$, then *the restriction of* $U_R(\nu, \check{z}) \in \mathscr{F}$ *to* ν_0 *is the value* $U_R^0(\nu, z)$ *of the restriction* U_R^0 *of* U_R *to* ν_0 at the point z, the projection of $\check{z}$ onto V.

Definition 3.4. Any additive subgroup $\mathscr{M}$ of $\mathscr{C}$ having the properties

$\mathscr{F} \subset \mathscr{M} \subset \mathscr{C}$, $\mathscr{M} = \overline{\mathscr{M}}$ (complex conjugate)

is called a *submodule of* $\mathscr{C}$. Multiplication in the algebra $\mathscr{C}$ makes $\mathscr{M}$ a module over the algebra $\mathscr{F}$.

In particular,

$u \in \mathscr{M}$ and $\varphi \in \mathbf{R}$ imply $ue^{\nu\varphi} \in \mathscr{M}$.

Let $\mathscr{N}$ be the submodule of $\mathscr{M}$ whose elements are

$$\sum_{j\in J} u_j(e^{\nu\varphi_j} - e^{\nu_0\varphi_j}), \qquad J \text{ finite}, \quad u_j \in \mathscr{M}, \quad \varphi_j \in \mathbf{R}.$$

Let us call the quotient

$$\mathscr{M} \to \mathscr{M}/\mathscr{N} = \mathscr{M}^0$$

the restriction of $\mathscr{M}$ *to* ν_0; $\mathscr{M}^0$ is a module over the algebra $\mathscr{F}^0$. Since $\mathscr{N} = \overline{\mathscr{N}}$, conjugation in $\mathscr{M}^0$ is induced by complex conjugation in $\mathscr{M}$.

Example 3.1. If $\mathscr{M} = \mathscr{F}$, then $\mathscr{M}^0 = \mathscr{F}^0$.

Example 3.2. Let $(1/|\nu|)^s\mathscr{F}, \ldots$ denote the images of $\mathscr{F}, \ldots$ under the multiplication

$$\mathscr{C} \ni \tilde{f} \to \frac{1}{|\nu|^s}\tilde{f} \in \mathscr{C}, \qquad s \in \mathbf{R}_+.$$

If $\mathscr{M} = (1/|\nu|^s)\mathscr{F}$, then $\mathscr{M}^0 = (1/|\nu|^s)\mathscr{F}^0$.

Definition 3.5 *of the scalar product* $(\cdot|\cdot)$. Let V and V' be two lagrangian manifolds in Z. Let $\mathscr{M}$ be a submodule of $\mathscr{C}$ (definition 3.4) such that

$$(\check{U}\check{|}\check{U}') \in \mathcal{M}, \quad \forall \check{U} \in \mathcal{F}_0(\check{V}), \quad \check{U}' \in \mathcal{F}_0(\check{V}'). \tag{3.8}$$

For any $U \in \mathcal{F}_0(V)$, $U' \in \mathcal{F}_0(V')$, there exist $\check{U} \in \mathcal{F}_0(\check{V})$, $\check{U}' \in \mathcal{F}_0(\check{V}')$ such that

$$U = \sum_{\gamma \in \pi_1[V]} \gamma \check{U}, \qquad U' = \sum_{\gamma' \in \pi_1[V']} \gamma' \check{U}'$$

according to lemma 3.1. By definition, the *scalar product of U and U′* is

$$(U \mid U') = (\check{U} \check{|} \check{U}')^0, \text{ the restriction of } (\check{U} \check{|} \check{U}') \in \mathcal{M} \text{ to } v_0. \tag{3.9}$$

Thus the scalar product *has values in* $\mathcal{M}^0$.

This definition makes sense by the following lemma.

LEMMA 3.2. $(\check{U}\check{|}\check{U}')^0$ depends only on U and U'.

Proof. Let

$$\sum_j b_j^* \circ b_j = 1$$

be a partition of unity (definition 1.4); we have

$$(\check{U} \check{|} \check{U}') = \sum_j (b_j \check{U} \check{|} b_j \check{U}');$$

it then suffices to prove that $(b_j \check{U} \check{|} b_j \check{U}')^0$ depends only on

$$\sum_{\gamma \in \pi_1[V]} \gamma b_j \check{U} = b_j U \text{ and } \sum_{\gamma' \in \pi_1[V']} \gamma' b_j \check{U}' = b_j U'.$$

By theorem 1.2, it suffices to prove the lemma when the following two conditions hold:

i. There exists a 2-frame R such that

$$\check{U}_R \in \mathcal{F}_0(\check{V} \backslash \check{\Sigma}_R, R), \; \check{U}'_R \in \mathcal{F}_0(\check{V}' \backslash \check{\Sigma}'_R, R).$$

ii. There exist two simply connected open sets, ω in V and ω' in V', that are projected injectively into X under R_X and contain the supports of U and U':

$$\operatorname{Supp} U \subset \omega \subset V, \qquad \operatorname{Supp} U' \subset \omega' \subset V'.$$

The connected components of the subset of $\check{V}$ projecting onto ω are the open sets $\gamma\check{\omega}$ in V, where $\gamma \in \pi_1[V]$ (see the proof of lemma 3.1). Given $x \in R_X \operatorname{Supp} U$, let

z be the unique point of ω such that $R_X z = x$,

$\check{z}$ be the unique point of $\check{\omega}$ projecting onto z in V.

By definition 2.3, §1,(1.9), and definition 5.1,

$$(\check{U} \,\check{|}\, \check{U}') = \tilde{\int_X} (\Pi_R \check{U}_R)(v, x) \cdot \overline{(\Pi_R \check{U}'_R)(v, x)} d^l x. \tag{3.10}$$

By definition (§1,theorem 2.1), if $x \in \operatorname{Supp}(\Pi_R \check{U}_R) = R_X \operatorname{Supp} U$, then

$$(\Pi_R \check{U}_R)(v, x) = \sum_{\gamma \in \pi_1[V]} \check{U}_R(v, \gamma^{-1} \check{z}),$$

that is, by (3.2)

$$(\Pi_R \check{U}_R)(v, x) = \sum_{\gamma \in \pi_1[V]} e^{(v_0 - v)c_\gamma} \gamma \check{U}_R(v, \check{z}).$$

We substitute this expression for $\Pi_R \check{U}_R$ and the analogous expression for $\Pi_R \check{U}'_R$ into (3.10). Using definition 3.4 of $\mathcal{N}$ and (3.7), we obtain the formula

$$(\check{U} \,\check{|}\, \check{U}') = \tilde{\int_X} U_R(v, \check{z}) \cdot \overline{U'_R(v, \check{z}')} d^l x \qquad \operatorname{mod} \mathcal{N} \tag{3.11}$$

under the assumptions (i) and (ii).

This formula (3.11) proves lemma 3.2 and is of use in proving the following theorem, part 1 of which is clear.

THEOREM 3.2. (Scalar product) 1°) *Let V and V' be two lagrangian manifolds in Z. The scalar product $(\cdot \,|\, \cdot)$ is a function of pairs U and U' of lagrangian functions on V and V' that is sesquilinear over $\mathcal{F}^0$.*

- *$(U \,|\, U')$ is defined when* $\operatorname{Supp} U \cap \operatorname{Supp} U'$ *is compact.*
- *$(U \,|\, U') \in \mathcal{M}^0$, a module over $\mathcal{F}^0$ depending on V and V'.*
- *$(U \,|\, U')$ depends only on the behavior of U and U' in a neighborhood of* $\operatorname{Supp} U \cap \operatorname{Supp} U'$.
- *$(U \,|\, U') = 0$ when* $\operatorname{Supp} U \cap \operatorname{Supp} U' = \varnothing$.
- *$(U \,|\, U')$ and $(U' \,|\, U)$ are conjugate.*
- *$(aU \,|\, U') = (U \,|\, a^* U')$ when a and a^* are adjoint lagrangian operators.*

2°) *Suppose $V = V'$. Then $(U \,|\, U') \in \mathcal{F}^0$* (algebra of formal numbers with vanishing phase):

$$0 \leqslant (U \,|\, U), \text{ equality implying } U = 0. \tag{3.12}$$

The norm of U,

$$0 \leqslant \|U\| = (U\,|\,U)^{1/2} \in \mathscr{F}^0, \tag{3.13}$$

satisfies the triangle and Schwarz inequalities:

$$\|U + U'\| \leqslant \|U\| + \|U'\|, \qquad |(U\,|\,U')| \leqslant \|U\| \cdot \|U'\| \text{ in } \mathscr{F}^0;$$

if the equality holds, then U *and* U' *are proportional*: *either* $U = 0$ *or there exists* $\alpha \in \mathscr{F}^0$ *such that* $U' = \alpha U$. *Let*

$$U_R^0(v, z) = \alpha_{R,0}(\check{z}) e^{v_0 \varphi_R(\check{z})} \mod \frac{1}{v} \tag{3.14}$$

$$= \left(\frac{\eta}{d^l x}\right)^{1/2} \beta_0(\check{z}) e^{v_0 \varphi_R(\check{z})} \mod \frac{1}{v}, \tag{3.15}$$

where $\check{z}$ *denotes any one of the points of* $\check{V}$ *projecting onto* z *in* V; (3.15) *assumes* V *is given a 2-orientation that defines the Maslov index* $m_R(\check{z})$ *and* $\arg d^l x$ (I,§3,corollary 3); *the lagrangian amplitude* β_0 *is independent of the choice of* R, *but depends on the choice of* η, *a regular measure on* V *such that* $\arg \eta = 0$. *Then, using analogous notation for* U' *and denoting* $R_X z$ *by* x,

$$\alpha_{R,0}(\check{z})\overline{\alpha'_{R,0}(\check{z})}\, d^l x = \beta_0(\check{z})\overline{\beta'_0(\check{z})}\eta \tag{3.16}$$

is a regular differential form on V, *which is independent of* R, *and*

$$(U\,|\,U') = \int_V \alpha_{R,0}(\check{z})\overline{\alpha'_{R,0}(\check{z})}\, d^l x = \int \beta_0(\check{z})\overline{\beta'_0(\check{z})}\eta \mod \frac{1}{v}. \tag{3.17}$$

3°) *Suppose* V *and* V' *are transverse. Then*

$$v^{l/2}(U\,|\,U') \in \mathscr{F}^\circ,$$

$$\left(\frac{|v|}{2\pi i}\right)^{l/2} (U\,|\,U')$$

$$= \sum_{z \in V \cap V'} [\operatorname{Hess}(\varphi - \varphi')]^{-1/2} \alpha_{R,0}(\check{z})\overline{\alpha'_{R,0}(\check{z}')} e^{v_0[\varphi_R(\check{z}) - \varphi_R(\check{z}')]} \mod \frac{1}{v} \tag{3.18}$$

$$= \sum_{z \in V \cap V'} \left|\frac{\eta_0 \wedge \eta_0}{d^{2l} z}\right|^{1/2} \beta_0(\check{z})\overline{\beta'_0(\check{z}')} e^{v_0[\psi(\check{z}) - \psi'(\check{z}')] - (\pi i/2) m(\lambda'_4, \lambda_4)} \mod \frac{1}{v}; \tag{3.19}$$

in (3.18),

• *φ and φ' are defined near Rz by*

$$\varphi(x) = \varphi_R(\check{R}_X^{-1}x \cap \check{V}), \qquad \varphi'(x) = \varphi'_R(\check{R}_X^{-1}x \cap \check{V}'),$$

• $\mathrm{Hess}(\varphi - \varphi') = \{\mathrm{Hess}_x[\varphi(x) - \varphi'(x)]\}_{x = R_X z}$;

in (3.19),

• *ψ is the lagrangian phase of V,*
• *$\lambda_4 \in \Lambda_4(Z)$ is the 2-oriented lagrangian plane tangent to $\check{V}$ at $\check{z}$,*
• *m is the Maslov index defined* mod 4 (I,§3,theorem 1),
• *$\eta_0 \wedge \eta'_0$ and $d^{2l}z$ are the measures on Z defined by* (5.11) *and* (5.15) (§1).

Proof of (3.16). Use §1,(5.9).

Proof of (3.17). Using a sufficiently fine partition of unity $\Sigma_j b_j^* \circ b_j = 1$, it suffices to prove (3.17) under the assumptions (i) and (ii) used in the proof of lemma 3.2, which imply (3.11). Since $V = V'$, formula (3.11) can be written

$$(U \mid U') = \int_X U_R^0(v, \check{z})\overline{U_R'^0(v, \check{z})}\, d^l x, \tag{3.20}$$

and hence proves (3.17).

Proof of (3.12). Using the same partition of unity, it suffices to prove (3.12) when (3.20) holds. Then (3.12) is evident, as well as the following more precise statement: If $U \neq 0$, then

$$(U \mid U) = \sum_{r=2s}^{\infty} \frac{\alpha_r}{v^r}, \text{ where } s \in \mathbf{N}, \quad \alpha_{2s} > 0.$$

Proof of (3.13). Use the preceding formula and part 2 of §1,theorem 2.3.

Proof of the Schwarz inequality. Suppose U and U' are not proportional. Then there exists $\alpha \in \mathscr{F}^0$ and $s \in \mathbf{N}$ such that

$$U = \alpha U' + \frac{1}{v^s} U'',$$

where the lagrangian amplitudes β'_0 and β''_0 of U' and U'' are not proportional. A classical calculation gives

$$\| U \|^2 \cdot \| U' \|^2 - |(U \mid U')|^2 = \frac{1}{|v|^{2s}} [\| U' \|^2 \cdot \| U'' \|^2 - |(U' \mid U'')|^2].$$

It then suffices to prove that the right-hand side is > 0 in $\mathscr{F}^0$. In other words, it suffices to prove that U and U' satisfy the strict Schwarz inequality when their lagrangian amplitudes β_0 and β'_0 are not proportional. Thus, from (3.17), we have

$$\|U\|^2\cdot\|U'\|^2 - |(U\,|\,U')|^2 = \int_V |\beta_0(z)|^2\eta\cdot\int_V |\beta'_0(z)|^2\eta - \left|\int_V \beta_0(z)\overline{\beta'_0(z)}\eta\right|^2 \mod\frac{1}{\nu}.$$

Hence, from the classical Schwarz inequality,

$$\|U\|^2\cdot\|U'\|^2 - |(U\,|\,U')|^2 = \rho_0 \mod\frac{1}{\nu}, \text{ where } 0 < \rho_0 \in \mathbf{R}.$$

Thus, by II,§1,theorem 2.3,

$$|(U\,|\,U')| < \|U\|\cdot\|U'\|.$$

Proof of the triangle inequality. Use the Schwarz inequality.

Proof of 3°). Using a partition of unity, we may assume that $(U\,|\,U')$ is expressed by (3.11). Then (3.18) follows from part 3 of theorem 5.1 (§1) and (3.19) of theorem 5.3 (§1).

Remark 3. Without assumption (i) (made in the proof of lemma 3.2), the integral in the right-hand side of (3.20) diverges (see theorem 2.2 on the structure of lagrangian functions). Thus *definition* 3.5 of $(\cdot\,|\,\cdot)$ *gives a meaning to this divergent integral.*

4. The Group $\mathrm{Sp}_2(Z)$

The group $\mathrm{Sp}_2(Z)$ of automorphisms of the 2-symplectic geometry of Z (see I,§3,4) clearly transforms lagrangian operators into lagrangian operators. The images of lagrangian functions under this group are lagrangian functions. The group leaves the scalar product invariant.

5. Lagrangian Distributions

Lagrangian distributions are defined by replacing functions by distributions (§1,7) in definitions 2.1, 3.1, and 3.2. Theorems 2.1, 2.2 (the β_{Rr} being distributions), 2.3, and 3.1 apply to distributions just as to functions.

Definition 2.3 and theorem 2.4,2°) apply to the scalar product $(\check{U}\,|\,\check{U}')$

of $\check{U}$ and $\check{U}'$, a lagrangian function and a distribution both defined on a lagrangian manifold $\check{V}$, when Supp $\check{U} \cap$ Supp $\check{U}'$ is compact.

Definition 3.5 and theorem 3.2,2°) apply to the scalar product $(U \mid U')$ of U and U', a lagrangian function and a distribution both defined on a lagrangian manifold V, when Supp $U \cap$ Supp U' is compact.

§3. Homogeneous Lagrangian Systems in One Unknown

0. Summary

The existence of an inverse a^{-1} of a lagrangian operator a under the assumption of theorem 1.1 of §2 shows that it is the *homogeneous* lagrangian systems in one (§3) or several (§4) unknowns whose study is nontrivial. §3 and §4 begin this study, which will be concluded in chapters III and IV with the special cases of the Schrödinger equation and the Dirac equation, the latter of which is a system in four unknowns.

1. Lagrangian Manifolds on Which Lagrangian Solutions of $aU = 0$ Are Defined

Let a be the lagrangian operator associated to a formal function

$a^0 : \Omega \to \mathscr{F}^0$ (Ω an open set in Z)

whose value at z is

$$a^0(v, z) = \sum_{r \in \mathbf{N}} \frac{1}{v^r} a_r^0(z).$$

THEOREM 1. *The support of a solution U of the equation $aU = 0$ is a lagrangian manifold V on which $a_0^0 = 0$.*

Proof. Consider a point in V at which $a_0^0 \neq 0$. Replace Ω by a neighborhood of this point on which $a_0^0 \neq 0$. Then a^{-1} exists on Ω and $aU = 0$ implies $U = 0$ on Ω.

We shall assume that a_0^0 *is real valued* and that the subset of Ω where $a_0^0 = 0$ *is a regular hypersurface* W given by the equation

$$W : H(z) = 0, \qquad H_z \neq 0.$$

V is therefore a submanifold of W such that

$$\dim V = l; \qquad d[z, dz] = 0 \text{ on } V.$$

V is studied by applying E. Cartan's theory of differential forms to the pfaffian form $[z, dz]$.

2. Review of E. Cartan's Theory of Pfaffian Forms[1]

We first recall some well-known results that are set forth, in particular, in the treatise of Y. Choquet-Bruhat [6].

Let ω be *a differential form* defined on Z that in section 2 will be any infinitely differentiable manifold of finite dimension. E. Cartan expresses ω locally using a minimal number of functions

$$f_1, \ldots, f_n : Z \to \mathbf{R}.$$

The number n of these functions is called the *rank* of ω. These functions are the first integrals (independent and maximal in number) of a completely integrable system of pfaffian equations called the *characteristic system* of ω. E. Cartan explicitly describes this system.

In other words, this characteristic system is equivalent to

$$df_1 = \cdots = df_n = 0;$$

there exists a differential form ϖ such that

$$\omega(z, dz) = \varpi(f(z), df(z)), \text{ where } f = (f_1, \ldots, f_n).$$

Let ω *be a pfaffian form* defined on Z (that is, a differential form of degree 1 in the differentials). Let i_ξ denote the *interior product* [6] with a tangent vector ξ of Z; the characteristic system of $d\omega$ may be written

$$i_\xi d\omega(z, dz) = 0 \qquad \forall \xi. \tag{2.1}$$

The rank of $d\omega$ is even; denote it by $2n$. Let

$$f = (f_1, \ldots, f_{2n})$$

be $2n$ independent first integrals of (2.1) locally. Then there exists a differential form ϖ' such that

$$d\omega(z, dz) = \varpi'(f(z), df(z)):$$

evidently $d\varpi' = 0$. Then $\varpi' = d\varpi$ locally. In other words, *locally there exist a pfaffian form* $\tilde{\varpi}$ of $2n$ *variables and a function* f_0 such that

[1] See Cartan [5].

$$\omega(z, dz) = df_0(z) + \varpi(f(z), df(z)), \text{ where } f = (f_1, \ldots, f_{2n}). \tag{2.2}$$

The rank of ω is evidently $2n$ or $2n + 1$ depending on whether f_0 is a composition with $f = (f_1, \ldots, f_{2n})$ or not. In the first case it is possible to choose ϖ such that $f_0 = 0$.

Let us state some less familiar facts with which E. Cartan [5] (chapter XII, "Les équations qui admettent un invariant intégral relatif") has supplemented this result; we recall their proofs.

LEMMA 2. Let ω be a pfaffian form; let $2n$ be the rank of $d\omega$. Locally there exist $2n + 1$ numerical functions $f_0, \ldots, f_{2n}$ such that

$$\omega(z, dz) = df_0(z) + \sum_{j=1}^{n} f_{2j-1}(z)df_{2j}(z); \tag{2.3}$$

$f_1, \ldots, f_{2n}$ are $2n$ independent first integrals of the characteristic system of $d\omega$.

Remark 2.1. $f_1, \ldots, f_{2n}$ is not an arbitrary sequence of first integrals of this characteristic system: see remark 2.2.

Proof. Let f_2 be a first integral of the characteristic system of $d\omega$. The restriction of $d\omega$ to the hypersurfaces $f_2 =$ const. has even rank $<2n$, so its rank is $\leqslant 2(n - 1)$. Let f_4 be a first integral of the characteristic system of this restricted form. Continuing in this fashion, the restriction of $d\omega$ to the manifolds

$$f_2 = \text{const.}, \quad \ldots, \quad f_{2j} = \text{const.}$$

has rank $\leqslant 2(n - j)$ and therefore vanishes for some value J of j such that $J \leqslant n$. In other words,

$$d\omega = 0 \text{ on the manifolds } f_2 = \text{const.}, \quad \ldots, \quad f_{2J} = \text{const.}$$

Then there exists a function f_0 such that

$$\omega = df_0 \bmod(df_2, \ldots, df_{2J});$$

that is,

$$\omega = df_0 + \sum_{j=1}^{J} f_{2j-1}\, df_{2j}.$$

But $d\omega$ has rank $2n$, so $J = n$ and $f_1, \ldots, f_{2n}$ are $2n$ independent first integrals of the characteristic system of $d\omega$.

E. Cartan has supplemented this lemma by using the notion of the Poisson bracket as follows.

Definition 2. Let ω be a pfaffian form that has been put in the form (2.3). Let g and g' be two first integrals of the characteristic system of $d\omega$. Each is then a composition:

$$z \mapsto g(z) = G[f_1(z), \ldots, f_{2n}(z)], \qquad z \mapsto g'(z) = G'[f_1(z), \ldots, f_{2n}(z)].$$

Then there exists a composition of the same type, called the *Poisson bracket of g and g'* and denoted (g, g'), such that

$$n(d\omega)^{n-1} \wedge dg \wedge dg' = (g, g')(d\omega)^n. \tag{2.4}$$

This bracket

$$(g, g') = -(g', g),$$

which is bilinear and independent of the choice of $f = \{f_1, \ldots, f_{2n}\}$, may be expressed as follows, using one such choice:

$$(g, g')(z) = \sum_{j=1}^{n} \left[\frac{\partial G}{\partial f_{2j-1}} \frac{\partial G'}{\partial f_{2j}} - \frac{\partial G'}{\partial f_{2j-1}} \frac{\partial G}{\partial f_{2j}} \right]_{f=f(z)}. \tag{2.5}$$

Hence every triple g, g', g'' of such first integrals satisfied the *Jacobi* identity:

$$(g, (g', g'')) + (g', (g'', g)) + (g'', (g, g')) = 0.$$

g and g' are said to be *in involution* when $(g, g') = 0$.

Remark 2.2. By (2.4), the pairs f_j, f_k $(1 \leqslant j \leqslant k)$ in the expression (2.3) for ω are in involution with the exception of the pairs f_{2j-1}, f_{2j}; these satisfy

$$(f_{2j-1}, f_{2j}) = 1.$$

Remark 2.3. Let $g_1, \ldots, g_q$ be independent first integrals of the characteristic system of $d\omega$. The restriction of ω to a manifold

$$g_1 = \text{const.}, \qquad \ldots, \qquad g_q = \text{const.}$$

reduces the rank of $d\omega$ by an even number that E. Cartan [5], chapter XII, section 124, has made explicit: it is $2(q - r)$ if all the Poisson brackets (g_j, g_k) $(j, k \in \{1, \ldots, q\})$ are compositions with $g_1, \ldots, g_q$ and if the exterior form

$$\sum_{j,k} (g_j, g_k)\xi_j \wedge \xi_k$$

in the variables $\xi_1, \ldots, \xi_q$ has rank $2r$ on this manifold.

In particular ($q = n, r = 0$), we obtain the following theorem, which will enable us to establish theorem 3.1. These theorems shed some light on the beginning of the study of the Schrödinger equation.

THEOREM 2. *Let ω be a pfaffian form; let $2n$ be the rank of $d\omega$.*

1°) *Locally, the characteristic system of $d\omega$ has n-tuples $\{f_2, f_4, \ldots, f_{2n}\}$ of independent first integrals that are pairwise in involution.*

2°) *Given one of these n-tuples, a quadrature locally gives $n + 1$ functions $f_0, f_1, f_3, \ldots, f_{2n-1}$ such that ω is expressed by* (2.3); *$f_1, f_2, f_3, \ldots, f_{2n}$ are independent first integrals of the characteristic system of $d\omega$.*

Proof of 1°). Lemma 2 and remark 2.2.

Proof of 2°). By assumption,

$$df_2 \wedge df_4 \wedge \cdots \wedge df_{2n} \neq 0, \qquad (d\omega)^{n-1} \wedge df_{2j} \wedge df_{2k} = 0 \qquad \forall j, k.$$

In other words, there exist differential forms Θ_j such that

$$(d\omega)^{n-1} = \sum_{j=1}^{n} \Theta_j \wedge df_2 \wedge \cdots \widehat{df_{2j}} \cdots \wedge df_{2n}$$

($\widehat{\ }$ suppresses the term that it covers). This relation expresses the existence of pfaffian forms θ_j such that

$$d\omega = \sum_{j=1}^{n} \theta_j \wedge df_{2j},$$

or equivalently, the condition

$$d\omega = 0 \mod(df_2, \ldots, df_{2n}),$$

that is, the existence of a function f_0 such that

$$\omega = df_0 \text{ for } f_2 = \text{const.}, \qquad \ldots, \qquad f_{2n} = \text{const.},$$

or the existence of $n + 1$ functions f_0, f_1, f_3, f_{2n-1} satisfying (2.3).

3. Lagrangian Manifolds in the Symplectic Space Z and in Its Hypersurfaces

Let

$$\omega(z, dz) = \tfrac{1}{2}[z, dz], \tag{3.1}$$

that is, in a frame R

$$\omega(z, dz) = \tfrac{1}{2}\langle p, dx\rangle - \tfrac{1}{2}\langle dp, x\rangle = \langle p, dx\rangle - \tfrac{1}{2}d\langle p, x\rangle,$$
$$\text{where } Rz = (x, p); \tag{3.2}$$

$d\omega$ evidently has rank $2l$; its characteristic system is $dz = 0$.

Lagrangian manifolds V in Z are manifolds of dimension l on which $d\omega = 0$. These are manifolds given by *local equations*

$$p = \varphi_{R,x}(x) \tag{3.3}$$

outside the apparent contour Σ_R of V, where φ_R is the phase of V in R; φ_R is an arbitrary function when V is arbitrary.

Given $x \in V$, it is possible to choose R such that $x \notin \Sigma_R$ and use the local equations (3.3) in a neighborhood of x.

But if we want to use a single frame, we must supplement the preceding result as follows (x^j and p_j will denote dual coordinates of x and p).

LEMMA 3.1. The lagrangian manifolds in Z that project under $R_X : z \mapsto x$ to a manifold in X of dimension $k \leqslant l$ given by equations

$$x^j = F_j(x^1, \ldots, x^k), \qquad k + 1 \leqslant j \leqslant l, \tag{3.4}$$

are the manifolds in V given by local equations (3.4) and

$$p_i = \frac{\partial\varphi_R(x^1, \ldots, x^k)}{\partial x^i} - \sum_{j=k+1}^{l} \frac{\partial F_j}{\partial x^i} p_j, \qquad 1 \leqslant i \leqslant k, \tag{3.5}$$

where φ_R is the phase of V in R and $x^1, \ldots, x^k, p_{k+1}, \ldots, p_l$ are local coordinates on V; φ_R is an arbitrary function of $x^1, \ldots, x^k$ when V is arbitrary.

If $k = l$, the equations of V reduce to (3.3).

If $k = 0$, V is a plane:

$$x = \text{const.}, \qquad p \text{ arbitrary}, \qquad \varphi_R = \text{const.}$$

Proof

$$\omega + \frac{1}{2}d\langle p, x\rangle = <p, dx> = \sum_{i=1}^{k}\left(p_i + \sum_{j=k+1}^{l}\frac{\partial F_j}{\partial x^i}p_j\right)dx^i.$$

Then $d\omega = 0$ on V if and only if there exists, on V, a function φ_R of $(x^1, \ldots, x^k)$ satisfying (3.5). The condition $\dim V = l$ is that $(x^1, \ldots x^k, p_{k+1}, \ldots, p_l)$ be independent on V. Obviously φ_R is the phase of V.

Notation. Given a function $f: Z \to \mathbf{R}$ and a frame R, f_R denotes the function $Z(l) \to \mathbf{R}$ such that $f_R(x, p) = f(z)$ for $Rz = (x, p)$; that is, $f_R \circ R = f$. Put

$$f_{R,x^k} = \frac{\partial f_R}{\partial x^k}, \qquad f_{R,x} = (f_{R,x^1}, \ldots, f_{R,x^l}).$$

W is a hypersurface in Z given by the equation

$$W: H(z) = 0, \qquad H_z \neq 0. \tag{3.6}$$

ω_W and f_W are the restrictions of ω and f to W.

LEMMA 3.2. (E. Cartan) *$d\omega_W$ has rank $2l - 2$; its characteristic system* is *Hamilton's system*

$$\frac{dx^j}{H_{R,p_j}(x, p)} = -\frac{dp_k}{H_{R,x^k}(x, p)} \text{ on } W \qquad \forall j, k \in \{1, \ldots, l\}, \tag{3.7}$$

where the vector $(H_{R,p}, -H_{R,x})$ is evidently tangent to W.

Proof. The condition

$$(d\omega_W)^{l-1} \wedge df_W = 0 \text{ on } W \tag{3.8}$$

may be written

$$\frac{(d\omega)^{l-1} \wedge df \wedge dH}{(d\omega)^l} = 0 \text{ on } W.$$

Then by (2.4) and (2.5), where we choose $f_{2j-1} = p_j$ and $f_{2j} = x^j$ in order to identify (2.3) and (3.2), condition (3.8) may be written

$$\sum_{j=1}^{l}(f_{R,p_j}H_{R,x^j} - H_{R,p_j}f_{R,x^j}) = 0 \text{ for } H_R = 0. \tag{3.9}$$

It does not hold identically; thus $(d\omega_W)^{l-1} \neq 0$. But $(d\omega_W)^l = 0$; thus the rank of $d\omega_W$ is $2(l-1)$. Hence the equivalent relations (3.8) and (3.9) mean that f_W is a first integral of the characteristic system of $d\omega_W$. Hence this system is Hamilton's system (3.7).

Remark 3. As local coordinates on W we shall use $2l - 2$ independent first integrals of Hamilton's system (3.7) and a function t such that (3.7) may be written

$$dx = H_{R,p}(x, p)dt, \qquad dp = -H_{R,x}(x, p)dt. \tag{3.10}$$

Definition 3.1. Given a function $g: Z \to \mathbf{R}$, define $g_z \in Z$ at the point z by

$$dg = [dz, g_z]; \tag{3.11}$$

in other words,

$$Rg_z = (g_{R,p}, -g_{R,x}). \tag{3.12}$$

Hamilton's system (3.10) may then be written

$$dz = H_z dt. \tag{3.13}$$

By (2.5), where we choose $f_{2j-1} = p_j$ and $f_{2j} = x^j$ in order to identify (2.3) and (3.2), the *Poisson bracket of g and g'* is

$$(g, g') = \langle g'_{R,x}, g_{R,p}\rangle - \langle g_{R,x}, g'_{R,p}\rangle = [g_z, g'_z]. \tag{3.14}$$

The tangent vector $\kappa = H_z$ of W is called the *characteristic vector*. The curves in W tangent to this vector at each point of W, that is, the solutions of Hamilton's system (3.7), are called *characteristic curves.*

The following classical theorem explicitly describes the lagrangian manifolds in W, that is, the resolution of the nonlinear first-order equation

$$H_R(x, \varphi_x) = 0,$$

in which the unknown is a function φ.

THEOREM 3.1. 1°) *Hamilton's system* (3.7) *has* $(l-1)$*-tuples*

$$h_1, \ldots, h_{l-1}$$

of independent first integrals that are pairwise in involution, that is,

$$(d\omega)^{l-2} \wedge dh_j \wedge dh_k = 0 \text{ on } W \qquad \forall j, k \in \{1, \ldots, l-1\}. \tag{3.15}$$

Given one of these $(l-1)$*-tuples, a quadrature locally defines* l *functions*

$$g_0, g_1, \ldots, g_{l-1}$$

such that

$$\omega = dg_0 + \sum_{j=1}^{l-1} g_j dh_j \text{ on } W. \tag{3.16}$$

$g_1, \ldots g_{l-1}, h_1, \ldots, h_{l-1}$ *constitute* $2(l-1)$ *independent first integrals of Hamilton's system* (3.7). *If* t *is defined by* (3.10), *then*

$$t, h_1, \ldots, h_{l-1}, g_1, \ldots, g_{l-1} \tag{3.17}$$

form local coordinates on W; g_0 *is a function of these coordinates.*

2°) *Lagrangian manifolds* V *in* W *are manifolds in* Z *that locally*

i. *are fibered by the characteristic curves*

t *arbitrary,*

$$h_1 = \text{const.}, \quad \ldots, \quad h_{l-1} = \text{const.},$$
$$g_1 = \text{const.}, \quad \ldots, \quad g_{l-1} = \text{const.},$$

ii. *have the lagrangian manifolds in the space* $Z(l-1)$ *with the coordinates*

$$x' = (h_1, \ldots, h_{l-1}), \qquad p' = (g_1, \ldots, g_{l-1}) \tag{3.18}$$

as a base for this fibration.

3°) *If the local coordinates* (3.17) *are used on* W, *we have the following local equations of a lagrangian manifold* V *in* W, *after suitably permuting the indices* $(1, \ldots, l-1)$, *choosing* $k \in \{0, \ldots, l-1\}$ *and choosing* $l-k$ *real numerical functions of* k *variables* $F_0, F_{k+1}, \ldots, F_{l-1}$:

$$V: \begin{cases} h_j = F_j(h_1, \ldots, h_k), & k+1 \leqslant j \leqslant l-1 \\ g_i = \dfrac{\partial F_0(h_1, \ldots, h_k)}{\partial h_i} - \displaystyle\sum_{j=k+1}^{l-1} \frac{\partial F_j}{\partial h_i} g_j, & 1 \leqslant i \leqslant k; \end{cases} \tag{3.19}$$

$t, h_1, \ldots, h_k, g_{k+1}, \ldots, g_{l-1}$ *are local coordinates on* V. *The lagrangian phase* ψ *of* V *is given by*

$$\psi(t, h_1, \ldots, h_k, g_{k+1}, \ldots, g_{l-1})$$
$$= g_0(t, h_1, \ldots, h_k, g_{k+1}, \ldots, g_{l-1}) + F_0(h_1, \ldots, h_k). \tag{3.20}$$

If $k = l - 1$, *the system* (3.19) *means*

$$g_i = \frac{\partial F_0(h_1, \ldots h_{l-1})}{\partial h_i}, 1 \leqslant i \leqslant l - 1. \tag{3.21}$$

If $k = 0$, *it means*

$$h = \text{const.}, \qquad g \text{ arbitrary}, \qquad F_0 = \text{const.} \tag{3.22}$$

Proof of 1°). Use theorem 2 and lemma 3.2.

Proof of 2°). By (3.16), the condition that $d\omega = 0$ on V is equivalent to the following: the mapping

$$V \ni z \mapsto (h_1, \ldots, h_{l-1}; g_1, \ldots, g_{l-1}) \in Z(l - 1) \tag{3.23}$$

sends V onto a manifold B in $Z(l - 1)$ satisfying $\Sigma_j dh_j \wedge dg_j = 0$. Then B has dimension $\leqslant l - 1$. A necessary and sufficient condition for $\dim V = l$ is that

i. $\dim B = l - 1$, that is, B be lagrangian;
ii. (3.23) map a characteristic curve of V onto each point of B.

Proof of 3°). Lemma 3.1, where

$$Z, x, p, \qquad V, \qquad d\varphi_R = \langle p, dx \rangle \text{ on } V,$$

are replaced by

$$Z(l - 1), x', p' \text{ [see (3.18)]}, \qquad B, \qquad dF_0 = \sum_{j=1}^{l-1} g_j dh_j \text{ on } B,$$

so that (3.16) gives

$\omega = dg_0 + dF_0$ on V. Now $\omega = d\psi$.

Section 4 will use the following definition.

Definition 3.2. Given a lagrangian manifold V in W, let us call a regular measure η on V that is invariant under the characteristic vector κ of W an *invariant measure on* V; in other words, an invariant measure on V is any differential form η defined on V that is homogeneous of degree l in the differentials and is annihilated by the Lie derivative $\mathscr{L}_\kappa$ in the direction of κ:

$$\mathscr{L}_\kappa \eta = 0. \tag{3.24}$$

Since, by definition,

$$\mathscr{L}_\kappa \eta = i_\kappa d\eta + d(i_\kappa \eta) \text{ and } d\eta = 0,$$

(3.24) means

$$d(i_\kappa \eta) = 0. \tag{3.25}$$

THEOREM 3.2. 1°) *Using the local coordinates* $x = R_X z$ *on* $V \backslash \Sigma_R$, *the condition that*

$$\eta(z, dz) = \chi_R(x) d^l x, \quad \text{where } \chi_R : V \backslash \Sigma_R \to \mathbf{R}, \tag{3.26}$$

be an invariant measure on V may be written

$$\left\langle \frac{\partial}{\partial x}, \chi_R(x) H_{R,p}(x, \varphi_{R,x}) \right\rangle = 0. \tag{3.27}$$

By (3.10), *this means that along the characteristics generating V*

$$\frac{d\chi_R}{dt} + \sum_j \frac{\partial H_{R,p_j}(x, \varphi_{R,x})}{\partial x^j} \chi_R = 0; \tag{3.28}$$

more explicitly,

$$\frac{d\chi_R}{dt} + \left[\sum_{j=1}^{l} H_{R,x^j p_j}(x, p) + \sum_{j=1}^{l} \sum_{k=1}^{l} H_{R,p_j p_k}(x, p) \varphi_{R,x^j x^k} \right]_{p = \varphi_{R,x}} \chi_R = 0. \tag{3.29}$$

2°) *Using the local coordinates* (3.17) *on V, the condition that*

$$\eta = \tau(t, h, g) dt \wedge d^{l-1} h \wedge d^{l-1} g, \quad \text{where } \tau : V \to \mathbf{R}, \tag{3.30}$$

be an invariant measure on V may be expressed in these words:

the function τ *is independent of* t. (3.31)

Proof of 1°). In the specified coordinates, the components of κ are

$$dx = H_{R,p}(x, \varphi_{R,x});$$

(3.27) reexpresses (3.25).

Proof of 2°). In the specified coordinates, the components of κ are

$(dt, dh, dg) = (1, 0, 0);$

(3.31) reexpresses (3.25).

4. Calculation of aU

The definition of a lagrangian operator, that is, formula (6.6) of §1, can be expressed more explicitly as follows.

Notation 4. Let a be a lagrangian operator associated to a formal function

$$a^0(\nu, \cdot) = \sum_{r \in \mathbf{N}} \frac{1}{\nu^r} a_r^0(\cdot)$$

defined on $\Omega \subset Z$. Let W be a hypersurface of Ω given by the equation

$$W: H(z) = 0, \qquad H_z \neq 0.$$

Let V be a lagrangian manifold in W. Let φ_R be its phase in a frame R. Let

$$\alpha: V \backslash \Sigma_R \to \mathbf{C}$$

be an infinitely differentiable function; $x = R_X z$ will serve as a local coordinate on $V \backslash \Sigma_R$.

Let us exclude the following case: the operator a vanishes on W; namely, by (6.6): $a^0(\nu, x, p)$ and all its derivatives vanish on W.

Let $\eta = \chi_R d^l x$ *be a positive invariant measure on* V (section 3).

THEOREM 4. 1°) *We have*

$$a_R^+\left(\nu, x, \frac{1}{\nu}\frac{\partial}{\partial x}\right)[\alpha(x)e^{\nu\varphi_R(x)}] = \sum_{r \in \mathbf{N}} \frac{1}{\nu r} e^{\nu\varphi_R(x)} L_r\left(x, \frac{\partial}{\partial x}\right)\alpha(x), \tag{4.1}$$

where L_r is a differential operator of order $\leqslant r$ depending on a, V, and R;

$$L_0(x) = a_0^0(x, \varphi_{R,x}). \tag{4.2}$$

2°) *If $a^0 = H$, then*

$$L_0 = 0, \qquad L_1\left(x, \frac{\partial}{\partial x}\right)\alpha = \chi_R^{1/2}\frac{d}{dt}(\alpha\chi_R^{-1/2}), \tag{4.3}$$

where d/dt is the derivative along the characteristic curves of W [see (3.10)], that is, the Lie derivative $\mathscr{L}_\kappa$.

3°) *More generally, suppose that* a_0^0 *vanishes to nth order on* W:

$$1 \leqslant n \leqslant \infty.$$

Then there exists $m \in \mathbf{N}$ *such that*

$$1 \leqslant m \leqslant n,$$

$$L_0 = L_1 = \cdots = L_{m-1} = 0, \tag{4.4}$$

$$L_m\left(x, \frac{\partial}{\partial x}\right)\alpha(x) = \chi_R^{1/2} M\left(z, \frac{d}{dt}\right)(\alpha\chi_R^{-1/2}) \text{ if } m \neq \infty.$$

The differential operator M *depends on* a *and* W *but is independent of* V *and* R. *It is not identically zero on* W. *Its order is* $\leqslant m$, *with equality holding only if* $m = n$.

In general, $m = 1$. *Hence, if* $n > 1$, *then* L_1 *is generally multiplication by a nonzero function.*

4°) *If* $a^0 = H$ *and if* H_R *is polynomial in* p *of degree* s *with principal part* $H_R^{(s)}$, *then*

$$L_r = 0 \text{ for } r > s,$$

$$L_s(x, p) = \left[\exp\frac{1}{2}\left\langle\frac{\partial}{\partial x}, \frac{\partial}{\partial p}\right\rangle\right] H_R^{(s)}(x, p). \tag{4.5}$$

Thus if $s = 2$, *formulas* (4.3) *and* (4.5) *explicitly describe* a.

Proof of 1°). Formula (6.6) of §1 is made explicit as follows. By formula (6.3) of §1,

$$a_R^+\left(v, x, \frac{1}{v}\frac{\partial}{\partial x}\right)[\alpha(x)e^{v\varphi_R(x)}] = \sum_{r\in\mathbf{N}} \frac{1}{v^r} e^{v\varphi_R(x)} l_r\left(v, x, \frac{\partial}{\partial x}\right)\alpha(x), \tag{4.6}$$

where l_r, a formal function of v, is the differential operator of order r:

$$l_r\left(v, x, \frac{\partial}{\partial x}\right)$$
$$= \sum_{k=0}^{r} \frac{1}{k!} \sum_{I,J_0,\ldots,J_k} \frac{2^{-|I|}}{I!J_0!\cdots J_k!} a^0_{Rx^Ip^{I+J_0+\cdots+J_k}}(v, x, \varphi_{R,x})\varphi_{R,x^{J_1}}\cdots\varphi_{R,x^{J_k}}\left(\frac{\partial}{\partial x}\right)^{J_0}; \tag{4.7}$$

the sum $\Sigma_{I,J_0,\ldots,J_k}$ extends over the collection of $(k + 2)$-tuples $(I, J_0, \ldots, J_k)$ of l-indices satisfying the condition

$$|I| + |J_0| + \ldots + |J_k| - k = r, \qquad 2 \leqslant |J_1|, \quad \ldots, \quad 2 \leqslant |J_k|; \tag{4.8}$$

this condition clearly implies

$$|I| + |J_0| + k \leqslant r, \qquad 2 \leqslant |J_k| \leqslant r + 1 \text{ for } k \geqslant 1.$$

Formulas (4.6) and (4.7) clearly imply (4.2) and (4.1), where L_r is independent of v, while l_r depends on v.

Proof of 2°). (4.2) gives $L_0 = 0$. We have $L_r = l_r$. For $r = 1$, condition (4.8) means

$$\begin{array}{llll} \text{either} & k = 0, & |I| = 0, & |J_0| = 1; \\ \text{or} & k = 0, & |I| = 1, & |J_0| = 0; \\ \text{or} & k = 1, & |I| = 0, & |J_0| = 0, \quad |J_1| = 2. \end{array}$$

Thus

$$L_1\left(x, \frac{\partial}{\partial x}\right)\alpha = \left\langle \frac{\partial \alpha}{\partial x}, H_{R,p}(x, \varphi_{R,x}) \right\rangle + \frac{1}{2}\left[\sum_{j=1}^{l} H_{R,x^j p_j}(x, p) + \sum_{j=1}^{l}\sum_{k=1}^{l} H_{R,p_j p_k}(x, p)\varphi_{R,x^j x^k}\right]_{p=\varphi_{R,x}} \alpha;$$

this relation is equivalent to $(4.3)_2$ by (3.10) and (3.29).

Proof of 3°). Let $k \in \mathbf{N}$; there exists a function ${}^0b^0 : \Omega \to \mathbf{C}$ such that

$$a_0^0 = {}^0b^0 \cdot H^{n_0},$$

where

- $n_0 = n$ and ${}^0b^0$ is not identically zero on W if $n \neq \infty$,
- $n_0 = k + 1$ and ${}^0b^0 = 0$ on W if $n = \infty$.

Since

$$a^0 \circ b^0 = a^0 \cdot b^0 \bmod \frac{1}{v} [\text{see §2, (1.5)}],$$

there exists a lagrangian operator c such that

$$a_R^+\left(v, x, \frac{1}{v}\frac{\partial}{\partial x}\right) - {}^0b_R^+\left(v, x, \frac{1}{v}\frac{\partial}{\partial x}\right) \circ \left[H_R^+\left(v, x, \frac{1}{v}\frac{\partial}{\partial x}\right)\right]^{n_0}$$
$$= \frac{1}{v} c_R^+\left(v, x, \frac{1}{v}\frac{\partial}{\partial x}\right).$$

If c^0 does not vanish on W, let ${}^1b^0 = c^0$, $n_1 = 0$. If c^0 vanishes on W, we apply the same reasoning to c as has been applied to a.

Proceeding by recursion, we obtain a decomposition of the *De Paris* type [7]:

$$a_R^+\left(v, x, \frac{1}{v}\frac{\partial}{\partial x}\right) = \sum_{j=0}^{h} \frac{1}{v^j}\, {}^jb_R^+\left(v, x, \frac{1}{v}\frac{\partial}{\partial x}\right) \circ \left[H_R^+\left(v, x, \frac{1}{v}\frac{\partial}{\partial x}\right)\right]^{n_j} \mod \frac{1}{v^{k+1}},$$

where

- $h \leqslant k, \qquad n_j \geqslant 0$
- $n_j = k + 1$ when ${}^jb^0 = 0$ on W.

By part 1 of the theorem,

$$\left[H_R^+\left(v, x, \frac{1}{v}\frac{\partial}{\partial x}\right)\right]^{n_j} [\alpha(x)e^{v\varphi_R(x)}] = 0 \mod \frac{1}{v^{n_j}}. \tag{4.9}$$

Let

$$m(k) = \inf_{j \in \{0, \cdots, k\}} (j + n_j);$$

if $m(k) > k$, we then have

$$a_R^+\left(v, x, \frac{1}{v}\frac{\partial}{\partial x}\right)[\alpha(x)e^{v\varphi_R(x)}] = 0 \mod \frac{1}{v^{k+1}}.$$

But we assumed that a does not vanish identically on W. Therefore we can choose k such that

$$m(k) \leqslant k.$$

We define $m = m(k)$; let J be the collection of j such that

$$j + n_j = m.$$

By 1°), ${}^jb_R^+(v, x, (1/v)\partial/\partial x)$ is $\mathrm{mod}(1/v)$ multiplication by the function ${}^jb^0$. Then, by (4.9), the De Paris decomposition gives

$$a_R^+\left(v, x, \frac{1}{v}\frac{\partial}{\partial x}\right)[\alpha(x)e^{v\varphi_R(x)}]$$
$$= \sum_{j \in J} \frac{1}{v^j}\, {}^jb^0(z)\left[H_R^+\left(v, x, \frac{1}{v}\frac{\partial}{\partial x}\right)\right]^{m-j} [\alpha(x)e^{v\varphi_R(x)}] \mod \frac{1}{v^{m+1}}$$

where

- $1 \leqslant m \leqslant n$,
- J is finite and nonempty,
- $0 \in J$ if and only if $m = n$,
- ${}^{j}b^0$ does not vanish on W.

Therefore, by $(4.3)_2$ the relations (4.4) hold with

$$L_m\alpha = \sum_{j\in J} {}^{j}b^0(z)\chi_R^{1/2}\left(\frac{d}{dt}\right)^{m-j}(\alpha\chi_R^{-1/2});$$

L_m is not zero; its order is $\leqslant m$; its order is m if and only if $m = n$.

In general, c^0 does not vanish on W, and therefore $n_1 = 0$, $m(k) = 1$, $m = 1$.

Proof of 4°). $a^0 = H$, so $l_r = L_r$. By (4.7), where $a^0 = H$, we can supplement condition (4.8) by the relation $|I| + |J_0| + \cdots + |J_k| \leqslant s$; that is, $k + r \leqslant s$.

If $r > s$, it cannot be satisfied; thus $L_r = 0$.

If $r = s$, it requires $k = 0$; (4.8) becomes

$$|I| + |J_0| = s$$

and (4.7) becomes

$$L_s(x, p) = \sum_I \frac{2^{-|I|}}{I!} H^{(s)}_{R,x^I p^I}(x, p),$$

the sum extending over the collection of l-indices I such that

$$|I| \leqslant s.$$

In other words,

$$L_s(x, p) = \sum_{j=0}^{s} \frac{2^{-j}}{j!}\left\langle\frac{\partial}{\partial x}, \frac{\partial}{\partial p}\right\rangle^{j} H^{(s)}_R(x, p),$$

where it is possible to replace $\Sigma_{j=0}^{s}$ by $\Sigma_{j=0}^{\infty}$, which gives (4.5).

5. Resolution of the Lagrangian Equation $aU = 0$

By theorem 1, solutions of this equation are defined on lagrangian submanifolds V of the manifold given by the equation $a_0^0 = 0$.

Notation 5. Assume that the equation

$$a_0^0 = 0$$

defines a hypersurface in Z. Let W be one of the parts of this hypersurface where it is regular. Let

$$U_R = \sum_{r\in\mathbf{N}} \frac{1}{\nu^r} \alpha_{R,r} e^{\nu\varphi_R}$$

be the expression in a frame R of a solution defined on a lagrangian manifold V in W. Notation 4 is kept.

By theorem 4, the condition that $aU = 0$ on $V\backslash\Sigma_R$ may be written

$$M\left(z, \frac{d}{dt}\right)[\chi_R^{-1/2}(x)\alpha_{R,r}(x)] + \chi_R^{-1/2} \sum_{s=1}^{r} L_{m+s}\left(x, \frac{\partial}{\partial x}\right)\alpha_{R,r-s}(x) = 0$$

$$\forall r \in \mathbf{N}; \qquad (5.1)_r$$

the first of these equations may be written

$$M\left(z, \frac{d}{dt}\right)\beta_0(z) = 0, \text{ where } \beta_0(z) = \chi_R^{-1/2}(x)\alpha_{R,0}(x) \qquad (5.1)_0$$

is the lagrangian amplitude of U, which is independent of R.

Theorem 2.2 of §2 supplements equations (5.1) as follows:

$$\chi_R^{-3r-1/2}\alpha_{R,r} \text{ is infinitely differentiable on } \check{V}, \qquad (5.2)$$

even at the points of $\check{\Sigma}_R$; recall that $\chi_R^{-1} = 0$ on Σ_R.

Remark 5.1. If a_0^0 vanishes to *nth* order on W, where $n > 1$, then, in general, the operator M is multiplication by a nonzero function $M: W \to \mathbf{C}$. Thus the equations (5.1) imply that the support of U is a lagrangian submanifold of the manifold in Z determined by the equations

$$H(z) = M(z) = 0.$$

The study of this case requires a generalization of sections 3 and 4: it is necessary to construct lagrangian submanifolds V of a given manifold W in Z when the codimension of W is >1; it is necessary to describe explicitly aU when U is defined on such a lagrangian manifold V. We shall not study this case and exclude it.

Let us show how conditions (5.1) and (5.2) enable us to solve the equation $aU = 0$ *using a single frame* R.

LEMMA 5. Let K be an arc of a characteristic generating V. Assume that the principal coefficient of M does not vanish on K.

1°) Let $z' \in K$. Let U' be a solution of the lagrangian equation $aU' = 0$ defined in a neighborhood of z' in V. Then there exists a neighborhood of K in V on which the equation $aU = 0$ has a unique solution U such that $U = U'$ in a neighborhood of z'.

2°) Let R be a frame such that Σ_R is transverse to K and χ_R^{-1} does not vanish to infinite order on $K \cap \Sigma_R$. Then U_R is the expression in R of a lagrangian solution U of the equation $aU = 0$ in a neighborhood of K if and only if U_R satisfies (5.1) and (5.2).

Notation used by the proof. V_K is a neighborhood of K in V of the form

$$V_K = B \times I,$$

where

- B is the ball $|b| \leqslant 1$ in $\mathbf{R}^{l-1}$,
- I is the segment $|t| \leqslant$ const. in $\mathbf{R}$;
- the segments $b \times I$ are the characteristics;
- $0 \times I$ is the given characteristic K, $z' = (0, 0)$;
- $\bigcup_j I_j = I$ is a finite covering of I, $V_j = B \times I_j$, j an integer.

Proof of 1°). We make a choice of V_K and V_j having the following properties:

- U' is defined on V_0;
- to each V_j there is associated a frame R_j such that $V_j \cap \Sigma_{R_j} = \emptyset$;
- I_j is the segment $j - 1 \leqslant t \leqslant j + 1$.

If U has been defined on $V_0 \cup V_1 \cup \cdots \cup V_{j-1}$ $(j > 0)$, then using (5.1) in the frame R_j makes it possible to extend successively the definitions of $\alpha_{R_j,0}, \ldots, \alpha_{R_j,r}, \ldots, U_{R_j}, U$ to V_j; these extensions are unique.

Proof of 2°). The expression U_R of a lagrangian function U, defined in a neighborhood of K, satisfies (5.1) and (5.2). It remains to prove the converse.

Let R be a frame, V_K a neighborhood of K, and

$$U_R = \sum_{r \in \mathbf{N}} \frac{1}{\nu^r} \alpha_{R,r} e^{\nu \varphi_R} \tag{5.3}$$

a formal function defined on $V_K \backslash \Sigma_R \cap V_K$ satisfying (5.1) and (5.2).

It has to be proved that, in a neighborhood of K, U_R is the expression of a lagrangian function.

Let $z' \in K$. By 1°), there exists a unique lagrangian function U defined in a neighborhood of K whose expression in R is U_R in some neighborhood of z'. It has to be proved that in some neighborhood of K, its expression in R is still U_R.

Since this expression, like U_R, satisfies (5.1) and (5.2), it suffices to prove the following uniqueness theorem: If the formal function (5.3) vanishes in a neighborhood of a point of K, then it vanishes in a neighborhood of K.

We choose V_K and V_j having the following properties:

- $U_R = 0$ on V_0;
- I_j is the segment $j - 1 \leqslant t \leqslant j$;
- $V_K \cap \Sigma_R = \bigcup_j B_j$, where $B_j = V_j \cap V_{j+1} = B \times j$.

Assume we have proved that

$$U_R = 0 \text{ on } V_0 \cup V_1 \cup \cdots \cup (V_j \backslash B_j), \qquad j > 0.$$

By (5.2), $\chi_R^{-1/2} \alpha_{R,r}$ has an infinite number of derivatives on B_j that all vanish; then (5.1) gives successively

$$\alpha_{R,0} = 0, \alpha_{R,1} = 0, \ldots, \alpha_{R,r} = 0, \ldots, U_R = 0 \text{ on } V_{j+1} \backslash B_{j+1}.$$

This lemma shows that *the differential operators M and L_r have very special properties*. The following is a consequence of part 2 of this lemma.

THEOREM 5. *Let W be a hypersurface in Z on which $a_0^0 = 0$; assume that the operator a is not zero and that the principal coefficient of M does not vanish. Let V be a lagrangian manifold in W. Let R be a frame. Let $\eta = \chi_R d^l x$ be an invariant positive measure on V; assume that χ_R^{-1} does not vanish to infinite order on Σ_R and that Σ_R is transverse to the characteristics of W generating V. Then*

$$\check{U}_R = \sum_{r \in \mathbf{N}} \frac{1}{\nu^r} \alpha_{R,r} e^{\nu \varphi_R}$$

is the expression in R of a lagrangian solution $\check{U}$ defined on the universal covering space $\check{V}$ of V if and only if U_R satisfies (5.1) *and* (5.2).

Remark 5.2. Lemma 5 and theorem 5 evidently apply to solutions $\check{U}$ of the equation

$$a\check{U} = 0 \mod \frac{1}{v^{m+s}}, \qquad s \in \dot{\mathbf{N}},$$

defined $\mathrm{mod}(1/v^s)$ on V.

6. Solutions of the Lagrangian Equation $aU = 0 \,\mathrm{mod}(1/v^2)$ with Positive Lagrangian Amplitude: Maslov's Quantization

Definition 6.1. We call an infinitely differentiable function

$H: \Omega \to \mathbf{R}$ (Ω an open subset of Z)

such that $H_z \neq 0$ on the hypersurface W given by the equation

W: $H(z) = 0$

a *hamiltonian.*

In classical mechanics, Hamilton's system (3.7) governs the movement of particles and, more generally, that of holonomic mechanics.

Let a be *the lagrangian operator associated to a hamiltonian* H; then it is *self-adjoint* (§1,6). Solving the equation

$$a\check{U} = 0 \mod \frac{1}{v^2}$$

amounts to finding *lagrangian manifolds* in W having *an invariant measure* that we assume is chosen >0.

Indeed, by (4.3), equation $(5.1)_0$ is written

$$\frac{d\beta_0}{dt} = 0$$

and means that the lagrangian amplitude of $\check{U}$ is constant on each of the characteristics generating V. In other words, $\beta_0\eta$ is invariant.

We require it to be $\geqslant 0$. In other words, we require that the lagrangian amplitude β_0 of $\check{U}$ satisfy the condition

$$\beta_0 \geqslant 0. \tag{6.1}$$

(Recall that this is the case in physics, where the amplitude should change more slowly than $e^{\nu\varphi_R}$ and not oscillate around the value 0.)

Definition 3.2 (§2) of lagrangian functions on V requires the datum of a purely imaginary number ν_0 that will be denoted by

$$\nu_0 = \frac{i}{\hbar}, \qquad \hbar > 0, \tag{6.2}$$

in chapter III because in quantum physics $2\pi\hbar$ is chosen to be Planck's constant.

Since

$$\alpha_{R,0} = \beta_0 \left(\frac{\eta}{d^l x}\right)^{1/2}, \text{ where } \beta_0 \geqslant 0,$$

the condition that U be lagrangian on V mod$(1/\nu)$ amounts to

$$\left(\frac{\eta}{d^l x}\right)^{1/2} e^{\nu_0 \varphi_R}$$

being defined (that is, uniform) on V. From the definitions of $(d^l x)^{1/2}$ and $\eta^{1/2}$ (I,§3,corollary 3 and II,§2,theorem 2.2), this condition is the following.

Definition 6.2. A lagrangian manifold V *satisfies Maslov's quantum condition* when the function

$$\frac{\nu_0}{2\pi i}\varphi_R - \tfrac{1}{4}m_R, \quad \text{where } \nu_0 = i/\hbar,$$

is defined mod 1 on V; this condition is independent of the choice of the frame R.

Remark 6. If V is oriented, in the euclidean sense, m_R is defined mod 2 on V. Then Maslov's quantum condition assigns to each period $\frac{1}{2}\oint_\gamma [z, dz]$ of ψ, that is, to each period $\oint_\gamma \langle p, dx \rangle$ of φ_R, one of the two values 0 or $\pi\hbar$ mod $2\pi\hbar$.

If V has a 2-orientation, m_R will be defined on V mod 4 and this condition requires that φ_R and ψ be defined on V mod $2\pi\hbar$. This will *not be* the case in the applications that are given in chapter III.

We have just proved the following theorem.

THEOREM 6. 1°) *Let a be the operator associated to a hamiltonian H. A lagrangian solution U of the equation*

$$aU = 0 \bmod \frac{1}{\nu^2} \tag{6.3}$$

with lagrangian amplitude $\geqslant 0$ *can be defined* $\operatorname{mod}(1/\nu)$ *on a manifold* V *if and only if* V *simultaneously has the following three properties:*

i. V *is a lagrangian manifold in the hypersurface* W: $H(z) = 0$;
ii. V *has an invariant positive measure (invariant under the characteristic vector of* H);
iii. V *satisfies Maslov's quantum condition.*

2°) *The datum of* V *having these three properties defines* U *up to a constant factor if and only if the invariant measure on* V *is unique, that is, if every function that is infinitely differentiable on* V *and constant on each of the characteristics generating* V *is constant on* V.

This theorem will be applied in III,§2.

7. Solution of Some Lagrangian Systems in One Unknown

Theorems 7.1 and 7.2 supplement theorem 6. They are applied in III,§1 and III,§3, respectively.

THEOREM 7.1. *Let* $a^{(j)}$ $(j = 1, \cdots, l)$ *be lagrangian operators associated to* l *hamiltonians*

$H^{(j)}: \Omega \to \mathbf{R}$ (Ω *an open subset of* Z)

that are independent and ***pairwise in involution***; *that is,*

$$dH^{(1)} \wedge \cdots \wedge dH^{(l)} \neq 0, (d\omega)^{l-2} \wedge dH^{(j)} \wedge dH^{(k)} = 0 \qquad \forall j, k, \tag{7.1}$$

where $\omega = \frac{1}{2}[z, dz]$.
1°) *The system*

$$a^{(1)}U = \cdots = a^{(l)}U = 0 \bmod \frac{1}{\nu^2} \tag{7.2}$$

has a lagrangian solution U *defined* $\operatorname{mod}(1/\nu)$ *on the connected manifold* V *if and only if* V *simultaneously satisfies the following two properties:*

i. V *is a connected component of the manifold in* Z *given by the equations*

$$V: H^{(1)}(z) = \cdots = H^{(l)}(z) = 0, \tag{7.3}$$

which imply that V *is lagrangian.*

ii. *V satisfies Maslov's quantum condition* (*definition* 6.2).
Then the lagrangian amplitude of U is constant when

$$\eta = \frac{d^{2l}z}{dH^{(1)} \wedge \cdots \wedge dH^{(l)}} \qquad \textit{(see remark 7)} \tag{7.4}$$

is chosen as the measure on V.

Thus, when V has been chosen, then U is defined mod$(1/\nu)$ *up to a constant numerical factor.*

2°) *The Lie derivatives $\mathscr{L}_{\kappa^j}$ in the direction of the characteristic vectors κ^j of the $H^{(j)}$ commute. If V is compact, then V is a torus whose translation group is generated by the infinitesimal transformations $\mathscr{L}_{\kappa^j}$.*

Remark 7. Recall that $d^{2l}z$ is defined by §1,(5.15); (7.4) means that η is the restriction to V of any form η_Z on Z such that

$$dH^{(1)} \wedge \cdots \wedge dH^{(l)} \wedge \eta_Z = d^{2l}z;$$

η is clearly independent of the choice of η_Z; η is *invariant* under each κ^j.

Proof of 1°). From theorem 1, the support of U belongs to one on the connected components V of the manifold given by the equations (7.3).

Conversely, let V be one on these components. The rank of $d\omega$ is $2l$ because its characteristic system is $dz = 0$. By theorem 2 (E. Cartan), locally there exist functions $g_0, \ldots, g_l$ such that

$$\omega = dg_0 + \sum_{j=1}^{l} g_j dH^{(j)};$$

then the restriction ω_V of ω to V satisfies

$$d\omega_V = 0.$$

Now dim $V = l$; thus V is a lagrangian manifold.

Moreover this result could be deduced from (3.22) of theorem 3.1, 3°).

Let R be a frame of Z; let $(x, p) = Rz$; the condition that $H^{(j)}$ and $H^{(k)}$ be in involution is expressed

$$\left\langle \frac{\partial H^{(j)}}{\partial x}, \frac{\partial H^{(k)}}{\partial p} \right\rangle - \left\langle \frac{\partial H^{(k)}}{\partial x}, \frac{\partial H^{(j)}}{\partial p} \right\rangle = 0,$$

that is,

$$\mathscr{L}_{\kappa^j} H^{(k)} = 0, \tag{7.5}$$

where $\mathscr{L}_{\kappa^j}$ is the Lie derivative in the characteristic direction of $H^{(j)}$:

$$\kappa^j = \left(\frac{\partial H^{(j)}}{\partial p}, -\frac{\partial H^{(j)}}{\partial x}\right).$$

From (7.5) follows

$$\mathscr{L}_{\kappa^j}\, dH^{(k)} = 0;$$

moreover the definition of the Lie derivative gives

$$\mathscr{L}_{\kappa^j}\, d^{2l} z = 0;$$

hence, by the definition (7.4) of η,

$$\mathscr{L}_{\kappa^j} \eta = 0 \qquad \forall j.$$

Then in view of theorem 4,2°), the condition that a lagrangian function $\check{U}$ defined on $\check{V}$ satisfy

$$a^{(1)}\check{U} = \cdots = a^{(l)}\check{U} = 0 \qquad \operatorname{mod}\frac{1}{\nu^2}$$

is that its lagrangian amplitude β_0 satisfy

$$\mathscr{L}_{\kappa^j}\beta_0 = 0 \qquad \forall j;$$

since the $H^{(j)}$ are independent, that is, since they satisfy $(7.1)_1$, this condition can be expressed as

β_0 is constant on V.

Then, by section 6, the condition that $\check{U}$ be lagrangian on V mod$(1/\nu)$ is equivalent to Maslov's quantum condition.

Proof of 2°). By §2,(1.1), the commutator of $a^{(j)}$ and $a^{(k)}$,

$$[a^{(j)}, a^{(k)}] = a^{(j)} \circ a^{(k)} - a^{(k)} \circ a^{(j)}$$

is the operator associated to the formal function

$$-2\left\{\sinh\frac{1}{2\nu}\left[\frac{\partial}{\partial z}, \frac{\partial}{\partial z'}\right] H^{(j)}(z) H^{(k)}(z')\right\}_{z=z'};$$

since $H^{(j)}$ and $H^{(k)}$ are in involution, this formal function vanishes

$\operatorname{mod}(1/v^2)$. It is an odd function of v, so it vanishes $\operatorname{mod}(1/v^3)$, from which follows

$$[a^{(j)}, a^{(k)}] = 0 \operatorname{mod} \frac{1}{v^3}. \tag{7.6}$$

From theorem 4,2°),

$$a_R^{(j)}(\alpha e^{v\varphi_R}) = \frac{1}{v} e^{v\varphi_R} \chi_R^{1/2} \ \mathscr{L}_{K^j}(\alpha \chi_R^{-1/2}) \operatorname{mod} \frac{1}{v^2} \qquad \forall \alpha.$$

so that

$$[a_R^{(j)}, a_R^{(k)}](\alpha e^{v\varphi_R}) = \frac{1}{v^2} e^{v\varphi_R} \chi_R^{1/2} [\mathscr{L}_{K^j}, \mathscr{L}_{K^k}](\alpha \chi_R^{-1/2}) \operatorname{mod} \frac{1}{v^3};$$

thus, by (7.6),

$$[\mathscr{L}_{K^j}, \mathscr{L}_{K^k}] = 0$$

Then, if V is compact, the infinitesimal transformations $\mathscr{L}_{K^1}, \cdots, \mathscr{L}_{K^l}$ generate an abelian group of homeomorphisms of V, whence 2°).

We supplement theorem 7.1 as follows.

THEOREM 7.2. *Let $a^{(j)}$ ($j = 1, \cdots, l$) be lagrangian operators that* ***commute with each other*** *and are equal* $\operatorname{mod}(1/v^2)$ *to operators associated to l independent hamiltonians $H^{(j)}$ [that is, satisfying $(7.1)_1$]; these hamiltonians are then in involution [that is, satisfy $(7.1)_2$]. Let us study the problem of defining* $\operatorname{mod}(1/v^{r+1})$, *on a connected manifold V, a lagrangian solution U of the system*

$$a^{(1)}U = \cdots = a^{(l)}U = 0 \operatorname{mod} \frac{1}{v^{r+2}}, \qquad r \geqslant 1. \tag{7.7}_r$$

Assume that conditions i) *and* ii) *of theorem 7.1 are satisfied; indeed they are necessary. Then this problem has a solution U if and only if, in addition, the following two conditions are satisfied:*

iii. *A solution of $(7.7)_{r-1}$ has to exist on V (we assume it is explicitly known).*
iv. *A function $\check{V} \to$* **C** *that is defined by integration of a closed pfaffian form on $V \backslash \Sigma_R$, and by the condition that it have polar singularities on Σ_R, has to be a function $V \to$* **C**. *(Knowledge of this function explicitly solves the problem.)*

When (i), (ii), (iii), *and* (iv) *are satisfied, then the solution* U *of* $(7.7)_r$ *is defined on* V *up to a factor that is a formal number with vanishing phase.*

Preliminary to the proof. Let R be a frame of Z; let V satisfy (i). We restate theorem4,1°),2°) as follows. Let $\beta: V\backslash\Sigma_R \to \mathbf{C}$; then

$$a_R^{(j)+}\left(\nu, x, \frac{1}{\nu}\frac{\partial}{\partial x}\right)(\chi_R^{1/2}(x)\,\beta(x)e^{\nu\varphi_R(x)})$$

$$= \chi_R^{1/2}(x)e^{\nu\varphi_R(x)}\sum_{r\in\mathbf{N}}\frac{1}{\nu^r}M_r^j\left(x, \frac{\partial}{\partial x}\right)\beta(x), \tag{7.8}$$

where M_r^j is a differential operator of order $\leqslant r$, depending on $a^{(j)}$ and V:

$M_0^j = 0,\ M_1^j = \mathscr{L}_{K^j}$.

The assumption that the operators $a^{(j)}$ commute obviously may be stated as follows:

$$\sum_{s=0}^{r}[M_{s+1}^j, M_{r-s+1}^k] = 0 \qquad \forall j, k, r. \tag{7.9}$$

Proof. Let U be a lagrangian function defined on V with lagrangian amplitude $\beta_0 = 1$. Let

$$U_R(\nu, x) = \chi_R^{1/2}e^{\nu\varphi_R(x)}\sum_{r\in\mathbf{N}}\frac{1}{\nu^r}\beta_r(x) \tag{7.10}$$

be its expression in the frame R, where $\beta_0 = 1$, $\beta_r: V\backslash\Sigma_R \to \mathbf{C}$. Suppose that U is given $\mathrm{mod}(1/\nu^r)$ and satisfies $(7.7)_{r-1}$: $\beta_0, \ldots, \beta_{r-1}$ are given and by (7.8) satisfy

$$\sum_{t=1}^{s}M_t^j\beta_{s-t} = 0 \text{ for } j = 1, \ldots, l, \quad s = 1, \ldots, r. \tag{7.11}$$

The condition that U satisfy $(7.7)_r$ is

$$\sum_{s=1}^{r+1}M_s^j\beta_{r+1-s} = 0 \qquad \forall j; \tag{7.12}$$

this condition is a system of l equations that defines

$M_1^j\beta_r \quad \forall j$; i.e., $\mathscr{L}_{K^j}\beta_r \quad \forall j$; i.e., $d\beta_r$,

by means of $\beta_0, \ldots, \beta_{r-1}$. Since the $M_1^j = \mathscr{L}_{K^j}$ commute, the condition of local integrability of this system, that is, the condition that the expression for $d\beta_r$ given by this system be a closed pfaffian form, is expressed by

$$M_1^j \circ \sum_{s=2}^{r+1} M_s^k \beta_{r+1-s} - M_1^k \circ \sum_{s=2}^{r+1} M_s^j \beta_{r+1-s} = 0 \qquad \forall j, k,$$

or, replacing s by $s + 1$, by

$$\sum_{s=1}^{r} [M_1^j, M_{s+1}^k]\beta_{r-s} + \sum_{s=1}^{r} [M_{s+1}^j, M_1^k]\beta_{r-s}$$

$$+ \sum_{s=1}^{r} (M_{s+1}^k \circ M_1^j - M_{s+1}^j \circ M_1^k)\beta_{r-s} = 0. \tag{7.13}$$

Now, with t and s replaced by $t + 1$ and $r - s + 1$, (7.11) implies

$$\sum_{s=1}^{r} M_{s+1}^k \circ M_1^j \beta_{r-s} + \sum_{s,t} M_{s+1}^k \circ M_{t+1}^j \beta_{r-s-t} = 0,$$

where $\Sigma_{s,t}$ signifies $\Sigma_{s=1}^{r-1} \Sigma_{t=1}^{r-s}$; that is, $\Sigma_{s,t}$ extends over the set of integer pairs (s, t) such that

$$1 \leqslant s, \qquad 1 \leqslant t, \qquad s + t \leqslant r.$$

In other words, (7.11) implies

$$\sum_{s=1}^{r} M_{s+1}^k \circ M_1^j \beta_{r-s} + \sum_{s=2}^{r} \sum_{t=1}^{s-1} M_{s-t+1}^k \circ M_{t+1}^j \beta_{r-s} = 0,$$

and similarly

$$\sum_{s=1}^{r} M_{s+1}^j \circ M_1^k \beta_{r-s} + \sum_{s=2}^{r} \sum_{t=1}^{s-1} M_{t+1}^j \circ M_{s-t+1}^k \beta_{r-s} = 0,$$

so that the condition (7.13) of local integrability may be written

$$\sum_{s=1}^{r} \sum_{t=0}^{s} [M_{t+1}^j, M_{s-t+1}^k]\beta_{r-s} = 0.$$

Thus it is satisfied by the assumption of commutivity (7.9).

Then the system (7.12) defines a function

$$\beta_r : \check{V} \backslash \check{\Sigma}_R \to \mathbf{C}$$

up to the addition of a function that is locally constant on $\check{V} \backslash \check{\Sigma}_R$. By this choice of β_r in (7.10), U_R locally becomes the expression in R of a lagrangian solution of $(7.7)_r$ on $\check{V} \backslash \check{\Sigma}_R$.

Then, by lemma 5,1°), which holds for systems mod$(1/v^{r+2})$, there exists a lagrangian solution of $(7.7)_r$ on $\check{V}$, whose expression in R is U_R, up to the addition to β_r of some function that is constant on each component of $\check{V}\backslash\check{\Sigma}_R$. By theorem 5, the addition of this function to β_r makes $\chi_R^{-3r}\beta_r$ infinitely differentiable on $\check{V}$; β_r is then defined up to the addition of a function that is constant on V, that is, a multiple of β_0.

By assumption (ii), Maslov's condition is satisfied; by section 6, U is lagrangian on V mod$(1/v^{r+1})$ if and only if β_r is a function: $V \to \mathbf{C}$. The theorem follows.

8. Lagrangian Distributions That Are Solutions of a Homogeneous Lagrangian System

Theorem 1 and lemma 5,1°) apply to solutions that are lagrangian distributions; condition (5.2) should then be stated as follows:

$\chi_R^{-3r-1/2}\alpha_{R,r}$ is a distribution on $\check{V}$.

Without trying to extend either lemma 5,2°) or theorem 5 to distributions, we merely remark that theorems 7.1 and 7.2 apply to solutions that are lagrangian distributions. Hence the following theorem holds.

THEOREM 8. (Regularity) *Under the assumption of theorem* 7.2, *every lagrangian distribution U that is a solution of the system* $(7.7)_r$ *is a lagrangian function* mod$(1/v^{r+1})$.

Conclusion

V. P. Maslov [10] *called the "solutions defined* mod $1/v$" *studied in sections* 6 *and* 7 *"asymptotics". But there is no reason for them to be equal* mod$(1/v)$ *to a lagrangian solution U of the equation* $aU = 0$ *(see theorem* 7.2 *and* III, §3), *and there is no reason for the expression* U_R *of one such solution U to be the asymptotic expansion of a solution of the differential operator or pseudodifferential operator that* a_R *can formally define. This is shown by the examples considered in chapter* III.

The most evident feature of this lagrangian analysis, which was motivated by the study of V. P. Maslov's treatise [10], *is that it is a new structure.*

It is formal. Therefore, in physics, this analysis could be reasonably applied only to the nonobservable quantities of quantum theory.

§4. Homogeneous Lagrangian Systems in Several Unknowns

Let us generalize the simplest results of §3: theorems 4, 5, 6, and 7.1.

1. Calculation of $\Sigma_{m=1}^{\mu} a_n^m U_m$

We shall extend a result of [8], section 8, and elucidate its proof; by doing this we shall generalize theorem 4 of §3.

Notation 1. Let a be a $\mu \times \mu$ matrix whose elements a_n^m $(m, n = 1, \ldots, \mu)$ are lagrangian operators associated to formal functions with vanishing phase; they are elements of a matrix

$$a^0 = \sum_{r \in \mathbf{N}} \frac{1}{\nu^r} a_r^0, \text{ where } a_r^0 : \Omega \to \mathbf{C}^{\mu^2}.$$

Let W be a hypersurface in Ω given by the equation

W: $H(z) = 0$, where $H_z \neq 0$ on W.

Let V be a lagrangian manifold in W; let φ_R be its phase in a frame R. Let

$$b(x, p) = a_0^0(z), \qquad c(x, p) = a_1^0(z) \text{ for } (x, p) = Rz. \tag{1.1}$$

Let $\alpha = \{\alpha_1, \ldots, \alpha_\mu\}$ be a vector whose components are infinitely differentiable functions

$$\alpha_m : V \backslash \Sigma_R \to \mathbf{C};$$

x will serve as a local coordinate on $V \backslash \Sigma_R$.
Let $\eta = \chi_R d^l x$ be a positive invariant measure on V (§3,3).

THEOREM 1. 1°) *We have*

$$a_R^+\left(\nu, x, \frac{1}{\nu}\frac{\partial}{\partial x}\right)[\alpha(x)e^{\nu\varphi_R(x)}] = \sum_{r \in \mathbf{N}} \frac{1}{\nu^r} e^{\nu\varphi_R(x)} L_r\left(x, \frac{\partial}{\partial x}\right)\alpha(x), \tag{1.2}$$

where L_r is a $\mu \times \mu$ matrix whose elements are differential operators of order $\leqslant r$; they depend on V, a, and R;

$$L_0(x) = b(x, \varphi_{R,x}). \tag{1.3}$$

2°) *Suppose* $\det b = H$. *There exist two nonzero μ-vectors f and g that are functions of $z \in Z$ such that on W*

$$bf = 0, \qquad {}^t bg = 0; \quad i.e.: \sum_{m=1}^{\mu} b_n^m f_m = 0, \qquad \sum_{n=1}^{\mu} g^n b_n^m = 0; \tag{1.4}$$

choose these (which is possible) so that on W,
$f_m g^n$ is the minor of b_n^m in the matrix b;
that is,

$$dH = \sum_{m,n} f_m g^n \, db_n^m \mod H. \tag{1.5}$$

Let

$$\langle g, u \rangle = \sum_{n=1}^{\mu} g^n u_n \text{ for any vector } u = (u_1, \ldots, u_\mu).$$

Evidently, by (1.3) *and* (1.4),

$$\langle g(x, \varphi_{R,x}), L_0(x)\alpha(x) \rangle = 0. \tag{1.6}$$

This formula is supplemented by the following, where γ is an arbitrary function $V \backslash \Sigma_R \to \mathbf{C}$, and where d/dt is the derivative along the characteristic curves of W, that is, the Lie derivative $\mathscr{L}_\kappa$:

$$\left\langle g(x, \varphi_{R,x}), L_1\left(x, \frac{\partial}{\partial x}\right)[\gamma(x) f(x, \varphi_{R,x})] \right\rangle$$
$$= \chi_R^{1/2} \frac{d}{dt}(\gamma \chi_R^{-1/2}) + J(x, \varphi_{R,x})\gamma(x). \tag{1.7}$$

Here $J(x, p)$ is defined on W by the formula

$$J = \frac{1}{2} \sum_{m,n} [(g^n, b_n^m) f_m + b_n^m (g^n, f_m) + g^n (b_n^m, f_m)] + \sum_{m,n} g^n c_n^m f_m \tag{1.8}$$

in which $(\cdot, \cdot)$ denotes the Poisson bracket, defined by §3, (3.14).

J has the following properties, all obvious except for the first:

Remark 1.1. J only depends on b and c and the restrictions f_W and g_W of f and g to W.

Remark 1.2. Multiplying f by $h: Z \to \mathbf{C} \backslash \{0\}$ multiplies g_W by h^{-1} and adds $d(\log h)/dt$ to J.

Remark 1.3. If the matrix b is *symmetric* and the matrix c is *antisymmetric*, then $J = 0$, since it is possible to choose $f = g$.

Remark 1.4. Suppose the matrices b and ic are *self-adjoint*. In particular, this is the case when the *matrix* a is self-adjoint, that is, when

$$a_n^m = (a_m^n)^* \qquad \forall m, n.$$

Then the values of J are purely imaginary, since it is possible to choose $g = \bar{f}$.

Proof of the theorem. Part 1 follows from §3, theorem 4. To prove part 2, let

$$a = a_R, \qquad \varphi = \varphi_R(x),$$

$$b_n^m = b_n^m(x, \varphi_x), \qquad b_{nx}^m = \left.\frac{\partial b_n^m(x,p)}{\partial x}\right|_{p=\varphi_x}, \qquad \ldots,$$

$$g_p^n = \left.\frac{\partial g^n(x,p)}{\partial p}\right|_{p=\varphi_x},$$

so that

$$\frac{\partial b_n^m}{\partial x^i} = b_{nx^i}^m + \sum_{j=1}^{l} b_{np_j}^m \varphi_{x^ix^j}, \qquad \ldots, \qquad \frac{\partial g^n}{\partial x^i} = g_{x^i}^n + \sum_{j=1}^{l} g_{p_j}^n \varphi_{x^ix^j}, \tag{1.9}$$

where x^i and p_j are the components of x and p ($i, j = 1, \ldots, l$). By the definition of a^+ in §1, (6.3)–(6.6),

$$a_n^{+m}\left(v, x, \frac{1}{v}\frac{\partial}{\partial x}\right)[\alpha_m e^{v\varphi}] = e^{v\varphi} b_n^m(x, \varphi_x)\alpha_m + \frac{e^{v\varphi}}{v}\left[\sum_j b_{np_j}^m \frac{\partial \alpha_m}{\partial x^j} + \left(\frac{1}{2}\sum_{i,j} b_{np_ip_j}^m \varphi_{x^ix^j} + \frac{1}{2}\sum_j b_{nx^jp_j}^m + c_n^m\right)\alpha_m\right] \mod \frac{1}{v^2};$$

hence, by (1.9) and the definition (1.2) of L_1,

$$\left\langle g, L_1\left(x, \frac{\partial}{\partial x}\right)\alpha\right\rangle = \sum_{j,m,n} g^n b_{np_j}^m \frac{\partial \alpha_m}{\partial x^j} + \sum_{m,n} g^n\left(\frac{1}{2}\sum_j \frac{\partial}{\partial x^j} b_{np_j}^m + c_n^m\right)\alpha_m.$$

Thus, by choosing $\alpha_m = \gamma f_m$ it follows from (1.5) that

$$\langle g, L_1(\gamma f)\rangle = \sum_j H_{p_j}\frac{\partial\gamma}{\partial x^j} + \left[\frac{1}{2}\sum_{j,m,n} g^n b^m_{np_j}\frac{\partial f_m}{\partial x^j} + \frac{1}{2}\sum_{j,m,n} g^n\frac{\partial}{\partial x^j}(b^m_{np_j}f_m) + \sum_{m,n} g^n c^m_n f_n\right]\gamma;$$

hence by (1.5) and the definition of d/dt on W,

$$\langle g, L_1(\gamma f)\rangle = \frac{d\gamma}{dt} + \frac{1}{2}\left(\sum_j \frac{\partial}{\partial x^j}H_{p_j}\right)\gamma + J\gamma, \tag{1.10}$$

where

$$J = \frac{1}{2}\sum_{j,m,n} g^n b^m_{np_j}\frac{\partial f_m}{\partial x^j} - \frac{1}{2}\sum_{j,m,n}\frac{\partial g^n}{\partial x^j}b^m_{np_j}f_m + \sum_{m,n} g^n c^m_n f_m. \tag{1.11}$$

Now, by §3, (3.28),

$$\sum_j \frac{\partial}{\partial x^j}H_{p_j} = -\frac{1}{\chi_R}\frac{d\chi_R}{dt};$$

hence (1.10) is equivalent to (1.7). Moreover, by (1.4),

$$\frac{\partial}{\partial x^j}\left(\sum_n g^n b^m_n\right) = \frac{\partial}{\partial x^j}\left(\sum_m b^m_n f_m\right) = 0;$$

hence (1.11) can be written

$$J = \frac{1}{2}\sum_{j,m,n}\begin{vmatrix} f_m & -b^m_n & g^n \\ \dfrac{\partial f_m}{\partial x^j} & \dfrac{\partial b^m_n}{\partial x^j} & \dfrac{\partial g^n}{\partial x^j} \\ f_{mp_j} & b^m_{np_j} & g^n_{p_j}\end{vmatrix} + \sum_{m,n} g^n c^m_n f_m.$$

By (1.9), this is

$$J = \frac{1}{2}\sum_{j,m,n}\begin{vmatrix} f_m & -b^m_n & g^n \\ f_{mx^j} & b^m_{nx^j} & g^n_{x^j} \\ f_{mp_j} & b^m_{np_j} & g^n_{p_j}\end{vmatrix} + \sum_{m,n} g^n c^m_n f_m, \tag{1.12}$$

because

$$\sum_{i,j,m,n} \begin{vmatrix} f_m & -b_n^m & g^n \\ f_{mp_j} & b_{np_j}^m & g_{p_j}^n \\ f_{mp_i} & b_{np_i}^m & g_{p_i}^n \end{vmatrix} \varphi_{x^i x^j} = 0,$$

since the above determinant is an antisymmetric function of (i, j). Clearly, (1.12) is equivalent to (1.8).

Proof of remark 1.1. (1.8) can be written

$$J = \frac{1}{2} \sum_{m,n} [(g^n, b_n^m f_m) + g^n(b_n^m, f_m)] + \sum_{m,n} g^n c_n^m f_m.$$

Now, by (1.4) there exist regular functions F_n such that

$$\sum_m b_n^m f_m = HF_n.$$

Hence, on W, where $H = 0$,

$$J = \frac{1}{2}\left[\sum_n F_n \frac{dg^n}{dt} + \sum_{m,n} g^n(b_n^m, f_m)\right] + \sum_{m,n} g^n c_n^m f_m;$$

thus J only depends on b, c, f, and the restriction of g to W.

2. Resolution of the Lagrangian System $aU = 0$ in Which the Zeros of $\det a_0^0$ Are Simple Zeros

Section 5 of §3 is easily extended in this case.

Notation 2. Notation 1 is kept. By theorem 1,1°), solutions of the system

$$\sum_{m=1}^{\mu} a_n^m U_m = 0 \tag{2.1}$$

are defined on lagrangian manifolds V in the hypersurface W given by the equation

$$W: H = 0, \text{ where } H = \det b \text{ (by assumption, } H_z \neq 0 \text{ on } W). \tag{2.2}$$

Let

$$U_R = \sum_{r \in \mathbf{N}} \frac{1}{\nu^r} \alpha_{R,r} e^{\nu\varphi_R}$$

be the expression, in a frame R, of a solution

$U = (U_1, \ldots, U_\mu)$

of the system (2.1), which will be written

$aU = 0;$

the U_m are lagrangian functions defined on $V \subset W$; the $\alpha_{R,r}$ are functions $\check{V}\backslash\check{\Sigma}_R \to \mathbf{C}^\mu$. In view of theorem 1, the condition $aU = 0$ on $V\backslash\Sigma_R$ may be written

$$L_0(x)\alpha_{R,r}(x) + \sum_{s=1}^{r} L_s\left(x, \frac{\partial}{\partial x}\right)\alpha_{R,r-s}(x) = 0 \qquad \forall r \in \mathbf{N}. \tag{2.3$_r$}$$

By (1.6), it implies

$$\left\langle g(x, \varphi_{R,x}), \sum_{s=1}^{r} L_s\left(x, \frac{\partial}{\partial x}\right)\alpha_{R,r-s}(x)\right\rangle = 0. \tag{2.4$_r$}$$

Let M_0 be one of the matrices $V \to \mathbf{C}^{\mu\times\mu}$ such that the relation

$L_0(x)\alpha(x) = \alpha'(x)$, where $\langle g(x, \varphi_{R,x}), \alpha'(x)\rangle = 0$ and $\alpha(x), \alpha'(x) \in \mathbf{C}^\mu$.

is equivalent to the existence of $\gamma\colon V\backslash\Sigma_R \to \mathbf{C}$ such that

$\alpha(x) + \gamma(x)f(x, \varphi_{R,x}) = M_0(x)\alpha'(x).$

Then, under condition $(2.4)_r$, equation $(2.3)_r$ may be written

$$\alpha_{R,r}(x) + \gamma_{R,r}(x)f(x, \varphi_{R,x}) + \sum_{s=1}^{r} M_0(x)L_s\left(x, \frac{\partial}{\partial x}\right)\alpha_{R,r-s}(x) = 0, \tag{2.5$_r$}$$

and, by (1.7), equation $(2.4)_{r+1}$ may be written

$$\begin{aligned}&\frac{d}{dt}(\gamma_{R,r}\chi_R^{-1/2}) + J\gamma_{R,r}\chi_R^{-1/2}\\ &+ \left\langle g(x, \varphi_{R,x}), \sum_{s=1}^{r} N_{s+1}\left(x, \frac{\partial}{\partial x}\right)\alpha_{R,r-s}(x)\right\rangle \chi_R^{-1/2} = 0,\end{aligned} \tag{2.6$_r$}$$

where

$$\begin{aligned}N_{s+1}\left(x, \frac{\partial}{\partial x}\right)\alpha(x) = {}& L_1\left(x, \frac{\partial}{\partial x}\right)\left[M_0(x)L_s\left(x, \frac{\partial}{\partial x}\right)\alpha(x)\right]\\ &- L_{s+1}\left(x, \frac{\partial}{\partial x}\right)\alpha(x).\end{aligned}$$

Theorem 2.2 in §2 supplements equations $(2.5)_r$ and $(2.6)_r$ as follows ($r \in \mathbf{N}$):

$\alpha_{R,r}\chi_R^{-3r-1/2}$ and thus $\gamma_{R,r}\chi_R^{-3r-1/2}$ are infinitely differentiable on $\check{V}$, $(2.7)_r$
even at the points of $\check{\Sigma}_R$; recall that $\chi_R^{-1} = 0$ on Σ_R.

Lemma 5 of §3 is easily extended. It shows that *the conditions* $(2.5)_r$, $(2.6)_r$, *and* $(2.7)_r$ *make it possible to solve the system* $aU = 0$ *using a single frame* R. More precisely (compare §3, theorems 5 and 6), the following theorem holds.

THEOREM 2.1. *Let V be a lagrangian manifold in the hypersurface W defined by* (2.2); *let $\check{V}$ be its universal covering space. Let R be a frame. Let $\eta = \chi_R d^l x$ be a positive invariant measure on V; assume that χ_R^{-1} does not vanish to infinite order on Σ_R; assume that Σ_R is transverse to the characteristics of W that generate V. Then*

$$U_R = \sum_{r \in \mathbf{N}} \frac{1}{\nu^r} \alpha_{R,r} e^{\nu\varphi_R}$$

is the expression in R of a lagrangian solution $U = (U_1, \ldots, U_\mu)$, defined on $\check{V}$ [or on V], of the system $aU = 0$ if and only if, for every $r \in \mathbf{N}$, the vectors $\alpha_{R,r}: \check{V}\backslash\check{\Sigma}_R \to \mathbf{C}^\mu$ and the functions $\gamma_{R,r}: \check{V}\backslash\check{\Sigma}_R \to \mathbf{C}$ satisfy the conditions:

$(2.5)_r$, $(2.6)_r$, $(2.7)_r$ [*and* $\gamma_{R,r} e^{\nu_0 \varphi_R}: V\backslash\Sigma_R \to \mathbf{C}$].

Remark 2. This theorem applies to solutions of the system

$$aU = 0 \ \mathrm{mod} \frac{1}{\nu^{s+1}}.$$

They are defined $\mathrm{mod}(1/\nu^{s+1})$ up to the addition of $(\alpha/\nu^s) f e^{\nu\varphi_R}$, where

$\alpha e^{\nu_0\varphi_R}: \check{V} \to \mathbf{C}^\mu$ [or $V \to \mathbf{C}^\mu$].

Evidently we have the following theorem.

THEOREM 2.2. [Reduction $\mathrm{mod}(1/\nu^2)$ of a system to an equation] *We keep the assumption of theorem 1, 2°), which defines H and J. The existence of a solution on $V \subset W$ [or on $\check{V}$] of the lagrangian system*

$$aU = 0 \ \mathrm{mod} \frac{1}{\nu^2}, \text{ that is, } \sum_m a_n^m U_m = 0 \ \mathrm{mod} \frac{1}{\nu^2}, \qquad (2.1)_2$$

is equivalent to the existence of a solution on V [*or* $\check{V}$] *of the lagrangian equation*

$$a'U' = 0 \bmod \frac{1}{\nu^2}, \tag{2.8}$$

where a' *is the lagrangian operator associated to the formal function*

$$H + \frac{1}{\nu} J.$$

To each solution U *of the system* $(2.1)_2$ *there corresponds a solution* U' *of the equation* (2.8) *such that*

$$U = U'f \bmod \frac{1}{\nu}.$$

In chapter IV, a reduction theorem analogous to the preceding one will be used in the special case of Dirac's equation.

3. A Special Lagrangian System $aU = 0$ in Which the Zeros of det a_0^0 Are Multiple Zeros

The following extension of theorem 7.1 of §3 is used in chapter IV. (Theorem 7.2 of §3 admits an analogous extension.)

THEOREM 3. *Let* $a^{(k)}$ $(k = 1, \ldots, l)$ *be* $\mu \times \mu$ *matrices whose elements are lagrangian operators. Assume* $a^{(k)}$ *is associated* $\bmod(1/\nu^2)$ *to the matrix*

$$H^{(k)}E + \frac{1}{\nu} J^{(k)},$$

where E *is the* $\mu \times \mu$ *identity matrix,* $H^{(k)}: \Omega \to \mathbf{C}$, *and* $J^{(k)}: \Omega \to \mathbf{C}^{\mu^2}$ *is a* $\mu \times \mu$ *matrix. Let* V *be the manifold given by the equations*

$$V: H^{(1)}(z) = \cdots = H^{(l)}(z) = 0. \tag{3.1}$$

Assume that the $a^{(k)}$ *commute* $\bmod(1/\nu^3)$ *and that*

$dH^{(1)} \wedge \cdots \wedge dH^{(l)} \neq 0$ *in a neighborhood of* V.

Then the following hold.

1°) *The hamiltonians* $H^{(k)}$ *are pairwise in involution:* V *is a lagrangian manifold; the measure on* V

$$\eta = \left.\frac{d^{2l}z}{dH^{(1)} \wedge \cdots \wedge dH^{(l)}}\right|_V$$

is invariant $\forall k$ *under the characteristic vector* $\kappa^{(k)}$ *of* $H^{(k)}$, *which is tangent to* V. *If* V *is compact, then* V *is a torus whose translation group is generated by the infinitesimal transformations* $\mathscr{L}_{\kappa^{(k)}}$.

2°) *Let* $U = (U_1, \ldots, U_\mu)$ *be a vector whose components* U_m *are lagrangian functions;* U *satisfies the lagrangian system*

$$a^{(k)}U = 0 \bmod \frac{1}{\nu^2} \qquad \forall k \tag{3.2}$$

if and only if U *is defined on* V *and the lagrangian amplitudes* $\beta = (\beta_1, \ldots, \beta_\mu)$ *of its components satisfy the system of first-order partial differential equations*

$$(H^{(k)}, \beta) + J^{(k)}\beta = 0, \text{ where } (H^{(k)}, \beta) = \mathscr{L}_{\kappa^{(k)}}\,\beta \tag{3.3}$$

and $(\cdot, \cdot)$ *is the Poisson bracket* (3.14) *of* §3. *This system is equivalent to a completely integrable system*

$$d\beta = \omega\beta, \quad \text{that is,} \quad d\beta_n = \sum_m \omega_n^m \beta_m, \tag{3.4}$$

where the elements ω_n^m *of the* $\mu \times \mu$ *matrix* ω *are pfaffian forms defined on* V. *In addition to* (3.3), β *has to satisfy the "quantum condition"*

$$\left(\frac{\eta}{d^l x}\right)^{1/2} \beta e^{\nu_0 \varphi_R} : V \backslash \Sigma_R \to \mathbf{C}^\mu. \tag{3.5}$$

Remark 3. The condition that (3.3) be completely integrable may be stated as follows:

$$d\omega = \omega \wedge \omega, \tag{3.6}$$

where $\omega \wedge \omega$ is the matrix whose elements are the $\sum_{h=1}^{\mu} \omega_n^h \wedge \omega_h^m$. Hence it is satisfied; it is equivalent to

$$(H^{(i)}, J^{(k)}) - (H^{(k)}, J^{(i)}) + J^{(i)}J^{(k)} - J^{(k)}J^{(i)} = 0 \qquad \forall i, k, \tag{3.7}$$

where by assumption $(H^{(i)}, H^{(k)}) = 0$, and by definition

$$(H^{(i)}, J^{(k)}) = \sum_{j=1}^{l} H^{(i)}_{p_j} J^{(k)}_{x^j} - \sum_{j=1}^{l} H^{(i)}_{x^j} J^{(k)}_{p_j}$$

in an arbitrary frame R; (3.6) or (3.7) is equivalent to the condition that the $a^{(j)}$ commute $\bmod(1/\nu^3)$.

Proof of 1°). Let $U_m(m = 1, \ldots, \mu)$ be lagrangian functions defined on V with lagrangian amplitudes β_m; let $U = (U_1, \ldots, U_\mu)$ and $\beta = (\beta_1, \ldots, \beta_\mu)$. By theorem 4 of §3,

$$a_R^{+(k)}\left(\nu, x, \frac{1}{\nu}\frac{\partial}{\partial x}\right)U = \frac{1}{\nu}\chi_R^{1/2}e^{\nu\varphi_R}L^{(k)}\left(x, \frac{\partial}{\partial x}\right)\beta \bmod \frac{1}{\nu^2}, \tag{3.8}$$

where $L^{(k)}$ is a $\mu \times \mu$ matrix whose elements are first order differential operators; the principal part of $L^{(k)}$ is $\mathscr{L}_{\kappa^{(k)}}E$, where $\mathscr{L}_{\kappa^{(k)}}$ is the Lie derivative in the direction of the characteristic vector $\kappa^{(k)}$ of $H^{(k)}$ and E is the $\mu \times \mu$ identity matrix. By assumption, the $a^{(k)}$ commute $\bmod(1/\nu^3)$. Thus the $L^{(k)}(x, \partial/\partial x)$ commute. Thus their principal parts $\mathscr{L}_{\kappa^{(k)}}$ commute. Therefore the $\kappa^{(k)}$ are tangent to the manifold V given by equation (3.1). [See the proof of theorem 7.1, 1°) of §3).]

Proof of 2°). By (3.8), the system (3.2) is equivalent to the system

$$L^{(k)}\left(x, \frac{\partial}{\partial x}\right)\beta = 0 \qquad \forall k; \tag{3.9}$$

since the $\mathscr{L}_{\kappa^{(k)}}$ commute, it is possible to find local coordinates $t_1, \ldots, t_l$ on V such that

$$\mathscr{L}_{\kappa^{(k)}} = \frac{\partial}{\partial t_k}. \tag{3.10}$$

By theorem 4 of section 3,

$$L^{(k)}\left(x, \frac{\partial}{\partial x}\right) = \frac{\partial}{\partial t_k} + J^{(k)}. \tag{3.11}$$

Thus system (3.9), in which the unknown function $\beta: \check{V} \to \mathbf{C}^\mu$ has to satisfy (3.5), is equivalent to the system

$$d\beta = \omega\beta,$$

on $\check{V}$, where

$$\omega = -\sum_{k=1}^{l} J^{(k)}\, dt_k \tag{3.12}$$

is a $\mu \times \mu$ matrix whose elements are pfaffian forms. Since the $a^{(k)}$ commute, the $L^{(k)}$ defined by (3.8) commute, which by (3.11) is expressed

$$\frac{\partial J^{(k)}}{\partial t_i} - \frac{\partial J^{(i)}}{\partial t_k} + J^{(i)}J^{(k)} - J^{(k)}J^{(i)} = 0, \tag{3.13}$$

or, by (3.12), $d\omega = \omega \wedge \omega$. This is the condition of complete integrability of the system (3.4). By (3.10), this condition is also expressed by (3.7).

III Schrödinger and Klein-Gordon Equations for One-Electron Atoms in a Magnetic Field

Introduction

Summary. The most interesting problems in the theory of linear and homogeneous partial differential equations are the eigenvalue problems. Their essential feature is that they have solutions only exceptionally. Examples of lagrangian problems having the same feature are given in this chapter. These problems assume

$$l = 3, \qquad Z(3) = X \oplus X^*, \qquad X = X^* = \mathbf{E}^3 \text{ (euclidean space)}.$$

They concern the lagrangian operator a associated to some convenient hamiltonian: this hamiltonian gives rise to a Hamilton system admitting two first integrals defined on $Z(3)$, namely, the length L and[1] one of the components M of the vector $x \wedge p$ in $\mathbf{E}^3$.

H may be the hamiltonian[2] of the nonrelativistic or relativistic electron under the simultaneous influence of the electric field of a stationary atomic nucleus and a constant magnetic field (Zeeman effect). Then H depends on a parameter: the energy level E of the electron. The operator a is the *Schrödinger* or (in the relativistic case) the *Klein-Gordon* operator. *The energy levels for which our lagrangian problems have a solution coincide with those defined by the problems that are classically studied regarding these operators.*

The advantage of the lagrangian point of view is its simplicity. By applying theorem 7.1 of II, §3, *these energy levels are obtained in* §1 *by a quadrature.* The latter is easily calculated in the Schrödinger and Klein-Gordon cases using the method of residues.

In §1, we determine *solutions defined* mod$(1/v)$ *on a compact lagrangian manifold* of the lagrangian system:

$$aU = (a_{L^2} - \text{const.})U = (a_M - \text{const.})U = 0 \text{ mod} \frac{1}{v^2}, \tag{1}$$

where a_{L^2} and a_M are the lagrangian operators associated to the first integrals L^2 and M. Here theorem 7.1 is applied. Three integers introduced by the Maslov quantization,

l, m, n such that $|m| \leqslant l < n$,

[1] The author uses $\cdot \wedge \cdot$ to denote the vector product in E^3. [Translator's note]
[2] Or the hamiltonian of the harmonic operator: see §3, remark 4.3.

characterize the equations having solutions (thus the energy levels) and also the solutions. These are Schrödinger's three quantum numbers. The lagrangian manifolds on which these solutions are defined are 3-dimensional *tori* $T(l, m, n)$ [see theorem 7.1, 2°)], given by the equations

$$T(l, m, n): H(x, p) = L^2(x, p) - \text{const.} = M - \text{const.} = 0.$$

These constants have the same values as in (1) and depend on (l, m, n).

In §2, we determine *solutions of the lagrangian equation*

$$aU = 0 \bmod \frac{1}{v^2} \tag{2}$$

that are defined $\bmod(1/v)$ *on a compact lagrangian manifold V and that have lagrangian amplitude* >0. This amounts to a formal problem analogous to the boundary-value problem whose study constitutes the classical study of the Schrödinger equation; the condition concerning behavior at infinity in the classical problem is now replaced by the condition that V is compact. Always, the condition of existence is the same: it is characterized by a triple of quantum integers; but the solution corresponding to such a tripel is not necessarily unique.

In §3, we determine *solutions defined on a compact lagrangian manifold* of the lagrangian system

$$aU = (a_{L^2} - \text{const.})U = (a_M - \text{const.})U = 0, \tag{3}$$

where the constants are formal numbers that are real $\bmod(1/v^2)$ and H *is the hamiltonian of the relativistic or nonrelativistic electron*; then a, a_{L^2}, *and* a_M *commute.* Here theorem 7.2 is applied. Solutions again are characterized by a triple of quantum integers (l, m, n). Solutions of problem (1) are, $\bmod(1/v)$, those of problem (3).

In §4, we recall *the problem classically posed regarding the Schrödinger and Klein-Gordon equations*: to find a function

$$u: \mathbf{E}^3 \to \mathbf{C},$$

that is square integrable—as is its gradient—and that satisfies the partial differential equation

$$au = 0. \tag{4}$$

In §4, we recall the resolution of this problem in order to show how it differs essentially from that of the preceding problems. *We observe that all these problems define the same energy levels, but we do not explain why this happens.*

The difficulties encountered in §2 and the length of the calculations used in §3 and §4 contrast with the simplicity of §1; §1 justifies the following conclusion.

CONCLUSION *Applying the Maslov quantization* (*II*,§3,6 *and* 7) *to an atom with one electron placed in a constant magnetic field gives the observable quantities the same values as does wave mechanics. However, the Maslov quantization is directly related to corpuscular mechanics, hence to the old quantum theory. Nevertheless it does not have the shortcomings of the old theory. It has a logical justification* (*chapters* I *and* II); *it does not require the determination of the nonquantized trajectories of the electron, but only the knowledge of the classical first integrals* L *and* M *of Hamilton's system, which defines these trajectories.*

The probabilistic interpretation of this quantization is the following: In the state defined by a choice of a triple of quantum integers (l, m, n), *the point* (x, p), *representing both the position* x *and the momentum* p *of the electron, belongs to a* 3-*dimensional torus* $T(l, m, n)$ *in the* 6-*dimensional space* $Z(3) = \mathbf{E}^3 \oplus \mathbf{E}^3$; *the probability that* (x, p) *belongs to a subset of* $T(l, m, n)$ *is defined by the invariant measure* η *on this torus* $T(l, m, n)$ [*see* §1,(3.16)].

Remark. Let

$$(l, m, n) \neq (l', m', n')$$

be two distinct triples of quantum integers. They define (§1) tori whose intersection is empty:

$$T(l, m, n) \cap T(l', m', n') = \varnothing.$$

Let U and U' be two lagrangian functions defined on $T(l, m, n)$ and $T(l', m', n')$, respectively. Then their scalar product is [II,§2,theorem 3.2,3°)]

$$(U \mid U') = 0.$$

Historical note. V. P. Maslov did not give this application of his quantization. He only studied the case of quantum numbers tending toward infinity, that is, the "correspondence principle" in quantum mechanics.

§1. A Hamiltonian H to Which Theorem 7.1 (Chapter II, §3) Applies Easily; the Energy Levels of a One-Electron Atom, with the Zeeman Effect

Theorem 7.1 of II,§3 assumes that l functions on $\Omega \subset Z(l)$ are given that are pairwise in involution. In chapter III, we choose $l = 3$ and choose a classical triple of such functions.

1. Four Functions Whose Pairs Are All in Involution on $\mathbf{E}^3 \oplus \mathbf{E}^3$ Except for One

Let $X = X^* = \mathbf{E}^3$ denote 3-dimensional euclidean space. Let us apply chapter II to

$$l = 3, \qquad Z(3) = \mathbf{E}^3 \oplus \mathbf{E}^3,$$

thus making a choice of a frame R_0 of Z: theorem 5 of II, §3 spares us the use of another frame.

But in $\mathbf{E}^3$ we use both a fixed orthonormal frame (I_1, I_2, I_3) and a moving orthonormal frame. Let

$$x \in X = \mathbf{E}^3, \qquad p \in X^* = \mathbf{E}^3;$$

let (x_1, x_2, x_3) and (p_1, p_2, p_3) be the coordinates of x and p in (I_1, I_2, I_3):

$$x = \sum_{j=1}^{3} x_j I_j, \qquad p = \sum_{j=1}^{3} p_j I_j.$$

Let us define five functions of (x, p):

$$R(x) = |x|, \quad P(p) = |p|, \quad Q(x, p) = \langle p, x \rangle, \quad L(x, p) = |x \wedge p|,$$
$$M(x, p) = x_1 p_2 - x_2 p_1 \text{ (third component of } x \wedge p); \tag{1.1}$$

they are connected by the obvious relations

$$L^2 + Q^2 = P^2 R^2, \qquad |M| \leqslant L, \quad 0 \leqslant P, \quad 0 \leqslant R. \tag{1.2}$$

Then the vector I_3 has a priviledged role: it will be, for example, the direction of the magnetic field producing the Zeeman effect.

In $\mathbf{E}^3 \oplus \mathbf{E}^3$, the characteristic system of $d\langle p, dx \rangle$ is

$$dx = dp = 0.$$

Every function $\mathbf{E}^3 \oplus \mathbf{E}^3 \to \mathbf{R}$ is a first integral of this system. Thus the

Poisson bracket $(\cdot,\cdot)$(definition 2 of II,§3) of two such functions is defined. By formula (3.14), of II,§3,

$$(L, M) = (L, Q) = (L, R) = (M, Q) = (M, R) = 0, \qquad (Q, R) = R. \tag{1.3}$$

From theorem 2 (*E. Cartan*) of II,§3, a quadrature locally defines four real numerical functions f_0, f_1, f_3, f_5 on $\mathbf{E}^3 \oplus \mathbf{E}^3$ such that

$$\langle p, dx \rangle = df_0 + f_1 dL + f_3 dM + f_5 dR. \tag{1.4}$$

Let us now *define the functions f_j explicitly*. Let (J_1, J_2, J_3) be the *orthonormal moving frame*, defined for $L \neq 0$, such that

$$x = RJ_1, \qquad x \wedge p = LJ_3, \tag{1.5}$$

which implies

$$p = QR^{-1}J_1 + LR^{-1}J_2, \qquad P \neq 0, \quad R \neq 0. \tag{1.6}$$

Let ω_1, ω_2, and ω_3 be the infinitesimal components of the displacement of that frame relative to itself (G. Darboux-E. Cartan); these are the pfaffian forms

$$\omega_1 = \langle J_3, dJ_2 \rangle = -\langle J_2, dJ_3 \rangle, \qquad \omega_2 = \langle J_1, dJ_3 \rangle,$$
$$\omega_3 = \langle J_2, dJ_1 \rangle$$

such that

$$dJ_1 = \omega_3 J_2 - \omega_2 J_3, \qquad dJ_2 = \omega_1 J_3 - \omega_3 J_1,$$
$$dJ_3 = \omega_2 J_1 - \omega_1 J_2. \tag{1.7}$$

Recall that exterior differentiation of these relations gives the equations

$$d\omega_1 = \omega_3 \wedge \omega_2, \qquad d\omega_2 = \omega_1 \wedge \omega_3, \qquad d\omega_3 = \omega_2 \wedge \omega_1 \tag{1.8}$$

(which are the structural equations of the orthogonal group).

By (1.7), differentiating $(1.5)_1$ gives

$$dx = (dR)J_1 + R\omega_3 J_2 - R\omega_2 J_3, \tag{1.9}$$

from which follows

$$\langle p, dx \rangle = QR^{-1} dR + L\omega_3 \tag{1.10}$$

by (1.6).

In order to transform this formula into a formula of the form (1.4), let us introduce the Euler angles Φ, Ψ, and Θ; these are the parameters of the frame (J_1, J_2, J_3) that are defined as follows for $J_3 \neq \pm I_3$, that is, $|M| < L$:

• Θ is the angle between I_3 and J_3; $0 < \Theta < \pi$;
• a rotation by Φ around I_3 transforms (I_1, I_2, I_3) into (I'_1, I'_2, I_3) such that

$$I'_2 \sin\Theta = I_3 \wedge J_3;$$

• a rotation by Θ around I'_2 transforms (I'_1, I'_2, I_3) into (I''_1, I'_2, J_3);
• a rotation by Ψ around J_3 transforms (I''_1, I'_2, J_3) into (J_1, J_2, J_3).

It is obvious that

$$M = L\cos\Theta, \tag{1.11}$$

Φ and Ψ are defined mod 2π. We have the classical formulas:

$$\begin{aligned} J_1 &= (\cos\Phi\cos\Psi\cos\Theta - \sin\Phi\sin\Psi)I_1 \\ &\quad + (\sin\Phi\cos\Psi\cos\Theta + \cos\Phi\sin\Psi)I_2 - \cos\Psi\sin\Theta\, I_3, \\ J_2 &= (-\cos\Phi\sin\Psi\cos\Theta - \sin\Phi\cos\Psi)I_1 \\ &\quad + (-\sin\Phi\sin\Psi\cos\Theta + \cos\Phi\cos\Psi)I_2 + \sin\Psi\sin\Theta\, I_3, \\ J_3 &= \cos\Phi\sin\Theta\, I_1 + \sin\Phi\sin\Theta\, I_2 + \cos\Theta\, I_3; \end{aligned} \tag{1.12}$$

$$\begin{aligned} \omega_1 &= -\cos\Psi\sin\Theta\, d\Phi + \sin\Psi\, d\Theta, \\ \omega_2 &= \sin\Psi\sin\Theta\, d\Phi + \cos\Psi\, d\Theta, \\ \omega_3 &= \cos\Theta\, d\Phi + d\Psi. \end{aligned} \tag{1.13}$$

By (1.11) and $(1.13)_3$, the explicit expression we sought for formula (1.10) is

$$\langle p, dx\rangle = Q\frac{dR}{R} + L\,d\Psi + M\,d\Phi. \tag{1.14}$$

This fundamental formula is of the form (1.4), *in agreement with E. Cartan's theorem*, which is our guide.

Complementary formulas. By (1.7), differentiating (1.6) gives

$$\begin{aligned} dp &= [d(QR^{-1}) - LR^{-1}\omega_3]J_1 + [d(LR^{-1}) + QR^{-1}\omega_3]J_2 \\ &\quad + [LR^{-1}\omega_1 - QR^{-1}\omega_2]J_3. \end{aligned} \tag{1.15}$$

By (1.9), $d^3x = dx_1 \wedge dx_2 \wedge dx_3$ is given by

$$d^3x = R^2(dR) \wedge \omega_2 \wedge \omega_3; \tag{1.16}$$

hence, by (1.15),

$$d^3x \wedge d^3p = L\frac{dR}{R} \wedge dQ \wedge dL \wedge \omega_1 \wedge \omega_2 \wedge \omega_3,$$

or, replacing the ω_j by their expressions (1.13) and eliminating Θ by means of (1.11),

$$d^3x \wedge d^3p = dL \wedge dM \wedge dQ \wedge \frac{dR}{R} \wedge d\Phi \wedge d\Psi. \tag{1.17}$$

Let Ω^6 be an *open subset* of $Z(3) = \mathbf{E}^3 \oplus \mathbf{E}^3$ on which

$$|M| < L;$$

we can use the coordinates

$$L, M, Q, R, \Phi, \Psi$$

on Ω^6, where Φ and Ψ are defined mod 2π.

Remark 1. $L \neq 0$ on Ω^6; thus by $(1.2)_1$, $P \neq 0$, $R \neq 0$.

LEMMA 1. Let V be a lagrangian manifold in Ω^6 ($\dim V = 3$). Let us replace each of the preceding functions and differential forms by its restriction to V.

1°) The *apparent contour* Σ_{R_0} of V is the surface in V *where*

$$\Sigma_{R_0}: dR \wedge \omega_2 \wedge \omega_3 = 0 \text{ on } V. \tag{1.18}$$

2°) In a neighborhood of a point of Σ_{R_0}, there exists a differential form ϖ on V of degree 3, that is nowhere zero, such that on Σ_{R_0}

$$dQ \wedge \omega_2 \wedge \omega_3/\varpi \geqslant 0, \qquad dL \wedge dR \wedge \omega_2/\varpi \geqslant 0,$$
$$dR \wedge \omega_3 \wedge \omega_1/\varpi \geqslant 0, \tag{1.19}$$

where the three left-hand-side functions do not vanish simultaneously. There exists a constant c such that, in a neighborhood of this point of V, the Maslov index m_{R_0} has the value

$$m_{R_0} = \begin{cases} c & \text{when } dR \wedge \omega_2 \wedge \omega_3/\varpi < 0 \\ 1 + c & \text{when } dR \wedge \omega_2 \wedge \omega_3/\varpi > 0. \end{cases} \tag{1.20}$$

Proof of 1°). Use expression (1.16) for d^3x.

Proof of 2°). Let us calculate the jump of m_{R_0} across Σ_{R_0} by applying theorem 3.2 of I,§3. By (1.9), (1.15), and (1.18), the components d_jx and d_jp of dx and dp in the frame (J_1, J_2, J_3) satisfy on V, at the points of Σ_{R_0},

$$\begin{aligned} d_1p \wedge d_2x \wedge d_3x &= R^2[d(QR^{-1}) - LR^{-1}\omega_3] \wedge \omega_2 \wedge \omega_3 \\ &= R\,dQ \wedge \omega_2 \wedge \omega_3, \\ d_1x \wedge d_2p \wedge d_3x &= -R\,dR \wedge [d(LR^{-1}) + QR^{-1}\omega_3] \wedge \omega_2 \\ &= dL \wedge dR \wedge \omega_2, \\ d_1x \wedge d_2x \wedge d_3p &= (dR) \wedge \omega_3 \wedge [L\omega_1 - Q\omega_2] \\ &= L\,dR \wedge \omega_3 \wedge \omega_1. \end{aligned}$$

The existence of ϖ satisfying (1.19) and (1.20) follows by this theorem (theorem 3.2, of I,§3).

2. Choice of a Hamiltonian H

Let

$$H : \Omega^6 \to \mathbf{R}$$

be an infinitely differentiable function in *involution with L and M*, that is, by (1.14), a composition with the functions L, M, Q, and R:

$$H(x, p) = H[L(x, p), M(x, p), Q(x, p), R(x)]. \tag{2.1}$$

$H[\cdot]$ is an infinitely differentiable function defined on an open subset Ω^4 of the space $\mathbf{R}^4$ with the coordinates L, M, Q, R. Assume that, on Ω^4

$$0 < R, \qquad |M| < L, \qquad (H_Q, H_R) \neq (0, 0) \text{ when } H = 0. \tag{2.2}$$

H_Q denotes the function $\partial H[L, M, Q, R]/\partial Q$.

By (1.3), *the three functions* H, L, and M of (x, p) are pairwise *in involution*, so that the results of II,§3 can be applied explicitly to the *lagrangian operator a associated to* H.

From theorem 2 (*E. Cartan*) of II,§3, a quadrature locally defines four real numerical functions on Ω^6,

$$g_0, g_1, g_2, g_3,$$

such that

$$\langle p, dx \rangle = dg_0 + g_1 dL + g_2 dM + g_3 dH. \tag{2.3}$$

We shall use this formula when $H = 0$; it is made explicit by lemma 2, assuming

$$H = 0.$$

According to (1.14), the quadrature is the definition of the function Ω by (2.8).

Notation. Let W be the hypersurface in Ω^6 given by the equation

$$W: H(x, p) = 0$$

(Compare II,§3). Let (L_0, M_0) be a pair of real numbers such that $|M_0| < L_0$; let $V[L_0, M_0]$ denote any connected component of the subset of W given by the equations

$$V[L_0, M_0]: H(x, p) = 0, \qquad L(x, p) = L_0, \qquad M(x, p) = M_0. \tag{2.4}$$

W is the union of the $V[L_0, M_0]$. By theorem 7.1, of II,§3, $V[L_0, M_0]$ is a lagrangian manifold. It is the topological product of

- the 2-dimensional torus with the coordinates Φ and Ψ mod 2π,
- a connected curve $\Gamma[L_0, M_0]$ in the open half-plane with the coordinates $Q, R > 0$.

The equation of this curve is

$$\Gamma[L_0, M_0]: H[L_0, M_0, Q, R] = 0. \tag{2.5}$$

By $(2.2)_3$, this curve does not have any singular points.

Let us define a real numerical function t of $[L, M, Q, R]$ on W up to the addition of a function of (L, M) by the condition

$$dt = \frac{dR}{RH_Q[L, M, Q, R]} = -\frac{dQ}{RH_R[L, M, Q, R]} \text{ on } \Gamma[L, M]; \tag{2.6}$$

when the curve $\Gamma[L, M]$ is a closed curve, then t is defined mod $c[L, M]$, where

$$c[L, M] = \oint_{\Gamma[L,M]} \frac{dR}{RH_Q} = -\oint_{\Gamma[L,M]} \frac{dQ}{RH_R}; \tag{2.7}$$

t is monotone on Γ; (L, M, t, Φ, Ψ) forms a system of local coordinates on W.

Let us define another real numerical function Ω on W up to the addition of a function of (L, M) by

$$d\Omega = Q\frac{dR}{R} \text{ on } \Gamma[L, M]; \tag{2.8}$$

when the curve $\Gamma[L, M]$ is a closed curve, then Ω is defined mod $2\pi N[L, M]$ where

$$N[L, M] = \frac{1}{2\pi}\oint_{\Gamma[L,M]} Q\frac{dR}{R} > 0. \tag{2.9}$$

Ω is not monotone on Γ. Let its differential be denoted by

$$d\Omega[L, M, t] = Q\frac{dR}{R} + \lambda[L, M, t]\,dL + \mu[L, M, t]\,dM. \tag{2.10}$$

The properties of the functions λ and μ thus defined will be specified in §2.

Now we can give the explicit expression of the restriction of (2.3) to W.

LEMMA 2. 1°) The restriction of the pfaffian form

$$\omega = \langle p, dx \rangle$$

to W is

$$\omega_W = d(\Omega + L\Psi + M\Phi) - (\lambda + \Psi)\,dL - (\mu + \Phi)\,dM. \tag{2.11}$$

2°) The characteristic system of $d\omega_W$ is Hamilton's system:

$$\frac{dx}{H_p(x, p)} = -\frac{dp}{H_x(x, p)}, \qquad H(x, p) = 0. \tag{2.12}$$

Its first integrals are compositions with the functions

$$L, M, \lambda + \Psi, \mu + \Phi; \tag{2.13}$$

L and M are in involution; $\lambda + \Psi$ and $\mu + \Phi$ are also in involution.

3°) On the solution curves of this system, that is, the characteristic curves of W, we have

$$dt = \frac{dx}{H_p(x, p)} = -\frac{dp}{H_x(x, p)} = \frac{dR}{RH_Q[L, M, R, Q]} = \frac{d\Psi}{H_L[\cdot]} = \frac{d\Phi}{H_M[\cdot]}$$

$$= -\frac{dQ}{RH_R[\cdot]}, \; dL = dM = 0. \tag{2.14}$$

4°) On W

$$\left.\frac{d^3x \wedge d^3p}{dH}\right|_W = dL \wedge dM \wedge dt \wedge d\Phi \wedge d\Psi. \tag{2.15}$$

Proof of 1°). Use the expression (1.14) for $\langle p, dx \rangle$ in Z and the expression (2.10) for $Q\,dR/R$ on W.

Proof of 2°). Lemma 3.2 of II,§3 (E. Cartan) proves that (2.12) is the characteristic system of $d\omega_W$ and that $d\omega_W$ has rank 4. Now, by (2.11),

$$d\omega_W = dL \wedge d(\lambda + \Psi) + dM \wedge d(\mu + \Phi);$$

thus the four functions (2.13) are first integrals of this system (see E. Cartan's definition of a characteristic system, II,§3,2). The pair (L, M) and $(\lambda + \Psi, \mu + \Phi)$ are in involution by remark 2.2 of II,§3.

Proof of 3°). The same lemma (E. Cartan) and the expression (1.14) for $\langle p, dx \rangle$ prove that the characteristic system of $d\omega_w$ is

$$\frac{dR}{RH_Q[L, M, Q, R]} = \frac{d\Psi}{H_L[\cdot]} = \frac{d\Phi}{H_M[\cdot]} = -\frac{dQ}{RH_R[\cdot]} = \frac{dL}{0} = \frac{dM}{0},$$
$$H = 0;$$

thus this system is equivalent to (2.12). Obviously

$$\frac{dx}{H_p(x, p)} = \frac{\langle x, dx \rangle}{\langle x, H_p(x, p) \rangle} = \frac{R\,dR}{\langle x, L_p \rangle H_L + \langle x, M_p \rangle H_M + \langle x, Q_p \rangle H_Q}.$$

Since $x \wedge (p + sx)$ is independent of the real variable s,

$$d(x \wedge p) = 0 \text{ for } dx = 0, \qquad dp \text{ parallel to } x;$$

therefore,

$$\langle x, L_p \rangle = \langle x, M_p \rangle = 0.$$

Moreover,

$$\langle x, Q_p \rangle = R^2.$$

The preceding five relations imply

$$\frac{dx}{H_p(x, p)} = \frac{dR}{RH_Q[L, M, Q, R]} = \frac{d\Psi}{H_L} = \frac{d\Phi}{H_M} = -\frac{dQ}{RH_R}, dL = dM = 0$$

and thus (2.14), by (2.6) and (2.12).

Proof of 4°). Formula (1.17) can be written

$$d^3x \wedge d^3p = dL \wedge dM \wedge dH \wedge \frac{dR}{RH_Q} \wedge d\Phi \wedge d\Psi,$$

where, for $H = 0$,

$$\frac{dR}{RH_Q} = dt \mod(dL, dM)$$

by (2.14); hence (2.15) holds, the following meaning being given to its left member:

$$\left.\frac{d^3x \wedge d^3p}{dH}\right|_W$$

denotes the restriction ϖ_W to W of any form ϖ of degree 5 such that $dH \wedge \varpi = d^3x \wedge d^3p$; ϖ_W is clearly independent of the choice of ϖ.

3. The Quantized Tori $T(l, m, n)$ Characterizing Solutions, Defined $\mathrm{mod}(1/\nu)$ on Compact Manifolds, of the Lagrangian System

$$aU = (a_{L^2} - L_0^2)U = (a_M - M_0)U = 0 \mod\frac{1}{\nu^2}; \tag{3.1}$$

a, a_{L^2}, and a_M denote *the lagrangian operators associated*, respectively, *to the hamiltonians*

$$H, L^2, M,$$

which are *in involution*; L_0 and M_0 are two real constants such that $|M_0| < L_0$.

From theorem 7.1 of II,§3, solutions of this system are *lagrangian functions U with constant lagrangian amplitude, defined* $\mathrm{mod}(1/\nu)$ *on those manifolds* $V[L_0, M_0]$ defined by (2.4), whether compact or not, *that satisfy Maslov's quantum condition* (II,§3, definition 6.2). $V[L_0, M_0]$ is chosen to be connected. The measure η_V on V, which is invariant under the characteristic vectors of H, $L^2 - L_0^2$, and $M - M_0$, and which serves to define the lagrangian amplitude, is

$$\eta_V = dt \wedge d\Phi \wedge d\Psi \tag{3.2}$$

from (2.15) and formula (7.4) of II,§3.

Recall the statement of Maslov's quantum condition: The function

$$\frac{1}{2\pi\hbar}\varphi_{R_0} - \frac{1}{4}m_{R_0} \quad \left(\text{where } \hbar = \frac{i}{\nu_0} \text{ is real}\right) \tag{3.3}$$

has to be uniform on V mod 1.

By (2.11), where $L = L_0$ and $M = M_0$, the phase φ_{R_0} of $V[L_0, M_0]$ (defined in I,§2,9 and I,§3,1) is

$$\varphi_{R_0} = \Omega + L_0\Psi + M_0\Phi. \tag{3.4}$$

Let us calculate the Maslov index of $V[L_0, M_0]$.

LEMMA 3. 1°) The *apparent contour* Σ_{R_0} of $V[L_0, M_0]$ is the union $\Sigma_1 \cup \Sigma_2$ of two surfaces in $V[L_0, M_0]$ given by the equations

$$\Sigma_1 : \Psi = 0 \bmod \pi; \qquad \Sigma_2 : H_Q[L, M, Q, R] = 0.$$

2°) The *Maslov index m of* $V[L_0, M_0]$ is

$$m_{R_0} = \left[\frac{1}{\pi}\Psi\right] - \left[\frac{1}{\pi}\arctan\frac{H_Q}{H_R}\right] \text{on } \check{V}[L_0, M_0] \tag{3.5}$$

up to the addition of an integer constant; $[\cdots]$ *denotes the integer part.*

Proof. Apply lemma 1. By (1.11) and (1.13),

$$\omega_1 = -\cos\Psi \sin\Theta\, d\Psi, \qquad \omega_2 = \sin\Psi\sin\Theta\, d\Phi,$$
$$L_0\omega_3 = L_0\, d\Psi + M_0\, d\Phi$$

on $V[L_0, M_0]$; hence, by (2.6),

$$dR \wedge \omega_2 \wedge \omega_3 = -RH_Q \sin\Psi\sin\Theta\, dt \wedge d\Psi \wedge d\Phi, \tag{3.6}$$

$$dQ \wedge \omega_2 \wedge \omega_3 = RH_R \sin\Psi\sin\Theta\, dt \wedge d\Psi \wedge d\Phi, \tag{3.7}$$

$$dR \wedge \omega_3 \wedge \omega_1 = -RH_Q \cos\Psi\sin\Theta\, dt \wedge d\Psi \wedge d\Phi, \tag{3.8}$$

where $R\sin\Theta \neq 0$ by $(2.2)_1$ and $(2.2)_2$.

By lemma 1,1°), Σ_{R_0} is given by the equation

$$H_Q \sin\Psi = 0,$$

from which follows part 1 of lemma 3.

In a neighborhood of a point of $\Sigma_1 \backslash \Sigma_1 \cup \Sigma_2$, $H_Q \cos \Psi \neq 0$; in lemma 1,2°), we can choose

$$\varpi = dR \wedge \omega_3 \wedge \omega_1; \qquad dR \wedge \omega_2 \wedge \omega_3/\varpi = \tan \Psi,$$

whence by this lemma

$$m_{R_0} = \left[\frac{1}{\pi}\Psi\right] + \text{const.} \tag{3.9}$$

By $(2.2)_3$, in a neighborhood of a point of $\Sigma_2 \backslash \Sigma_1 \cup \Sigma_2$, $H_R \sin \Psi \neq 0$; in lemma 1,2°), we can choose

$$\varpi = dQ \wedge \omega_2 \wedge \omega_3; \qquad dR \wedge \omega_2 \wedge \omega_3/\varpi = -H_Q/H_R,$$

where

$$m_{R_0} = \text{const.} - \left[\frac{1}{\pi}\arctan\frac{H_Q}{H_R}\right]. \tag{3.10}$$

The global expression for m_{R_0} follows from the two local expressions (3.9) and (3.10).

By (3.4) and (3.5), *Maslov's quantum condition* may be formulated as follows: The function

$$\frac{1}{\hbar}\frac{\Omega}{2\pi} + \frac{1}{4\pi}\arctan\frac{H_Q}{H_R} + \left(\frac{L_0}{\hbar} - \frac{1}{2}\right)\frac{\Psi}{2\pi} + \frac{M_0}{\hbar}\frac{\Phi}{2\pi}$$

is defined mod 1 on $V[L_0, M_0]$.

If $V[L_0, M_0]$ is *compact*, that is, if the curve $\Gamma[L_0, M_0]$ is a closed curve, by (2.9) this condition is that

$$\frac{1}{\hbar}N[L_0, M_0] + \frac{1}{2}, \qquad \frac{1}{\hbar}L_0 - \frac{1}{2}, \qquad \frac{1}{\hbar}M_0$$

be three integers. Denote them by

$$n - l, \quad l, \quad m$$

in order to recover the classical notation of quantum physics. Since $N > 0$ and $|M_0| < L_0$, we have

$$|m| \leqslant l < n.$$

If $V[L_0, M_0]$ *is not compact*, Maslov's quantum condition is that

$$\frac{1}{\hbar}L_0 - \frac{1}{2} = l, \qquad \frac{1}{\hbar}M_0 = m$$

be two integers such that $|m| \leqslant l$.

Thus the conclusion of this section is the following theorem.

THEOREM 3. *The connected manifolds on which the solutions of the lagrangian system* (3.1) *are defined are*

1°) *The* **compact** *manifolds* $V[L_0, M_0]$ *defined by* (2.4) *such that there exist three integers*

$$l, \quad m, \quad n$$

satisfying the conditions

$$|m| \leqslant l < n, \tag{3.11}$$

$$L_0 = \hbar(l + \tfrac{1}{2}), \qquad M_0 = \hbar m, \qquad L_0 + N[L_0, M_0] = \hbar n; \tag{3.12}$$

thus $V[L_0, M_0]$ *is a torus, which will be denoted by* $T(l, m, n)$*; the coordinates at a point of this torus are*

$$t \bmod c[L_0, M_0] \text{ defined by (2.6)–(2.7)}, \qquad \Psi \bmod 2\pi, \qquad \Phi \bmod 2\pi. \tag{3.13}$$

2°) *The* **noncompact** *manifolds* $V[L_0, M_0]$ *defined by* (2.4) *such that there exist two integers*

$$l, \quad m$$

satisfying the conditions

$$|m| \leqslant l, \tag{3.14}$$

$$L_0 = \hbar(l + \tfrac{1}{2}), \qquad M_0 = \hbar m; \tag{3.15}$$

thus $V[L_0, M_0]$ *is a product of*

- *a 2-dimensional torus with the coordinates* $\Psi \bmod 2\pi$, $\Phi \bmod 2\pi$,
- *a line, half-line, or segment with coordinate* t.

We provide $V[L_0, M_0]$, *whether compact or noncompact, with the following measure, which is invariant under the characteristic vectors of* H, $L^2 - L_0$, *and* $M - M_0$:

$$\eta_V = dt \wedge d\Phi \wedge d\Psi; \tag{3.16}$$

then the solutions of (3.1) *defined on* $V[L_0, M_0]$ *are lagrangian functions with constant lagrangian amplitude.*

Remark 3.1. From now on, we shall limit ourselves to the study of compact lagrangian manifolds (for example, that of the electron belonging to an atom).

Remark 3.2. The characteristic vectors of $L^2 - L_0^2$ and $M - M_0$ are

$$\kappa_{L^2-L_0^2}: dL = dM = dt = 0, \qquad d\Psi = 2L_0, \qquad d\Phi = 0, \tag{3.17}$$

$$\kappa_{M-M_0}: dL = dM = dt = 0, \qquad d\Psi = 0, \qquad d\Phi = 1, \tag{3.18}$$

respectively. It is immediately verified that these vectors leave η_V invariant.

Proof of (3.17). It follows from $(1.2)_1$, then $(1.5)_1$ and (1.6), and finally $(1.5)_1$ and (1.12) that $\kappa_{L^2-L_0^2}$ is the vector tangent to $V[L_0, M_0]$ such that

$$dx = 2LL_p = 2(R^2p - Qx) = 2LRJ_2 = 2L\frac{\partial x(L, M, t, \Psi, \Phi)}{\partial \Psi}.$$

This proves (3.17).

Proof of (3.18). It follows from (1.1), then $(1.5)_1$ and (1.12), that κ_{M-M_0} is the vector tangent to $V[L_0, M_0]$ such that

$$dx = (-x_2, x_1, 0) = \frac{\partial x(L, M, t, \Psi, \Phi)}{\partial \Phi}$$

Remark 3.3. In agreement with theorem 7.1,2°) of II,§3, when the manifold $V[L_0, M_0]$ is *compact*, hence a *torus*, the characteristic vector of H [see (2.14)],

$$\kappa: dL = dM = 0, \qquad dt = 1, \qquad d\Psi = H_L, \qquad d\Phi = H_M, \tag{3.19}$$

and those of $L^2 - L_0^2$ and $M - M_0$ generate a *group of translations of* $V[L_0, M_0]$, namely, the translations in the following coordinates [see (2.7) and III,§2,(1.5) and III,§2,(2.3)]:

$$t \mod c[L_0, M_0],$$

$$\Psi + \lambda[L_0, M_0, t] - \frac{N_L[L_0, M_0]}{c[L_0, M_0]}t \mod 2\pi,$$

$$\Phi + \mu - \frac{N_M}{c}t \mod 2\pi.$$

4. Examples: The Schrödinger And Klein-Gordon Operators

Let us choose

$$H(x, p) = \frac{1}{2}\left[P^2 - \frac{K[R, M]}{R^2}\right], \tag{4.1}$$

where $K:\mathbf{R}_+ \oplus \mathbf{R} \to \mathbf{R}$ is a given function. In other words,

$$H[L, M, Q, R] = \frac{1}{2R^2}[L^2 + Q^2 - K[R, M]]. \tag{4.2}$$

Theorem 3 can be applied immediately.

The condition that (2.4) define at least one compact hypersurface $V[L_0, M_0]$, which is a torus, is that the function

$$\mathbf{R}_+ \ni R \mapsto K[R, M_0] - L_0^2 \in \mathbf{R}$$

be positive between two consecutive zeros R_1 and R_2 $(0 < R_1 < R_2)$. Then (2.9) defines

$$N[L_0, M_0] = \frac{1}{\pi}\int_{R_1}^{R_2} \sqrt{K[R, M_0] - L_0^2}\,\frac{dR}{R} > 0. \tag{4.3}$$

Remark 4.1. if K is *an affine function of* M, then, by (4.1),

$$\left\langle \frac{\partial}{\partial x}, \frac{\partial}{\partial p}\right\rangle H(x, p) = 0, \tag{4.4}$$

and consequently [II,§1,definition 6.2 and II,§1,(6.3)], the expression in R_0 of *the operator associated to* H is

$$a = \frac{1}{2\nu^2}\Delta - \frac{1}{2R^2}K\left[R, \frac{1}{\nu}\left(x_1\frac{\partial}{\partial x_2} - x_2\frac{\partial}{\partial x_1}\right)\right], \tag{4.5}$$

where

- $\Delta = \Sigma_{j=1}^3 \partial^2/\partial x_j^2$,

- the multiplication by any function of R commutes with the operator

$$\frac{1}{\nu}\left(x_1\frac{\partial}{\partial x_2} - x_2\frac{\partial}{\partial x_1}\right).$$

Example 4. We call the operator associated to the hamiltonian

$$H(x, p) = \frac{1}{2}\left[P^2 + A(M) - \frac{2B(M)}{R} + \frac{C(M)}{R^2}\right], \tag{4.6}$$

where A, B, and $C:\mathbf{R} \to \mathbf{R}$ are some given functions, *the Schrödinger–Klein-Gordon operator.* Let

$$A_0 = A(M_0), \qquad B_0 = B(M_0), \qquad C_0 = C(M_0).$$

The condition that there exist at least one compact manifold $V[L_0, M_0]$ defined by (2.4) may be expressed as follows:

$$A_0 > 0, \qquad L_0^2 + C_0 > 0, \qquad B_0 > \sqrt{A_0}\sqrt{L_0^2 + C_0}. \tag{4.7}$$

When it is satisfied, $V[L_0, M_0]$ is unique, and by (4.3),

$$N[L_0, M_0] = \frac{1}{2\pi}\oint\sqrt{-A_0 R^2 + 2B_0 R - L_0^2 - C_0}\frac{dR}{R},$$

where the integral is calculated along a cycle of $\mathbf{C}$ enclosing the cut $[R_1, R_2]$; taking residues at $R = \infty$ and $R = 0$ gives

$$N[L_0, M_0] = \frac{B_0}{\sqrt{A_0}} - \sqrt{L_0^2 + C_0} > 0. \tag{4.8}$$

The statement of theorem 3 becomes the following.

THEOREM 4.1. *Let a be the Schrödinger–Klein-Gordon operator, that is, the operator associated to the hamiltonian* (4.6). *Let a_{L^2} and a_M be the operators associated to the hamiltonians L^2 and M. Then the lagrangian system*

$$aU = (a_{L^2} - L_0^2)U = (a_M - M_0)U = 0 \bmod \frac{1}{\nu^2} \tag{4.9}$$

has solutions defined on a ***compact*** *manifold if and only if there exists a triple of integers* (l, m, n) *such that*

$$L_0 = \hbar(l + \tfrac{1}{2}), \qquad M_0 = \hbar m, \tag{4.10}$$

$$\begin{gathered} |m| \leqslant l < n, \qquad (l + \tfrac{1}{2})^2 + C_0\hbar^{-2} > 0, \\ A_0 = B_0^2\hbar^{-2}[n - l - \tfrac{1}{2} + \sqrt{(l + \tfrac{1}{2})^2 + C_0\hbar^{-2}}]^{-2}. \end{gathered} \tag{4.11}$$

If this condition is satisfied, this compact manifold is unique: it is the torus $T(l, m, n)$ given by the equations

$$T(l, m, n): H(x, p) = L(x, p) - L_0 = M(x, p) - M_0 = 0. \tag{4.12}$$

We provide this torus with the invariant measure (3.16); *then the solutions of* (4.9) *defined* $\mathrm{mod}(1/v)$ *on* $T(l, m, n)$ *are the lagrangian functions with constant lagrangian amplitude.*

Notation. (Related to physics and used only at the end of this section) In $\mathbf{E}^3$, we choose an electric potential $\mathscr{A}_0 : \mathbf{E}^3 \to \mathbf{R}$ and a magnetic potential vector $(\mathscr{A}_1, \mathscr{A}_2, \mathscr{A}_3) : \mathbf{E}^3 \to \mathbf{E}^3$ satisfying the physical law

$$\sum_{j=1}^{3} \frac{\partial \mathscr{A}_j}{\partial x_j} = 0; \tag{4.13}$$

they are time independent.

The trajectories in $\mathbf{E}^3$ of the relativistic or nonrelativistic electron of mass μ and electric charge $-\varepsilon < 0$ are the solutions of Hamilton's system, defined by the hamiltonian given by

$$H(x, p) = \frac{1}{2\mu} \sum_{j=1}^{3} \left[p_j + \frac{\varepsilon}{c} \mathscr{A}_j(x) \right]^2 - E - \varepsilon \mathscr{A}_0(x) \tag{4.14}$$

in the *nonrelativistic* case, and by

$$H(x, p) = \frac{1}{2\mu} \sum_{j=1}^{3} \left[p_j + \frac{\varepsilon}{c} \mathscr{A}_j(x) \right]^2 - \frac{1}{2\mu c^2} [E + \varepsilon \mathscr{A}_0(x)]^2 + \tfrac{1}{2} \mu c^2 \tag{4.15}$$

in the *relativistic* case. Here

- c is the speed of light,
- E is a constant, which is the energy of the electron whose position x and momentum p satisfies $H(x, p) = 0$ (energy including the rest mass μc^2 in the relativistic case).

By (4.13)

$$\left\langle \frac{\partial}{\partial x}, \frac{\partial}{\partial p} \right\rangle H(x, p) = 0.$$

Then by II,§1,(6.3) and II,§1,definition 6.2, in the nonrelativistic case the operator associated to H is the *Schrödinger* operator

$$a = \frac{1}{2\mu} \left[\frac{1}{v^2} \Delta + \frac{2\varepsilon}{c} \sum_j \mathscr{A}_j(x) \frac{1}{v} \frac{\partial}{\partial x_j} + \frac{\varepsilon^2}{c^2} \sum_j \mathscr{A}_j^2(x) \right] - E - \varepsilon \mathscr{A}_0(x), \tag{4.16}$$

and in the relativistic case it is the *Klein-Gordon* operator

$$a = \frac{1}{2\mu}\left[\frac{1}{v^2}\Delta + \frac{2\varepsilon}{c^2}\sum_j \mathscr{A}_j \frac{1}{v}\frac{\partial}{\partial x_j} + \frac{\varepsilon^2}{c^2}\sum_j \mathscr{A}_j^2\right] - \frac{1}{2\mu c^2}[E + \varepsilon\mathscr{A}_0]^2 + \frac{\mu c^2}{2}. \tag{4.17}$$

Let us choose these hamiltonians and operators as follows: $\mathscr{A}_0$ is the electric potential of a nucleus with atomic number Z, placed at the origin; the magnetic field is constant, parallel to the third coordinate axis, and of intensity $\mathscr{H}$. In other words,

$$\mathscr{A}_0(x) = \frac{\varepsilon Z}{R}, \qquad \mathscr{A}_1(x) = -\tfrac{1}{2}\mathscr{H}x_2, \qquad \mathscr{A}_2(x) = \tfrac{1}{2}\mathscr{H}x_1, \qquad \mathscr{A}_3 = 0. \tag{4.18}$$

Let us neglect the $\mathscr{H}^2$ terms, as is done in physics; the hamiltonians (4.14) and (4.15) of the electron become hamiltonians of the form (4.6)

$$H(x, p) = \frac{1}{2\mu}\left[P^2 + \frac{\varepsilon}{c}\mathscr{H}M - 2\mu E - \frac{2\mu\varepsilon^2 Z}{R}\right] \tag{4.19}$$

in the nonrelativistic case, and

$$H(x, p) = \frac{1}{2\mu}\left[P^2 + \frac{\varepsilon}{c}\mathscr{H}M + \mu^2 c^2 - \frac{E^2}{c^2} - \frac{2\varepsilon^2 ZE}{c^2 R} - \frac{\varepsilon^4 Z^2}{c^2 R^2}\right] \tag{4.20}$$

in the relativistic case.

The *Schrödinger* operator becomes

$$a = \frac{1}{2\mu}\left[\frac{1}{v^2}\Delta + \frac{\varepsilon\mathscr{H}}{cv}\left(x_1\frac{\partial}{\partial x_2} - x_2\frac{\partial}{\partial x_1}\right) - 2\mu E - \frac{2\mu\varepsilon^2 Z}{R}\right]; \tag{4.21}$$

the *Klein-Gordon* operator becomes

$$a = \frac{1}{2\mu}\left[\frac{1}{v^2} + \frac{\varepsilon\mathscr{H}}{cv}\left(x_1\frac{\partial}{\partial x_2} - x_2\frac{\partial}{\partial x_1}\right) + \mu^2 c^2 - \frac{E^2}{c^2} - \frac{2\varepsilon^2 ZE}{c^2 R} - \frac{\varepsilon^4 Z^2}{c^2 R^2}\right]. \tag{4.22}$$

Relation $(4.11)_3$ defines E as a function of (l, m, n); in the course of this calculation, two constants, which are classical in physics, appear.

$\alpha = \dfrac{\varepsilon^2}{\hbar c} \simeq \dfrac{1}{137}$, the dimensionless *fine-structure constant*,

$$\beta = \frac{\varepsilon\hbar}{2\mu c}, \text{ the } \textit{Bohr magneton} \ (\beta\mathscr{H} \text{ has the dimensions of energy}),$$

as well as the function given by

$$F(n, k) = \frac{1}{\sqrt{1 + \left(\dfrac{\alpha Z}{n - k + \sqrt{k^2 - \alpha^2 Z^2}}\right)^2}}. \tag{4.23}$$

The statement of theorem 4.1 becomes the following.

THEOREM 4.2. *Let a be the Schrödinger operator* (4.21) *or the Klein-Gordon operator* (4.22), *H being* (4.19) *or* (4.20). *Let a_{L^2} and a_M be the operators associated to L^2 and M. For certain values of the constants E, L_0, and M_0, the lagrangian system*

$$aU = (a_{L^2} - L_0^2)U = (a_M - M_0)U = 0 \bmod \frac{1}{\nu^2} \tag{4.9}$$

has solutions defined $\bmod(1/\nu)$ *on* ***compact*** *manifolds. These values of E, L_0, and M_0 are expressed as functions of a triple of integers (l, m, n) such that*

$$|m| \leqslant l < n \text{ and, in the Klein-Gordon case, } l + \tfrac{1}{2} \geqslant \alpha Z;$$

these expressions for E, L_0, and M_0 are

$$E = -\mu c^2 \frac{\alpha^2 Z^2}{2n^2} + \beta\mathscr{H}m \text{ in the Schrödinger case,} \tag{4.24}$$

$$E^2 = \mu c^2(\mu c^2 + 2\beta\mathscr{H}m)F^2(n, l + \tfrac{1}{2}) \text{ in the Klein-Gordon case,} \tag{4.25}$$

$$L_0 = \hbar(l + \tfrac{1}{2}), \qquad M_0 = \hbar m. \tag{4.26}$$

These solutions of (4.9) *are defined on the tori* (4.12). *Let us provide these tori with the invariant measure* (3.16); *then these solutions are lagrangian functions, defined on these tori, with constant lagrangian amplitude.*

Remark 4.2. Physicists, neglecting β^2 and $\beta\alpha^2$, simplify (4.25) as follows:

$$\pm E \simeq \mu c^2 F(n, l + \tfrac{1}{2}) + \beta\mathscr{H}m.$$

The minus sign concerns antimatter ($\mu < 0$).

Remark 4.3. The energy levels (4.24) and (4.25) are those given by the study of the partial differential equation

$au = 0,$

whose unknown $u:\mathbf{E}^3\backslash\{0\} \to \mathbf{C}$ and its gradient have to be square integrable (III,§4).

§2. The Lagrangian Equation $aU = 0 \bmod(1/\nu^2)$ (a Associated to H, U Having Lagrangian Amplitude $\geqslant 0$ Defined on a Compact V)

0. Introduction

Summary. In §1, we studied problem (1) (introduction to chapter III), that is, a system of three lagrangian equations. In §2, we study *the first of these three equations* and, more precisely, problem (2) (introduction to chapter III).

Each solution of problem (1) is a solution of problem (2), by theorem 3 of §1. In the exceptional case of a hamiltonian H independent of M (for example, Schrödinger and Klein-Gordon without a magnetic field), a rotation in $\mathbf{E}^3$ acting simultaneously on x and p obviously transforms solutions of problem (1) into solutions of problem (2) that are no longer solutions of problem (1). Without considering this exceptional case, theorem 1 of §2 constructs solutions of problem (2) that are not solutions of problem (1): *a solution of problem* (2) *defined on a torus* $T(l, m, n)$ (theorem 3 of §1) *is not necessarily unique up to a constant multiplicative factor.*

On the other hand, theorem 2 shows that even if H depends on some parameters, *these tori* $T(l, m, n)$ *are, in general, the only compact lagrangian manifolds* V *on which solutions of problem* (2) are defined. Moreover, theorem 3.1 specifies that in the Schrödinger and Klein-Gordon cases problems (1) and (2) define *the same energy levels*: the classical levels.

Remark. Theorem 1 of §2 does not have any physical significance: it takes into account the condition that some numbers, which in the Schrödinger and Klein-Gordon cases are measurements of physical quantities, take rational values.

CONCLUSION. Let us call a problem that when posed for the Schrödinger–Klein-Gordon equation has *the essential features of the classical problem* (4) (introduction to chapter III) a *well-posed problem. Problem* (1) *is well-posed by* theorem 4.2 of §1. From the preceding remark, the main result of §2 is that *problem* (2) *is not well-posed.*

1. Solutions of the Equation $aU = 0 \mod(1/\nu^2)$ with Lagrangian Amplitude $\geqslant 0$ Defined on the Tori $V[L_0, M_0]$

In section 2 of §1 we chose H and defined the manifolds

$$W: H = 0, \qquad V[L_0, M_0]: H = L - L_0 = M - M_0 = 0.$$

Any compact lagrangian manifold V in W is a union of characteristics K of H (whose paramter t varies from $-\infty$ to $+\infty$); V is thus a union of compact closures $\bar{K}$ of such characteristics.

Properties of a characteristic K of H with compact closure $\bar{K}$. By (2.13) of §1, such a characteristic K stays on a torus $V[L_0, M_0]$ and is given by the equations

$$K: \Psi - \Psi_0 + \lambda[L_0, M_0, t] = \Phi - \Phi_0 + \mu[L_0, M_0, t] = 0 \quad (\Phi_0, \Psi_0 \text{ constants}) \tag{1.1}$$

on this torus. Recall that the coordinates of a point of $V[L_0, M_0]$,

$$(\Psi, \Phi, t),$$

are defined

$$(\text{mod } 2\pi, \text{ mod } 2\pi, \text{ mod } c_0), \text{ where } c_0 = c[L_0, M_0].$$

More explicitly, let $\mathbf{R}^3$ be given the coordinates (Ψ, Φ, t) and let $\mathbf{Z}^3$ be the additive group of triples of integers (ξ, η, ζ) acting on $\mathbf{R}^3$ as follows:

$$\mathbf{Z}^3 \ni (\xi, \eta, \zeta): (\Psi, \Phi, t) \mapsto (\Psi + 2\pi\xi, \Phi + 2\pi\eta, t + c_0\zeta);$$

the quotient of $\mathbf{R}^3$ *by this group* $\mathbf{Z}^3$ is $V[L_0, M_0]$: there exists a natural mapping

$$\mathbf{R}^3 \to \mathbf{R}^3/\mathbf{Z}^3 = V[L_0, M_0]. \tag{1.2}$$

Given a function

$$F: (L, M, t) \to F[L, M, t] \in \mathbf{R},$$

let

$$\Delta_t F[L, M, t] = F[L, M, t + c[L, M]] - F[L, M, t].$$

Obviously

$$\Delta_t R = \Delta_t Q = 0. \tag{1.3}$$

By the definitions (2.8) of Ω and (2.9) of N in §1,

$$\Delta_t\Omega[L, M, t] = 2\pi N[L, M]; \tag{1.4}$$

then, by (1.3) and §1,definition (2.10) of λ and μ,

$$2\pi\, dN[L, M] = \Delta_t\lambda[L, M, t]\, dL + \Delta_t\mu[L, M, t]\, dM;$$

that is,

$$\Delta_t\lambda = 2\pi N_L, \qquad \Delta_t\mu = 2\pi N_M. \tag{1.5}$$

Let

$$N_{L_0} = N_L[L_0, M_0], \qquad N_{M_0} = N_M[L_0, M_0].$$

We can now describe $\bar{K}$.

LEMMA 1.1. 1°) *Assume* N_{L_0} *and* N_{M_0} *are rational*:

$$N_{L_0} = -\frac{L_1}{N_1}, \qquad N_{M_0} = -\frac{M_1}{N_1},$$

$$\textit{where } (L_1, M_1 N_1) \in \mathbf{Z}^3, \quad \text{G.C.D.}(L_1, M_1, N_1) = 1 \tag{1.6}$$

(that is, L_1, M_1, and N_1 are integers with greatest common divisor 1). Then $K = \bar{K}$ is *a closed curve* given by the equations (1.1).

More precisely, the equations (1.1) define an open curve $\tilde{\mathbf{R}}$ (that is, a curve homeomorphic to $\mathbf{R}$) in $\mathbf{R}^3$. By (1.5) and (1.6), the subgroup $\tilde{\mathbf{Z}}$ of $\mathbf{Z}^3$ generated by (L_1, M_1, N_1) leaves $\tilde{\mathbf{R}}$ invariant; we have

$$K = \bar{K} = \tilde{\mathbf{R}}/\tilde{\mathbf{Z}}. \tag{1.7}$$

2°) Assume that N_{L_0} *and* N_{M_0} *are connected by a unique affine relation*

$$L_1 N_{L_0} + M_1 N_{M_0} = N_1 \tag{1.8}$$

where $(L_1, M_1, N_1) \in \mathbf{Z}^3$, G.C.D.$(L_1, M_1, N_1) = 1$.

Then $\bar{K}$ *is the* 2-*dimensional torus* T^2 defined in $V[L_0, M_0]$ by the equation

$$L_1\{\Psi - \Psi_0 + \lambda[L_0, M_0, t]\} + M_1\{\Phi - \Phi_0 + \mu[L_0, M_0, t]\} = 0. \tag{1.9}$$

More precisely, equation (1.9) defines a surface $\tilde{\mathbf{R}}^2$ in $\mathbf{R}^3$ homeomorphic to $\mathbf{R}^2$. By (1.5) and (1.8), the subgroup $\tilde{\mathbf{Z}}^2$ of $\mathbf{Z}^3$ given by the equation

$$\tilde{\mathbf{Z}}^2 : L_1\xi + M_1\eta + N_1\zeta = 0 \tag{1.10}$$

acts on $\tilde{\mathbf{R}}^2$. We have

$$T^2 = \tilde{\mathbf{R}}^2/\tilde{\mathbf{Z}}^2. \tag{1.11}$$

In order to describe the generators of $\tilde{\mathbf{Z}}^2$, let

$$L_2 = \text{G.C.D.}(M_1, N_1), \qquad M_2 = \text{G.C.D.}(L_1, N_1),$$
$$N_2 = \text{G.C.D.}(L_1, M_1). \tag{1.12}$$

M_2 and N_2 divide L_1; G.C.D.$(M_2, N_2) = 1$ by $(1.8)_3$; then there exists an integer L_3 and similarly integers M_3 and N_3 such that

$$L_1 = L_3 M_2 N_2, \qquad M_1 = L_2 M_3 N_2, \qquad N_1 = L_2 M_2 N_3. \tag{1.13}$$

$\tilde{\mathbf{Z}}^2$ is generated by its three elements

$$(0, M_2N_3, -M_3N_2), \qquad (-L_2N_3, 0, L_3N_2), \qquad (L_2M_3, -L_3M_2, 0), \tag{1.14}$$

which evidently are connected by the relation

$$L_3(0, M_2N_3, -M_3N_2) + M_3(-L_2N_3, 0, L_3N_2)$$
$$+ N_3(L_2M_3, -L_3M_2, 0) = 0. \tag{1.15}$$

3°) Suppose *there is no affine relation with integer coefficients connecting* N_{L_0} *and* N_{M_0}. Then $\bar{K}$ *is the torus* $V[L_0, M_0]$.

Proof. The subset of $\mathbf{R}^3$ whose natural image in $V[L_0, M_0]$ is K is defined by the condition

$$\left(\frac{\Psi - \Psi_0 + \lambda[L_0, M_0, t]}{2\pi}, \frac{\Phi - \Phi_0 + \mu[L_0, M_0, t]}{2\pi}\right) \in G,$$

where G is the image of $\mathbf{Z}^3$ in the additive group $\mathbf{R}^2$ under the morphism

$$\mathbf{Z}^3 \ni (\xi, \eta, \zeta) \mapsto (\xi + N_{L_0}\zeta, \eta + N_{M_0}\zeta) \in \mathbf{R}^2.$$

Then $\bar{K}$ is the natural image in $V[L_0, M_0]$ of the closed subset of $\mathbf{R}^3$ defined by the condition

$$\left(\frac{\Psi - \Psi_0 + \lambda[L_0, M_0, t]}{2\pi}, \frac{\Phi - \Phi_0 + \mu[L_0, M_0, t]}{2\pi}\right) \in \bar{G}, \tag{1.16}$$

where $\bar{G}$ is the closure of G; $\bar{G}$ *is a closed subgroup of* $\mathbf{R}^2$.

Three cases occur: (1°) G is discrete, (2°) $\dim \bar{G} = 1$, (3°) $\bar{G} = \mathbf{R}^2$.

1°) *G is discrete*, that is, $\bar{G} = G$. This is the case if and only if N_{L_0} *and* N_{M_0} *are rational* (use the rapidity of convergence to an irrational number of its rational approximations given by its continued fraction development), that is, if and only if (1.6) holds. Then $\bar{K} = K$; the elements of the subgroup $\tilde{\mathbf{Z}}$ of $\mathbf{Z}^3$ leaving invariant the curve $\tilde{\mathbf{R}} \subset \mathbf{R}^3$ given by the equations (1.1) are the (ξ, η, ζ) in $\mathbf{Z}^3$ such that

$$N_1\xi = L_1\zeta, \qquad N_1\eta = M_1\zeta.$$

By $(1.6)_4$, N_1 divides ζ. Thus $\tilde{\mathbf{Z}}$ is generated by $(L_1, M_1, N_1) \in \mathbf{Z}^3$.

2°) $\dim \bar{G} = 1$. Then $\bar{G}$ is the set of $(\theta, \tau) \in \mathbf{R}^2$ satisfying a condition of the form

$$L_1\theta + M_1\tau \in \mathbf{Z}, \quad \text{where } L_1, M_1 \in \mathbf{R}; \tag{1.17}$$

by the definition of G, $\bar{G}$ is the subgroup (1.17) of $\mathbf{R}^2$ if and only if (1.8) is satisfied and G is not discrete, that is, by (1°), if and only if N_{L_0} and N_{M_0} are connected by a unique affine relation with integer coefficients, which is (1.8).

Assuming this hypothesis, by definition (1.16) of $\bar{K}$ and definition (1.17) of $\bar{G}$, $\bar{K}$ is the image under (1.2) of the manifold $\tilde{\mathbf{R}}^2$ in $\mathbf{R}^3$ given by equation (1.9); obviously (1.10) defines the subgroup $\tilde{\mathbf{Z}}^2$ of $\mathbf{Z}^3$ that leaves $\tilde{\mathbf{R}}^2$ invariant; and (1.10) implies by $(1.8)_3$ and (1.12) that

$$L_2, M_2, \text{ and } N_2 \text{ divide } \xi, \eta, \text{ and } \zeta, \text{ respectively.} \tag{1.18}$$

By (1.13), the three elements (1.14) are contained in $\tilde{\mathbf{Z}}^2$. On one hand, $(1.14)_1$ generates the subgroup of $\tilde{\mathbf{Z}}^2$ given by the equation

$$\xi = 0,$$

because G.C.D.$(M_2N_3, M_3N_2) = 1$ by $(1.12)_1$, $(1.13)_2$, and $(1.13)_3$. Thus G.C.D.$(N_3, M_3) = 1$, which implies that the values taken by ξ in the subgroup of $\tilde{\mathbf{Z}}^2$ generated by the elements $(1.14)_2$ and $(1.14)_3$ are all the multiples of L_2. Thus, by (1.18), the three elements (1.14) of $\tilde{\mathbf{Z}}^2$ generate $\tilde{\mathbf{Z}}^2$.

3°) $\bar{G} = \mathbf{R}^2$. By (1°) and (2°), $\bar{G} = \mathbf{R}^2$ if and only if N_{L_0} and N_{M_0} are not connected by any affine relation with integer coefficients. If $\bar{G} = \mathbf{R}^2$, then $\bar{K}$ is the image of $\mathbf{R}^3$ under (1.2); thus $\bar{K} = V[L_0, M_0]$.

Invariant measures on $V[L_0, M_0]$. Recall that $V[L_0, M_0]$ has a measure >0 that is invariant under the characteristic vector κ of H [(3.2) of §1]:

$$\eta_V = dt \wedge d\Phi \wedge d\Psi.$$

Every measure on $V[L_0, M_0]$ that is invariant under κ is the product of η_V and a function $V[L_0, M_0] \to \mathbf{R}$ that is invariant under κ, that is, constant on the closures $\bar{K}$ of the characteristics K of H staying on $V[L_0, M_0]$.

Then the following lemma is an obvious consequence of lemma 1.1.

LEMMA 1.2. 1°) *Suppose that* N_{L_0} *and* N_{M_0} *are the rational numbers* (1.6). Then the characteristics of H staying on $V[L_0, M_0]$ are the closed curves given by the equations

$$K(c_1, c_2)\colon \Psi + \lambda[L_0, M_0, t] = c_1, \qquad \Phi + \mu[L_0, M_0, t] = c_2;$$

e_1 and c_2 are constants defined mod

$$2\pi\frac{M_2}{N_1} = \frac{2\pi}{L_2 N_3}, \qquad 2\pi\frac{L_2}{N_1} = \frac{2\pi}{M_2 N_3},$$

respectively, where

$$L_2 = \text{G.C.D.}(M_1, N_1), \qquad M_2 = \text{G.C.D.}(L_1, N_1),$$
$$N_3 = N_1/L_2 M_2 \in \mathbf{Z}.$$

The *measures on* $V[L_0, M_0]$ *that are invariant* under the characteristic vector κ of H are given by

$$F(\Psi + \lambda[L_0, M_0, t], \Phi + \mu[L_0, M_0, t])\eta_V, \tag{1.19}$$

where $F(\cdot,\cdot)$ *is an arbitrary function with periods* $2\pi(M_2/N_1)$ *and* $2\pi(L_2/N_1)$ in the first and second arguments, respectively.

2°) Suppose that N_{L_0} *and* N_{M_0} *are connected by the unique affine relation* (1.8). Then the closures $\bar{K}$ of the characteristics K of H staying on $V[L_0, M_0]$ are the tori given by the equations

$$T^2(c_0)\colon L_1\{\Psi + \lambda[L_0, M_0, t] + M_1\{\Phi + \mu[L_0, M_0, t]\} = c_0, \tag{1.20}$$

where c_0 is a constant defined mod 2π. The *measures on* $V[L_0, M_0]$ *that are invariant* under the characteristic vector κ of H are given by

$$F[L_1(\Psi + \lambda) + M_1(\Phi + \mu)]\eta_V, \tag{1.21}$$

where $F[\cdot]$ *is an arbitrary function with period* 2π.

3°) Suppose that *there is no affine relation with integer coefficients connecting* N_{L_0} *and* N_{M_0}. Then the closure $\bar{K}$ of each characteristic K of H

staying on $V[L_0, M_0]$ *is* $V[L_0, M_0]$. Every *measure on* $V[L_0, M_0]$ *that is invariant* under the characteristic vector κ of H is given by

$$\text{const.}\, \eta_V. \tag{1.22}$$

The following theorem is an easy consequence.

THEOREM 1. 1°) *The tori* $V[L_0, M_0]$ *on which a lagrangian solution* U *of the equation*

$$aU = 0 \bmod \frac{1}{v^2} \qquad (a \text{ is the operator associated to } H)$$

with lagrangian amplitude $\geqslant 0$ *may be defined are defined by the condition* (3.11), (3.12) *of* §1: *there exist three integers*

l, m, n

such that

$$|m| \leqslant l < n,$$

$$L_0 = \hbar(l + \tfrac{1}{2}), \qquad M_0 = \hbar m, \qquad L_0 + N[L_0, M_0] = \hbar n.$$

2°) *A necessary and sufficient condition for the solution* U *defined on such a torus to be unique up to a constant factor is that the derivatives of* N,

$$N_L[L_0, M_0], \qquad N_M[L_0, M_0],$$

are not connected by any affine relation with integer coefficients.

Proof. By theorem 6 of II,§3, the condition that there exist such a solution on $V[L_0, M_0]$ is Maslov's quantum condition. The condition that it be unique is the condition that the invariant measure η_V on $V[L_0, M_0]$ be unique (up to a constant factor). In section 3 of §1, Maslov's quantum condition is formulated as (3.11)–(3.12). The condition that the invariant measure be unique is given in lemma 1.2.

2. Compact Lagrangian Manifolds V, Other Than the Tori $V[L_0, M_0]$, on Which Solutions of the Equation $aU = 0 \bmod(1/v^2)$ with Lagrangian Amplitude $\geqslant 0$ Exist

We shall show that such manifolds V exist only exceptionally.

The calculation of their Maslov index (lemmas 2.3 and 2.4) uses the following properties.

Other properties of the characteristics K of H with compact closures. Differentiating the definition (2.10) (§1) of λ and μ gives

$$\frac{1}{R} dQ \wedge dR + d\lambda \wedge dL + d\mu \wedge dM = 0,$$

where L, M, Q, R are functions of (L, M, t) satisfying $H[L, M, Q, R] = 0$; then

$$H_L dL + H_M dM + H_Q dQ + H_R dR = 0,$$

whence, by elimination of dQ:

$$\left[\frac{H_L}{RH_Q} dR + d\lambda\right] \wedge dL + \left[\frac{H_M}{RH_Q} dR + d\mu\right] \wedge dM = 0;$$

thus there exist three real numerical functions ρ, σ, and τ of (L, M, t), defined when $H_Q \neq 0$, such that

$$d\lambda = -\frac{H_L}{RH_Q} dR + \rho\, dL + \sigma\, dM,$$
$$d\mu = -\frac{H_M}{RH_Q} dR + \sigma\, dL + \tau\, dM. \tag{2.1}$$

By the expression (1.5) for $\Delta_t \lambda$ and $\Delta_t \mu$ and definition (2.6) of t in §1, these relations imply

$$\Delta_t \rho = 2\pi N_{L^2}, \qquad \Delta_t \sigma = 2\pi N_{LM}, \qquad \Delta_t \tau = 2\pi N_{M^2}, \tag{2.2}$$

$$\lambda_t[L, M, t] = -H_L[L, M, Q, R], \qquad \mu_t = -H_M. \tag{2.3}$$

Let us describe explicitly *the singular part of* ρ, σ, τ, and $\rho\tau - \sigma^2$ when $H_Q = 0$; by (2.1),

$$\rho = \frac{R_L H_L}{RH_Q} + \lambda_L, \qquad \sigma = \frac{R_M H_L}{RH_Q} + \lambda_M = \frac{R_L H_M}{RH_Q} + \mu_L,$$
$$\tau = \frac{R_M H_M}{RH_Q} + \mu_M; \tag{2.4}$$

thus

$$\rho\tau - \sigma^2 = \frac{1}{RH_Q}[R_L H_L \mu_M - R_M H_L \mu_L - R_L H_M \lambda_M + R_M H_M \lambda_L]$$
$$+ \lambda_L \mu_M - \lambda_M \mu_L.$$

Since $H[L, M, Q, R] = 0$,

$$H_R R_L + H_L + H_Q Q_L = H_R R_M + H_M + H_Q Q_M = 0.$$

When $H_R \neq 0$, these relations make it possible to eliminate R_L and R_M from the preceding expressions for ρ, σ, τ, and $\rho\tau - \sigma^2$; by assumption (2.2) of §1,

$$H_R \neq 0 \text{ when } H_Q = 0;$$

hence:

LEMMA 2.1. The functions

$$\rho + \frac{H_L^2}{RH_Q H_R}, \qquad \sigma + \frac{H_L H_M}{RH_Q H_R}, \qquad \tau + \frac{H_M^2}{RH_Q H_R},$$
$$\rho\tau - \sigma^2 + \frac{1}{RH_Q H_R}[H_L^2 \mu_M - H_L H_M(\mu_L + \lambda_M) + H_M^2 \lambda_L] \tag{2.5}$$

are *bounded* in a neighborhood of the points where $H_Q = 0$.

Properties of compact lagrangian manifolds V in W. V is generated by characteristics of H with compact closure, which thus stay on tori $V[L_0, M_0]$; then the functions

$$L, M, N = N[L, M], \qquad c = c[L, M]$$

are defined on V.

Let V_2 be the open subset of V where $dL \wedge dM \neq 0$ if it is not empty. When $dL \wedge dM = 0$ on V, let V_1 be the open subset of V where $(dL, dM) \neq 0$ if it is not empty.[3] Then V_1 and V_2 are lagrangian manifolds, not necessarily compact, that are generated by characteristics K of H with compact closures $\bar{K}$; they contain these closures.

When V_2 and V_1 do not exist, *then V is one of the tori $V[L_0, M_0]$*. By the following lemma, V_1 and V_2 only exist if the graph

$$N : (L, M) \mapsto N[L, M]$$

contains a rectilinear segment with rational direction.

Thus the consequence of theorem 2 stated in the introduction (§2,0) follows from this lemma.

[3] This notation means that dL and dM are not simultaneously zero. [Translator's note]

LEMMA 2.2. 1°) *On* V_2, *N is an affine function of* (L, M). More precisely,

$$L_1\,dL + M_1\,dM + N_1\,dN = 0, \tag{2.6}$$

where $(L_1, M_1, N_1) \in \mathbf{Z}^3$, G.C.D.$(L_1, M_1, N_1) = 1$.

V_2 is defined in W by the datum of a function F of two variables and by the equations

$$\Psi + \lambda[L, M, t] + F_L[L, M] = \Phi + \mu[L, M, t] + F_M[L, M] = 0. \tag{2.7}$$

More precisely, in the space $\mathbf{R}^5$ with the coordinates (L, M, Ψ, Φ, t), equations (2.7) define a 3-dimensional manifold $\tilde{V}_2$. By (1.5), where

$$N_L = -L_1/N_1, \qquad N_M = -M_1/N_1,$$

the elements (ξ, η, ζ) of the subgroup $\tilde{\mathbf{Z}}$ of $\mathbf{Z}^3$ generated by $(L_1, M_1, N_1) \in \mathbf{Z}^3$ act on $\tilde{V}_2$ as follows:

$$(\xi, \eta, \zeta): (L, M, \Psi, \Phi, t) \mapsto (L, M, \Psi + 2\pi\xi, \Phi + 2\pi\eta, t + c[L, M]\zeta). \tag{2.8}$$

We have

$$V_2 = \tilde{V}_2/\tilde{\mathbf{Z}}. \tag{2.9}$$

V_2 has a *measure* that is *invariant* under the characteristic vector of H, given by

$$\eta_V = dL \wedge dM \wedge dt.$$

2°) *On* V_1, *L, M, and N are affine functions of a single variable s.* More precisely,

$$\frac{dL}{L_1} = \frac{dM}{M_1} = \frac{dN}{N_1} = ds \tag{2.10}$$

where $(L_1, M_1) \neq 0$, $(L_1, M_1, N_1) \in \mathbf{Z}^3$, G.C.D.$(L_1, M_1, N_1) = 1$.

V_1 is defined in W by the datum of three functions of s—the affine functions L and M satisfying (2.10) and F—and by the equations

$$\begin{aligned} &L = L(s), \qquad M = M(s), \\ &L_1\{\Psi + \lambda[L, M, t]\} + M_1\{\Phi + \mu[L, M, t]\} + F_s(s) = 0. \end{aligned} \tag{2.11}$$

More precisely, equations (2.11) define a 3-dimensional manifold $\tilde{V}_1$ in $\mathbf{R}^5$

on which the subgroup $\tilde{\mathbf{Z}}^2$ of $\mathbf{Z}^3$ given by equation (1.10) acts according to (2.8). Recall that this subgroup is generated by its three elements (1.14). We have

$$V_1 = \tilde{V}_1/\tilde{\mathbf{Z}}^2. \tag{2.12}$$

V_1 has a *measure* that is *invariant* under the characteristic vector of H, given by

$$\eta_V = (M_1 \, d\Psi - L_1 \, d\Phi) \wedge ds \wedge dt.$$

3°) The *phases* of V_1 and V_2 in the frame R_0 (section 1 of §1) are given by

$$\varphi_{R_0} = \Omega + L\Psi + M\Phi + F. \tag{2.13}$$

Remark 2.1. (2.8) and (2.10) in §1 define

- Ω up to the addition of a function F of (L, M)
- λ and μ up to the addition of its derivatives F_L and F_M.

Given V_1 or V_2, it is possible to choose Ω such that

$$F = 0$$

in (2.7), (2.11), and (2.13).

In order to quantize V, we shall use only the following consequence of (2.13) and the definitions (2.7) of $c[L, M]$ and (2.9) of N in §1: choosing F to be zero, the function

$$\varphi_{R_0} - 2\pi N \frac{t}{c[L, M]} - L\Psi - M\Phi \tag{2.14}$$

is defined on V_1 and on V_2.

Preliminary to the proof. Formula (2.11) of §1 shows that theorem 3.1 of II,§3 applies with

$$l = 3, \qquad h_1 = L, \qquad h_2 = M,$$

$$g_0 = \Omega + L\Psi + M\Phi, \qquad g_1 = -\Psi - \lambda, \qquad g_2 = -\Phi - \mu,$$

the phase φ_{R_0} replacing the lagrangian phase.

Hence any lagrangian manifold V in W is given locally by equations of one of the four following types:

$$\Psi + \lambda + F_L[L, M] = \Phi + \mu + F_M = 0; \tag{2.15$_1$}$$

$$M = f(L), \qquad \Psi + \lambda + f_L(L)(\Phi + \mu) + F_L(L) = 0; \tag{2.15$_2$}$$

the result of the permutation of (L, Ψ), (M, Φ) in $(2.15)_2$; $(2.15)_3$

$$L = \text{const.}, \qquad M = \text{const.}; \tag{2.15$_4$}$$

F and f are functions of one or two variables. The phase φ_{R_0} of V is expressed by (2.13), where $F = 0$ in the fourth case.

Proof of 1°). Locally, V_2 is given by equations of the form $(2.15)_1$. Hence V_2 is generated by characteristics K staying on disjoint tori $V[L_0, M_0]$. Their closures $\bar{K}$ are disjoint and are contained in V_2; $\dim V_2 = 3$, so that

$$\dim \bar{K} = 1.$$

Then by lemma 1.1 the values of N_L and N_M on V are rational numbers. Thus the functions N_L and N_M are constant on V and satisfy (1.6), which expresses (2.6), whence 1°) and (2.13).

Proof of 2°). Locally, V_1 is given by equations of the form $(2.15)_2$ or $(2.15)_3$, that is, equations of the form

$$\begin{aligned} &L = L(s), \qquad M = M(s), \\ &L_s(s)(\Psi + \lambda) + M_s(s)(\Phi + \mu) + F_s(s) = 0, \end{aligned} \tag{2.16}$$

where $(L_s, M_s) \neq 0$. For each value of s, let $T(s)$ be the manifold in W given by equations (2.16): $\dim T(s) = 2$.

Since V_1 is generated by characteristics given by the equations

$$L = \text{const.}, \qquad M = \text{const.}, \qquad \Psi + \lambda = \text{const.}, \qquad \Phi + \mu = \text{const.}$$

and contains their closures, V_1 contains the closures $\overline{T(s)}$ of the $T(s)$.

Let us determine the $\overline{T(s)}$.

By expression (1.5) for $\Delta_t \lambda$ and $\Delta_t \mu$ and since $L_s N_L + M_s N_M = N_s$, $T(s)$ is the image in W of the set of $(\Psi, \Phi, t) \in \mathbf{R}^3$ such that

$$L_s \frac{\Psi + \lambda[L, M, t]}{2\pi} + M_s \frac{\Phi + \mu[L, M, t]}{2\pi} + \frac{1}{2\pi} F_s \in G(s),$$

where $G(s)$ is the image of $\mathbf{Z}^3$ in the additive group $\mathbf{R}$ under the morphism

$$\mathbf{Z}^3 \ni (\xi, \eta, \zeta) \mapsto L_s \xi + M_s \eta + N_s \zeta.$$

Then $\overline{T(s)}$ is the image in W of the set of $(\Psi, \Phi, t) \in \mathbf{R}^3$ such that

$$L_s \frac{\Psi + \lambda[L, M, t]}{2\pi} + M_s \frac{\Phi + \mu[L, M, t]}{2\pi} + \frac{1}{2\pi} F_s \in \overline{G(s)},$$

where $\overline{G(s)}$ is the closure of $G(s)$ and thus a closed subgroup of $\mathbf{R}$.

$\overline{G(s)} = \mathbf{R}$, and thus $\overline{T(s)}$ is the 3-dimensional torus $V[L(s), M(s)]$ unless $G(s)$ is discrete, that is, unless there exists $(L_1, M_1, N_1) \in \mathbf{Z}^3$ such that

$$\frac{L_s}{L_1} = \frac{M_s}{M_1} = \frac{N_s}{N_1}, \qquad \text{G.C.D.}(L_1, M_1, N_1) = 1.$$

Since $\dim V_1 = 3$, this must be the case for every s. The functions M_s/L_s and N_s/L_s have rational values; thus they are constants. The integers L_1, M_1, and N_1 are independent of s; and s can be chosen so as to arrive at (2.10); thus 2°) and (2.13) hold.

Quantization of V. Let us impose Maslov's quantum condition on V. In order to express this condition, let us calculate the *Maslov index* m_{R_0} of V_2 and of V_1 using lemma 1 of §1.

We make $F = 0$ in lemma 2.2 (see remark 2.1).

LEMMA 2.3. 1°) The functions ρ, σ, τ are defined on V_2. Let $\bar{\mathbf{R}}$ be the one-point compactification of $\mathbf{R}$. Then the function $F : V_2 \to \bar{\mathbf{R}}$ given by

$$F(L, M, t) = \frac{\rho L^2 + 2\sigma LM + \tau M^2}{L(L^2 - M^2)(\rho\tau - \sigma^2)} \in \bar{\mathbf{R}} \tag{2.17}$$

is defined on $V_2 \backslash \Sigma''$, Σ'' being the subset of V_2, where

$$\Sigma'' : \frac{\rho}{M^2} = -\frac{\sigma}{LM} = \frac{\tau}{L^2}. \tag{2.18}$$

2°) *If* $V_2 \backslash \Sigma''$ is connected, then

$$m_{R_0} = \left[\frac{1}{\pi}\Psi + \frac{1}{\pi} \arctan F(L, M, t)\right] ([\cdot] \text{ the integer part of}) \tag{2.19}$$

on its universal covering space.

3°) The function

$$m_{R_0} - \frac{\Psi}{\pi} - 2\frac{t}{c[L, M]} \tag{2.20}$$

is defined (that is, uniform) on V_2.

Remark 2.2. Assume a function

$N:(L, M) \mapsto N[L, M]$ satisfying (2.6)

is given. Choose H satisfying (2.9) of §1. If this choice is *generic*, then

$\dim \Sigma'' = 1, \qquad V_2 \backslash \Sigma''$ is connected.

Proof of 1°). Recall that (L, M, t) are local coordinates on V_2. By (2.2) and (2.6),

$$\Delta_t \rho = \Delta_t \sigma = \Delta_t \tau = 0;$$

1°) follows.

Proof of 2°). *The apparent contour of* V_2. Formulas (1.11) and (1.13) in §1 give the relation

$$\frac{L}{\sin \Theta} \omega_2 \wedge \omega_3$$

$$= L \sin \Psi \, d\Phi \wedge d\Psi + \frac{\cos \Psi}{L^2 - M^2} (M\,dL - L\,dM) \wedge (L\,d\Psi + M\,d\Phi) \tag{2.21}$$

on W. Differentiating (2.7), where $F = 0$, gives

$$d\Psi + \rho\,dL + \sigma\,dM = d\Phi + \sigma\,dL + \tau\,dM = 0 \mod dR$$

on V_2, from definition (2.1) of ρ, σ, and τ. By (2.6) of §1,

$$dR = RH_Q\,dt \mod(dL, dM),$$

hence

$$\frac{L}{R \sin \Theta} dR \wedge \omega_2 \wedge \omega_3 = G(L, M, t, \Psi)\, dL \wedge dM \wedge dt \tag{2.22}$$

where G denotes the function

$$G = -H_Q \left[L(\rho\tau - \sigma^2) \sin \Psi + \frac{\rho L^2 + 2\sigma LM + \tau M^2}{L^2 - M^2} \cos \Psi \right],$$

which is regular on V_2 by lemma 2.1. Then by lemma 1 of §1, the apparent contour Σ_{R_0} of V_2 is

$$\Sigma_{R_0} = \Sigma' \cup \Sigma'',$$

where Σ'' is defined by (2.18) and Σ' is the surface in $V_2\backslash\Sigma''$ given by the equation

$$\Sigma' : \tan\Psi + F(L, M, t) = 0, \tag{2.23}$$

where F is defined by (2.17).

Calculation of m_{R_0}. Formulas (1.11) and (1.13) in §1 give

$$\frac{L}{\sin\Theta}\omega_3 \wedge \omega_1$$

$$= L\cos\Psi\, d\Phi \wedge d\Psi - \frac{\sin\Psi}{L^2 - M^2}(M\,dL - L\,dM) \wedge (L\,d\Psi + M\,d\Phi) \tag{2.24}$$

on W, from which, by the same calculations used to deduce (2.22) from (2.21),

$$\frac{L}{R\sin\Theta}dR \wedge \omega_3 \wedge \omega_1 = G_\Psi(L, M, t, \Psi)\, dL \wedge dM \wedge dt. \tag{2.25}$$

Relations (2.22), (2.25) and lemma 1 of §1 give

$m_R = \text{const. for } G/G_\Psi < 0,$

$m_R = 1 + \text{const. for } G/G_\Psi > 0$

in a neighborhood of a point of Σ'. This result obviously holds for any function G vanishing to first order on Σ', for example, the function

$$G = \frac{1}{\pi}\Psi + \frac{1}{\pi}\arctan F(L, M, t) \mod 1.$$

Thus, on $V_2\backslash\Sigma''$, m_{R_0} is expressed locally by (2.19), up to an additive constant; 2°) follows.

Proof of 3°). First suppose that H and Ω are generic (remark 2.2). Then $V_2\backslash\Sigma''$ is connected and

$$LH_L + MH_M \neq 0 \quad \text{for } H_Q = 0,$$

which implies by lemma 2.1 that the function

$$f = \rho L^2 + 2\sigma LM + \tau M^2 : V_2\backslash\Sigma'' \to \bar{\mathbf{R}}$$

is defined. By (2.19), the function

$$m_{R_0} - \frac{1}{\pi}\Psi - \frac{1}{\pi}\arctan F \text{ is defined on } V_2\backslash\Sigma''; \tag{2.26}$$

by the definition of F, where $|M| < L$ from assumption (2.2) of §1, we have

$$F/f < 0$$

in a neighborhood of the points where $F = 0$; $F = 0$ is equivalent to $f = 0$, so that

$$\arctan F + \arctan f \text{ is defined on } V_2\backslash\Sigma''; \tag{2.27}$$

we see that

$$\arctan f + \arctan\frac{1}{f} = \text{const.}; \tag{2.28}$$

by lemma 2.1, $1/f = 0$ is equivalent to $H_Q = 0$ and

$$\frac{H_Q}{H_R} f < 0$$

in a neighborhood of the points where $H_Q = 0$; thus

$$\arctan\frac{1}{f} + \arctan\frac{H_Q}{H_R} \text{ is defined on } V_2\backslash\Sigma''; \tag{2.29}$$

now, from the orientation of $\Gamma[(2.9)$ of §1$]$,

$$\arctan\frac{H_Q}{H_R} + 2\pi\frac{t}{c[L, M]} \text{ is defined on } \Gamma[L, M]. \tag{2.30}$$

The formulas (2.26)–(2.30) prove 3°) for H generic; 3°) follows.

LEMMA 2.4. 1°) Define a constant N_0 and a function r on V_1 by the relations

$$L_1 M - M_1 L = N_0, \tag{2.31}$$

$$\Psi + \lambda + M_1 r = \Phi + \mu - L_1 r = 0. \tag{2.32}$$

The functions (r, s, t) are coordinates of the points of V_1. A function $F : V_1 \to \bar{\mathbf{R}}$ is defined by the formula

$$F(r, s, t) = \frac{N_0^2}{L(L^2 - M^2)(\rho L_1^2 + 2\sigma L_1 M_1 + \tau M_1^2)} \tag{2.33}$$

when V_1 is generic.

2°) If V_1 is generic ($N_0 \neq 0$; $H_{L^2}L_1^2 + 2H_{LM}L_1M_1 + M_{M^2}M_1^2 \neq 0$ for $H_Q = 0$), then

$$m_{R_0} = \left[\frac{1}{\pi}\Psi + \arctan F(r, s, t)\right] ([\cdot] \text{ is the integer part}). \tag{2.34}$$

3°) The function (2.20) is defined on V_1.

Proof of 1°). Formula $(2.11)_2$, where $F = 0$ (remark 2.1), justifies the definition (2.32) of r. By (2.2) and (2.10),

$$\begin{aligned}\Delta_t(\rho L_1^2 + 2\sigma L_1 M_1 + \tau M_1^2) &= 2\pi[L_1^2 N_{L^2} + 2L_1 M_1 N_{LM} + M_1^2 N_{M^2}] \\ &= 2\pi\frac{d^2N}{ds^2} = 0.\end{aligned}$$

Proof of 2°). *The apparent contour of* V_1. Differentiating (2.32) gives

$$\begin{aligned}d\Psi + (\rho L_1 + \sigma M_1)ds + M_1\,dr &= d\Phi + (\sigma L_1 + \tau M_1)ds - L_1 dr \\ &= 0 \bmod dR\end{aligned}$$

by (2.10) and definition (2.1) of ρ, σ, τ. Whence by (2.21)

$$\frac{L}{R\sin\Theta}dR \wedge \omega_2 \wedge \omega_3 = G(s, t, \Psi)ds \wedge dr \wedge dt, \tag{2.35}$$

where

$$G(s, t, \Psi) = H_Q\left[L(\rho L_1^2 + 2\sigma L_1 M_1 + \tau M_1^2)\sin\Psi + \frac{N_0^2}{L^2 - M^2}\cos\Psi\right],$$

since $dR = RH_Q\,dt \bmod(dL, dM)$, that is, mod ds. By 1°) and lemma 2.1, the function $G: V_1 \to \bar{\mathbf{R}}$ is defined and regular on V_1. Thus, by lemma 1 of §1, the apparent contour Σ_{R_0} of V_1 is given by the equation

$$\Sigma_{R_0}: \tan\Psi + F(r, s, t) = 0.$$

Calculation of m_{R_0}. The same calculation used to deduce (2.35) from (2.21) enables us to deduce the formula

$$\frac{L}{R\sin\Theta}dR \wedge \omega_3 \wedge \omega_1 = G_\Psi(s, t, \Psi)ds \wedge dr \wedge dt$$

from (2.24). Applying lemma 1 of §1 as is done in lemma 2.4, it follows that m_{R_0} can be expressed by (2.34).

Proof of 3°). Suppose that V_1 is generic. By (2.34),

$$m_{R_0} - \frac{1}{\pi}\Psi - \frac{1}{\pi}\arctan F(r, s, t) \text{ is defined on } V_1. \tag{2.36}$$

From (2.33), where $|M| < L$ by assumption (2.2) of §1, and from lemma 2.1, $F = 0$ is equivalent to $H_Q = 0$, and

$$F\frac{H_R}{H_Q} < 0$$

in a neighborhood of the points of V_1 where $H_Q = 0$; thus

$$\arctan F + \arctan\frac{H_Q}{H_R} \text{ is defined on } V_1. \tag{2.37}$$

Formulas (2.30), (2.36), and (2.37) prove that the function (2.20) is defined on generic V_1, hence on every V_1.

LEMMA 2.5. 1°) *V_2 satisfies Maslov's quantum condition* if and only if, on V_2, the functions L, M, and N are connected by a relation

$$L_1\left(L - \frac{\hbar}{2}\right) + M_1 M + N_1\left(N + \frac{\hbar}{2}\right) = \hbar N_0, \tag{2.38}$$

where

$$L_1, M_1, N_1, N_0 \in \mathbf{Z}, \qquad N_1 \neq 0, \qquad \text{G.C.D.}(L_1, M_1, N_1) = 1. \tag{2.39}$$

2°) *V_1 satisfies Maslov's quantum condition* if and only if, on V_1, the functions L, M, and N are connected by the three relations

$$(L_1, M_1, N_1) \wedge \left(L - \frac{\hbar}{2}, M, N + \frac{\hbar}{2}\right) = \hbar(L_0, M_0, N_0), \tag{2.40}$$

where $\wedge$ denotes the vector product in $\mathbf{E}^3$ and

$$\begin{gathered} L_1, M_1, N_1, L_0, M_0, N_0 \in \mathbf{Z}, \qquad L_1^2 + M_1^2 \neq 0, \\ \text{G.C.D.}(L_1, M_1, N_1) = 1, \qquad L_0 L_1 + M_0 M_1 + N_0 N_1 = 0. \end{gathered} \tag{2.41}$$

Thus only two of the three relations (2.40) are independent.

Preliminary to the proof. A manifold V satisfies Maslov's quantum condition (II,§2, definition 6.2) if and only if the function

$$\frac{1}{2\pi\hbar}\varphi_R - \frac{1}{4}m_R \text{ is defined mod 1 on } V.$$

Then by (2.14) and (2.20), V_1 or V_2 satisfies this condition if and only if the function

$$\left(\frac{L}{\hbar}+\frac{1}{2}\right)\frac{\Psi}{2\pi}+\frac{M}{\hbar}\frac{\Phi}{2\pi}+\left(\frac{N}{\hbar}+\frac{1}{2}\right)\frac{t}{c[L, M]} \tag{2.42}$$

is defined mod 1 on V_1 or V_2.

Proof of 1°). By lemma 2.2,1°) the function (2.42) is defined on $\tilde{V}_2$, $V_2 = \tilde{V}_2/\tilde{\mathbf{Z}}$, where $\tilde{\mathbf{Z}}$ is generated by

$$(L_1, M_1, N_1)$$

and acts on $\tilde{V}_2$ according to (2.8). Thus Maslov's quantum condition is

$$\left(\frac{L}{\hbar}+\frac{1}{2}\right)L_1+\frac{M}{\hbar}M_1+\left(\frac{N}{\hbar}+\frac{1}{2}\right)N_1 \in \mathbf{Z};$$

whence 1°).

Proof of 2°). By lemma 2.2,2°) the function (2.42) is defined on $\tilde{V}_1$, $V_1 = \tilde{V}_1/\tilde{\mathbf{Z}}^2$, where $\tilde{\mathbf{Z}}^2$ is generated by

$$(0, M_2N_3, -M_3N_2), \qquad (-L_2N_3, 0, L_3N_2), \qquad (L_2M_3, -L_3M_2, 0),$$

$L_2, \ldots, N_3$ being defined by (1.12) and (1.13), and $\tilde{\mathbf{Z}}^2$ acts on V_1 according to (2.8). Thus Maslov's quantum condition is

$$\frac{M}{\hbar}M_2N_3 - \left(\frac{N}{\hbar}+\frac{1}{2}\right)M_3N_2 \in \mathbf{Z},$$

$$-\left(\frac{L}{\hbar}+\frac{1}{2}\right)L_2N_3+\left(\frac{N}{\hbar}+\frac{1}{2}\right)L_3N_2 \in \mathbf{Z},$$

$$\left(\frac{L}{\hbar}+\frac{1}{2}\right)L_2M_3 - \frac{M}{\hbar}L_3M_2 \in \mathbf{Z}.$$

By (1.13), this quantum condition is equivalent to the condition (2.40)–(2.41) supplemented by the following one: L_0, M_0, and N_0 are multiples of L_2, M_2, and N_2, respectively. Now

$L_2 = \text{G.C.D.}(M_1, N_1), \ldots;$

thus this last condition is a result of (2.41); 2°) follows.

THEOREM 2. *Assume that the lagrangian equation*

$$aU = 0 \bmod \frac{1}{\nu^2} \quad (a \text{ the operator associated to } H)$$

has a solution U that has a lagrangian amplitude $\geqslant 0$ *and is defined* $\bmod (1/\nu)$ *on some compact lagrangian manifold V other than the tori* $V[L_0, M_0]$ *studied in theorem* 1; *then the graph of the function*

$N: (L, M) \mapsto N[L, M]$ [*see* III,§1,(2.9)]

contains a rectilinear segment given by a plückerian equation

$$(L_1, M_1, N_1) \wedge \left(L - \frac{\hbar}{2}, M, N + \frac{\hbar}{2}\right) = \hbar(L_0, M_0, N_0) \tag{2.40}$$

such that

$$\begin{aligned} &L_1, M_1, N_1, L_0, M_0, N_0 \in \mathbf{Z}, \qquad L_1^2 + M_1^2 \neq 0, \\ &\text{G.C.D.}(L_1, M_1, N_1) = 1, \qquad L_0L_1 + M_0M_1 + N_0N_1 = 0. \end{aligned} \tag{2.41}$$

COROLLARY. *The condition*

$N_0' \in \mathbf{Z}$

is necessary and sufficient for there to exist such a segment in a planar domain belonging to the graph of N and satisfying an equation

$$L_1'\left(L - \frac{\hbar}{2}\right) + M_1'M + N_1'\left(N + \frac{\hbar}{2}\right) = \hbar N_0', \tag{2.43}$$

where

$$\begin{aligned} &L_1', M_1', N_1' \in \mathbf{Z}, \qquad L_1'^2 + M_1'^2 \neq 0, \\ &\text{G.C.D.}(L_1', M_1', N_1') = 1, \qquad N_0' \in \mathbf{R}. \end{aligned} \tag{2.44}$$

Remark 2.3. This corollary makes it easier to apply the theorem (see section 3).

Proof of the theorem. By II,§3,theorem 6, V must satisfy Maslov's quantum condition. Since V is not one of the tori $V[L_0, M_0]$, V must contain a lagrangian manifold V_2 or V_1 satisfying this quantum condition. Then by lemma 2.5 the graph of the function N contains

- either a rectilinear segment with the equations (2.40)–(2.41) or
- a planar domain with the equation (2.38)–(2.39) and hence such a segment.

Proof of the corollary. The condition $N_0' \in \mathbf{Z}$ evidently emplies that such a segment exists. Conversely, assume that in the space $\mathbf{R}^3$ with coordinates (L, M, N), the plane with the equations (2.43)–(2.44) contains a line with the equations (2.40)–(2.41). This assumption is expressed by the four relations

$$L_1 L_1' + M_1 M_1' + N_1 N_1' = 0,$$

$$N_0' = \frac{M_1' N_0 - N_1' M_0}{L_1} = \frac{N_1' L_0 - L_1' N_0}{M_1} = \frac{L_1' M_0 - M_1' L_0}{N_1}$$

(two of the last three result from the other two). These relations imply $N_0' \in \mathbf{Z}$ since G.C.D.$(L_1, M_1, N_1) = 1$.

3. Example: The Schrödinger–Klein–Gordon Operator

Let us choose a to be the operator associated to the hamiltonian H defined by (4.6) of §1, where *A is assumed to be an affine function of M, B and C are assumed to be constant, and $B > 0$.*

In §1, we studied the system

$$aU = (a_{L^2} - L_0^2)U = (a_{M^2} - M_0)U = 0 \mod \frac{1}{\nu^2}$$

and recovered the classical energy levels. In studying the single equation

$$aU = 0 \mod \frac{1}{\nu^2}$$

we shall find again the same condition for existence, thus the same *classical energy levels.*

THEOREM 3.1. *The lagrangian equation*

$$aU = 0 \mod \frac{1}{\nu^2} \tag{3.1}$$

has a solution U with **lagrangian amplitude** $\geqslant 0$ *defined* $\bmod(1/\nu)$ *on a* **compact** *manifold V if and only if there exists a triple of integers* (l, m, n) *satisfying condition* (4.11) *of* §1,*theorem* 4.1.

Remark 3. Under this condition, neither the unicity of V nor the unicity of U up to a constant factor is assured.

Proof. Under condition (4.11) of §1, the existence of such a solution of (3.1) is assured by theorem 1 and also by theorem 4.1 of §1.

By theorem 1, there exists such a solution of (3.1) defined on a torus $V[L_0, M_0]$ only if this condition is assumed.

By theorem 2 and formula (4.8) of §1, which gives the value of the function N as

$$N[L, M] = \frac{B}{\sqrt{A(M)}} - \sqrt{L^2 + C}, \tag{3.2}$$

the existence of such a solution on a manifold V other than a torus $V[L_0, M_0]$ requires that *the graph of N contains a line segment*. Thus it requires that one of the three following cases occur.

First case: $A(M) = A_0$ is independent of M; $C = 0$.
Then by (3.2) the graph of N is the plane with the equation

$$L + N = \frac{B}{\sqrt{A_0}}.$$

Theorem 2 and its corollary require the existence of an integer n such that

$$\frac{B}{\sqrt{A_0}} = \hbar n.$$

Then condition (4.11) of §1,theorem 4.1 is satisfied since it is independent of l for $C = 0$ and independent of m for $A(M)$ independent of M.

Second case: $C = 0$; A depends on M.
By (3.2), the only lines contained in the graph of N are given by the equations

$$M = M_0, N + L = \frac{B}{\sqrt{A_0}}, \text{ where } M_0 = \text{const.}, \qquad A_0 = A(M_0). \tag{3.3}$$

Their plückerian equations are then

$$(1, 0, -1) \wedge \left(L - \frac{\hbar}{2}, M, N + \frac{\hbar}{2}\right) = \left(M_0, -\frac{B}{\sqrt{A_0}}, M_0\right).$$

Theorem 2 requires the existence of integers m and n such that

$$M_0 = \hbar m, \qquad \frac{B}{\sqrt{A_0}} = \hbar n. \tag{3.4}$$

From (3.3) and (3.4) it follows that

$n > |m|$ because $N > 0$ and $L > |M|$ by assumption [see (2.2) of §1].

Thus condition (4.11) of §1 is satisfied.

Third case: $A = A_0$ is independent of M; $C \neq 0$.
By (3.2), the only lines contained in the graph of N are given by the equations

$$L = L_0, \qquad N = \frac{B}{\sqrt{A_0}} - \sqrt{L_0^2 + C},$$

where $L_0 = \text{const.}, \qquad L_0^2 + C > 0.$

Their plückerian equations are

$$(0, 1, 0) \wedge \left(L - \frac{\hbar}{2}, M, N + \frac{\hbar}{2}\right) = \left(\frac{B}{\sqrt{A_0}} - \sqrt{L_0^2 + C} + \frac{\hbar}{2}, 0, \frac{\hbar}{2} - L_0\right).$$

Theorem 2 requires the existence of integers l and n such that

$$\frac{B}{\sqrt{A_0}} + L_0 - \sqrt{L_0^2 + C} = \hbar n, \; L_0 = \hbar(l + \tfrac{1}{2});$$

$0 \leqslant l$ because $0 < L_0$; $l < n$ because $N > 0$ by assumption.

Thus condition (4.11) of §1 is satisfied.

THEOREM 3.2. *Choose a to be the Klein-Gordon operator* (4.22) *of* §1. *Suppose the magnetic field $\mathscr{H} \neq 0$. Then the tori $T(l, m, n)$ defined by* (4.12) *of* §1 *are the only compact manifolds on which there exists a lagrangian solution U of the equation*

$$aU = 0 \mod \frac{1}{\nu^2}$$

with lagrangian amplitude $\geqslant 0$.

Proof. The proof of theorem 3.1 shows that the necessary condition stated in theorem 2 can not be satisfied when a is the Klein-Gordon operator ($C \neq 0$) and $\mathscr{H} \neq 0$ (A depends on M).

Conclusion

We do not pursue this difficult study of the equation $aU = 0 \bmod(1/\nu^2)$. In particular, we do not describe the lagrangian manifolds other than the tori $T(l, m, n)$ on which there exist solutions of the Klein-Gordon equation when $\mathscr{H} = 0$ or solutions of the Schrödinger equation.

§3. The Lagrangian System $aU = (a_M - \text{const.})U = (a_{L^2} - \text{const.})U = 0$ When a Is the Schrödinger–Klein-Gordon Operator

0. Introduction

In §3, we study the lagrangian system that was solved $\bmod(1/\nu^2)$ in §1.

In section 1, we determine the condition under which theorem 7.2 of II, §3 applies.

In sections 2, 3, and 4, we apply this theorem under assumptions that become more and more strict. These assumptions finally amount to the assumption that a is the Schrödinger–Klein-Gordon operator. Existence theorem 4.1 is finally obtained.

Remark 0. *A Voros* orally pointed out that these properties of the Schrödinger and Klein-Gordon equations extend *to the case where the electric potential is any positive-valued function of the variable* R, if the energy level E is not constrained to be a real number, and if it can be taken to be any *formal number with vanishing phase.*

1. Commutivity of the Operators a, a_{L^2}, and a_M Associated to the Hamiltonians H (§1, Section 2), L^2, and M (§1, Section 1)

We want to determine when theorem 7.2 of II,§3 applies to these operators, that is, when they commute.

LEMMA 1. 1°) a_M and a (thus, in particular, a_M and a_{L^2}) commute.

2°) a_{L^2} and a commute if and only if

$$H_{M^2Q} = H_{M^2R} = 0 \qquad \forall L, M, Q, R. \tag{1.1}$$

Proof. Let a and a' be the lagrangian operators associated to two hamiltonians H and H'. By formula (1.1) of II,§2, their commutator

$$a \circ a' - a' \circ a$$

is associated to the formal function given by

$$-2 \sum_{r \in \mathbf{N}} \frac{1}{(2r+1)!} \frac{1}{(2\nu)^{2r+1}} \times \left[\left\langle \frac{\partial}{\partial x}, \frac{\partial}{\partial p'} \right\rangle - \left\langle \frac{\partial}{\partial x'}, \frac{\partial}{\partial p} \right\rangle\right]^{2r+1} H(x, p) H'(x', p') \bigg|_{\substack{x'=x \\ p'=p}} . \tag{1.2}$$

Suppose H and H' are in involution, which is the case for H(§1,2), L^2, and M(§1,1) by (1.3) of §1. Then the first term of (1.2) is zero. If H' is a polynomial of degree 2 in (x', p'), then all of the other terms evidently are zero; thus a and a' commute and part 1 of the lemma follows. Suppose that H' is a polynomial that is homogeneous of degree 4 in (x', p'). Then by (1.2), $a \circ a' - a' \circ a$ is associated to H''/ν^3 where H'' is the hamiltonian given by

$$H''(x, p) = -\frac{1}{24} \left[\left\langle \frac{\partial}{\partial x}, \frac{\partial}{\partial p'} \right\rangle - \left\langle \frac{\partial}{\partial x'}, \frac{\partial}{\partial p} \right\rangle\right]^3 H(x, p) H'(x', p') \bigg|_{\substack{x'=x \\ p'=p}} .$$

Two applications of Taylor's formula show that the term of $H'(x' + y, p' + q)$ that is homogeneous of degree 1 in (x', p') and degree 3 in (y, q) is

$$\frac{1}{6} \left[\left\langle q, \frac{\partial}{\partial p'} \right\rangle + \left\langle y, \frac{\partial}{\partial x'} \right\rangle\right]^3 H'(x', p') = \left[\left\langle p', \frac{\partial}{\partial q} \right\rangle + \left\langle x', \frac{\partial}{\partial y} \right\rangle\right] H'(y, q);$$

thus, if H' is homogeneous of degree 2 in each of its two variables, then

$$H''(x, p) = -\frac{1}{4} \left[\left\langle p, H'_{p'}\left(\frac{\partial}{\partial p}, \frac{\partial}{\partial x}\right) \right\rangle - \left\langle x, H'_{x'}\left(\frac{\partial}{\partial p}, \frac{\partial}{\partial x}\right) \right\rangle\right] H(x, p).$$

In particular, if $H' = L^2$, that is, $H'(x', p') = |x' \wedge p'|^2$, then

$$H''(x, p) = \frac{1}{2} \left\langle p \wedge \frac{\partial}{\partial p} + x \wedge \frac{\partial}{\partial x}, \frac{\partial}{\partial p} \wedge \frac{\partial}{\partial x} \right\rangle H(x, p). \tag{1.3}$$

In this formula, for each j the pair of operators

$$\left(p \wedge \frac{\partial}{\partial p} + x \wedge \frac{\partial}{\partial x}\right)_j \quad \text{and} \quad \left(\frac{\partial}{\partial p} \wedge \frac{\partial}{\partial x}\right)_j \quad \text{commutes.}$$

The operator

$$\left(p \wedge \frac{\partial}{\partial p} + x \wedge \frac{\partial}{\partial x}\right)_j$$

is an infinitesimal rotation acting on x and p; thus it annihilates L, Q, and R. Obviously,

$$\left(p \wedge \frac{\partial}{\partial p} + x \wedge \frac{\partial}{\partial x}\right) M(x, p) = x_3 p - p_3 x.$$

Suppose H is a composition with L, M, Q, and R [§1,(2.1)]. Then (1.3) becomes

$$H''(x, p) = \frac{1}{2}\left\langle \frac{\partial}{\partial p} \wedge \frac{\partial}{\partial x}, H_M x_3 p - H_M p_3 x \right\rangle.$$

Let

$$\mathscr{X}F = \begin{vmatrix} x_1 & x_2 & x_3 \\ p_1 & p_2 & p_3 \\ F_{x_1} & F_{x_2} & F_{x_3} \end{vmatrix}, \qquad \mathscr{P}F = \begin{vmatrix} x_1 & x_2 & x_3 \\ p_1 & p_2 & p_3 \\ F_{p_1} & F_{p_2} & F_{p_3} \end{vmatrix}$$

for any function F of (x, p). The preceding expression for H'' may be written

$$H''(x, p) = \frac{1}{2}\frac{\partial}{\partial x_3}\mathscr{P}H_M - \frac{1}{2}\frac{\partial}{\partial p_3}\mathscr{X}H_M.$$

Now, the linear differential operators $\mathscr{X}$ and $\mathscr{P}$ evidently annihilate P^2, Q, and R^2, hence L^2, by (1.2) of §1; moreover,

$$\mathscr{X}M = -\frac{1}{2}\frac{\partial L^2}{\partial x_3}, \qquad \mathscr{P}M = -\frac{1}{2}\frac{\partial L^2}{\partial p_3}.$$

Thus, letting $\mathscr{D}F$ denote the functional determinant

$$\mathscr{D}F = \frac{\partial L^2}{\partial x_3}\frac{\partial F}{\partial p_3} - \frac{\partial L^2}{\partial p_3}\frac{\partial F}{\partial x_3},$$

the expression for H'' becomes

$$H''(x, p) = \tfrac{1}{4}\mathscr{D}H_{M^2}.$$

Now, the linear differential operator $\mathscr{D}$ evidently annihilates L and M; moreover,

$$\tfrac{1}{2}\mathscr{D}Q = P^2x_3^2 - R^2p_3^2 \quad \text{and} \quad \frac{R}{2}\mathscr{D}R = Qx_3^2 - R^2p_3x_3;$$

thus the preceding expression for H'' becomes

$$H''(x, p) = \tfrac{1}{2}\left(P^2H_{M^2Q} + \frac{Q}{R}H_{M^2R}\right)x_3^2 - \frac{R}{2}H_{M^2R}p_3x_3 - \tfrac{1}{2}R^2H_{M^2Q}p_3^2. \tag{1.4}$$

By §1.1, p_3/x_3 is independent of (L, M, Q, R). Thus the condition

$$H''(x, p) = 0 \qquad \forall x, p$$

is equivalent to (1.1), which proves part 2 of the lemma.

2. Case of an Operator a Commuting with a_{L^2} and a_M

Suppose (1.1) holds. Lemma 1 proves that theorem 7.2 of II,§3 applies to the lagrangian system

$$aU = (a_{L^2} - c_L)U = (a_M - c_M)U = 0 \ \text{mod}\frac{1}{\nu^{r+2}}, \qquad r \geqslant 1, \qquad (2.1)_r$$

where c_L and c_M are two formal numbers with vanishing phase such that

$$c_L - L_0^2 = c_M - M_0 = 0 \ \text{mod}\frac{1}{\nu^2}. \tag{2.2}$$

The lagrangian operators a_{L^2} and a_M have the following expressions $a_{L^2}^+$ and a_M^+ in R_0 by I,§1,3 and the formula

$$\left[\exp\frac{1}{2\nu}\left\langle\frac{\partial}{\partial x}, \frac{\partial}{\partial p}\right\rangle\right]L^2(x, p) = \left[1 + \frac{1}{2\nu}\left\langle\frac{\partial}{\partial x}, \frac{\partial}{\partial p}\right\rangle + \frac{1}{8\nu^2}\left\langle\frac{\partial}{\partial x}, \frac{\partial}{\partial p}\right\rangle^2\right] \cdot(P^2R^2 - Q^2)$$

$$= P^2R^2 - Q^2 - \frac{2}{\nu}Q - \frac{3}{2\nu^2};$$

namely

$$a_{L^2}^+\left(\nu, x, \frac{1}{\nu}\frac{\partial}{\partial x}\right) = \frac{1}{\nu^2}\left(\Delta_0 - \frac{3}{2}\right), \tag{2.3}$$

where[4]

$$\Delta_0 = R^2\Delta - \sum_{j,k} x_j x_k \frac{\partial^2}{\partial x_j \partial x_k} - 2\sum_j x_j \frac{\partial}{\partial x_j} \tag{2.4}$$

is the spherical laplacian (2.24), which acts on the restrictions of functions to spheres $R = \text{const.}$;

$$a_M^+\left(v, x, \frac{1}{v}\frac{\partial}{\partial x}\right) = \frac{1}{v}\left(x_1\frac{\partial}{\partial x_2} - x_2\frac{\partial}{\partial x_1}\right),$$

which is an infinitesimal rotation.

Let us require that the unknown U be defined $\mathrm{mod}(1/v^{r+1})$ on a *compact* lagrangian manifold V. By theorem 3 of §1,

i. V is necessarily one of the tori

$$V = V[L_0, M_0] = T(l, m, n) : H = L^2 - L_0^2 = M - M_0 = 0,$$

which part 1) of this theorem defines.

ii. The lagrangian amplitude β_0 of U is necessarily constant.

If $\beta_0 = 0$, $(2.1)_r$ reduces to $(2.1)_{r-1}$: hence we impose the condition

$$\beta_0 = \text{const.} \neq 0. \tag{2.5}$$

The problem defined by the system $(2.1)_r$, condition (2.5), and the condition that V be *compact* will be called *problem* $(2.1)_r$.

Notation. The invariant measure η_V is given by (3.16) of theorem 3 of §1; d^3x by (1.16) and (3.6) of §1; $\arg d^3x = \pi m_{R_0}$ by definition [(3.4) of I,§3]; m_{R_0} is given by (3.5) of §1; $\arg \eta_V = 0$ by definition; hence the value of the function χ, defined and used in II,§3, and the value of its argument are, respectively,

$$\chi = \frac{\eta_V}{d^3x} = [R^3 H_Q[L_0, M_0, Q, R] \sin\Psi \sin\Theta]^{-1}, \text{ where } \Theta = \text{const.};$$

$$\arg\chi = -\arg H_Q - \arg\sin\Psi, \tag{2.6}$$

where $\arg H_Q = -\pi[(1/\pi)\arctan(H_Q/H_R)]$, $\arg\sin\Psi = \pi[\Psi/\pi]$.

(Recall that $[\cdot]$ is the integer part.)

The apparent contour of V is

[4] As usual: $\Delta = \Sigma_j(\partial/\partial x_j)^2$.

$$\Sigma_{R_0}: H_Q \sin \Psi = 0; \text{ thus } \chi: V \backslash \Sigma_{R_0} \to \mathbf{R}. \tag{2.7}$$

Let U be a lagrangian function on V with lagrangian amplitude

$\beta_0 = \text{const.} \neq 0.$

Its expression in R_0 will be denoted

$$U_{R_0}(v) = \sqrt{\chi}\beta(v)e^{v\varphi_{R_0}}, \text{ where } \beta(v) = \sum_{s\in\mathbf{N}} \frac{\beta_s}{v_s}. \tag{2.8}$$

By the structure theorem 2.2 and definition 3.2 of lagrangian functions in II,§2, the function

$\chi^{-3s}\beta_s: V \to \mathbf{C}$

is regular even on Σ_{R_0}.

Let D_H, D_L, and D_M be operators such that

$$\begin{aligned} [aU]_{R_0} &= \frac{\sqrt{\chi}}{v} e^{v\varphi_{R_0}} D_H\beta, \\ [(a_{L^2} - L_0^2)U]_{R_0} &= \frac{\sqrt{\chi}}{v} e^{v\varphi_{R_0}} D_L\beta, \\ [(a_M - M_0)U]_{R_0} &= \frac{\sqrt{\chi}}{v} e^{v\varphi_{R_0}} D_M\beta. \end{aligned} \tag{2.9}$$

D_H, D_L, and D_M *commute* since a, a_{L^2}, and a_M commute.

Let us use the local coordinates (R, Ψ, Φ) on $V\backslash\Sigma_{R_0}$.

LEMMA 2.1. 1°) With this choice of coordinates,

$$D_M = \frac{\partial}{\partial\Phi}. \tag{2.10}$$

2°) If U is a solution of the equation

$$(a_M - c_M)U = 0 \bmod \frac{1}{v^{r+2}}, \tag{2.11}$$

where c_M is a formal number with vanishing phase such that

$$c_M = M_0 \bmod \frac{1}{v^2},$$

then

$$c_M = M_0 \mod \frac{1}{v^{r+2}};$$

β depends, $\mathrm{mod}(1/v^{r+1})$, only on the coordinates (R, Ψ).

Proof of 1°). Let us calculate D_M by using theorem 4 of II,§3: in this theorem, the hamiltonian $M - M_0$ is substituted for H. By (3.18) of §1, the characteristics of this hamiltonian are given by the equations

$$dt = d\Psi = 0; \text{ that is, } dR = d\Psi = 0.$$

The parameter of these characteristics is Φ, which is substituted for t in this theorem. We obtain (2.10).

Proof of 2°). For $r = 0$, 2°) is obvious. Proceeding by induction on r, we can assume 2°) is true when r is replaced by $r - 1$. Then

$$c_M = M_0 + \frac{M_{r+1}}{v^{r+1}}, \qquad M_{r+1} \in \mathbf{C}.$$

By (2.9) and (2.10), (2.11) is equivalent to

$$\frac{\partial \beta_r}{\partial \Phi} = M_{r+1}\beta_0, \text{ where } \beta_0 = \text{const.} \neq 0.$$

Now, β_r is a function of Φ having period 2π; thus

$$M_{r+1} = 0, \qquad \frac{\partial \beta_r}{\partial \Phi} = 0,$$

from which 2°) follows.

Notation. From now on, we assume that *β depends only on the variables* (R, Ψ). By (2.10), D_H and D_{R_0} commute with $\partial/\partial\Phi$ and thus act on functions of (R, Ψ).

LEMMA 2.2. 1°) In the local coordinates (R, Ψ, Φ),

$$D_L\beta = 2L_0\frac{d\tau}{d\Psi}\left[\frac{\partial}{\partial \tau} - \frac{1}{v}F\right]\beta - \frac{5}{4v}\beta, \tag{2.12}$$

where τ is the variable

$$\tau = \cot \Psi$$

and F *is the operator with polynomial coefficients* given by

$$F\beta = F_1\beta + \frac{\partial}{\partial\tau}\left[F_2\frac{\partial\beta}{\partial\tau}\right],$$

F_1 and F_2 being the polynomials in τ given by

$$F_1(\tau) = \frac{5M_0^2\tau^2 + L_0^2 + M_0^2}{8L_0(L_0^2 - M_0^2)}, \qquad F_2(\tau) = \frac{(M_0^2\tau^2 + L_0^2)(\tau^2 + 1)}{2L_0(L_0^2 - M_0^2)}.$$

2°) Let U be a lagrangian function that is defined on the torus $V[L_0, M_0]$, has a lagrangian amplitude $\beta_0 \neq 0$, and is a solution of the system

$$(a_{L^2} - c_L)U = (a_M - M_0)U = 0 \mod \frac{1}{\nu^{r+2}}, \tag{2.13}$$

where c_L is a formal number, with vanishing phase, such that

$$c_L = L_0^2 \mod \frac{1}{\nu^2}.$$

Let Σ be the set of points on the curve $\Gamma[L_0, M_0]$ [(2.5) of §1], where

$\Sigma: H_Q = 0.$

Then

$$c_L = L_0^2 - \frac{5}{4\nu^2} \mod \frac{1}{\nu^{r+2}} \tag{2.14}$$

and

$$\beta(\nu, R, \Psi) = g(\nu, R)f(\nu, \tau) \mod \frac{1}{\nu^{r+1}}, \tag{2.15}$$

g being some formal function with vanishing phase defined on

$\Gamma[L_0, M_0]\backslash\Sigma$

and f being defined as follows.

3°) There exists a unique formal function f defined on $\mathbf{R}$, of the form

$$f(\nu, \tau) = \sum_{s\in\mathbf{N}} \frac{1}{\nu^s} f_s(\tau), \text{ where } \tau \in \mathbf{R}, \tag{2.16}$$

such that

$$\frac{df}{d\tau} = \frac{1}{\nu} Ff, \tag{2.17}$$

$f_0 = 1$, f_s is a real polynomial in τ, of degree 3s, having the parity of s, (2.18)

and

$$f\bar{f} = 1 + \frac{1}{\nu} F_2 \left(\bar{f} \frac{df}{d\tau} - f \frac{d\bar{f}}{d\tau} \right) \tag{2.19}$$

[$\bar{f}$ is the complex conjugate of f; (2.19) expresses f_{2s} using $f_1, \ldots, f_{2s-1}$]. Any solution of (2.17) $\mathrm{mod}(1/\nu^r)$ is, $\mathrm{mod}(1/\nu^r)$, the product of f and a formal number with vanishing phase.

Remark 2. Let U' be the formal function of x that is homogeneous of degree 0, is defined for

$M_0 R < L_0 \sqrt{x_1^2 + x_2^2}$,

and is given by

$$U'(\nu, x) = \frac{f(\nu, \Psi)}{\sqrt{\sin \Psi}} e^{\nu(L_0 \Psi + M_0 \Phi)}; \tag{2.20}$$

U' satisfies

$$\frac{1}{\nu}\left(x_1 \frac{\partial}{\partial x_2} - x_2 \frac{\partial}{\partial x_1} \right) U' = M_0 U', \qquad \frac{1}{\nu^2} \Delta U' = \frac{1}{R^2}\left(L_0^2 + \frac{1}{4\nu^2} \right) U'. \tag{2.21}$$

Proof of 1°). Let us calculate D_L using theorem 4 of II,§3: in this theorem the hamiltonian $L^2 - L_0^2$ is substituted for H. By (3.17) of §1, the characteristics of this hamiltonian are given by the equations

$dt = d\Phi = 0$; that is, $dR = d\Phi = 0$.

The parameter of these characteristics is $\Psi/2L_0$, which is substituted for t in this theorem. The right-hand side of formula $(4.5)_2$ in this theorem has to be replaced by

$$\left[\exp \frac{1}{2} \left\langle \frac{\partial}{\partial x}, \frac{\partial}{\partial p} \right\rangle \right] L^2(x, p) = \left[1 + \frac{1}{2} \left\langle \frac{\partial}{\partial x}, \frac{\partial}{\partial p} \right\rangle + \frac{1}{8} \left\langle \frac{\partial}{\partial x}, \frac{\partial}{\partial p} \right\rangle^2 \right]$$
$$(P^2R^2 - Q^2) = P^2R^2 - Q^2 - 2Q - \frac{3}{2}.$$

This theorem gives

$$D_L\beta = 2L_0\frac{\partial\beta}{\partial\Psi} + \frac{1}{v}\left[\chi^{-1/2}\Delta_0(\beta\chi^{1/2}) - \frac{3}{2}\beta\right], \tag{2.22}$$

where Δ_0 is defined by (2.4) in terms of the coordinates (x_1, x_2, x_3).

By (1.5) and (1.12) of §1, we have

$$R\frac{\partial}{\partial R} = \sum_{j=1}^{3} x_j\frac{\partial}{\partial x_j}; \text{ thus } R^2\frac{\partial^2}{\partial R^2} = \sum_{j,k} x_j x_k \frac{\partial^2}{\partial x_j \partial x_k} \tag{2.23}$$

using the coordinates (R, Ψ, Φ) in the left-hand sides and the coordinates (x_1, x_2, x_3) in the right-hand sides. Then definition (2.4) of Δ_0 is formulated as

$$\Delta_0 = R^2\Delta - R^2\frac{\partial^2}{\partial R^2} - 2R\frac{\partial}{\partial R}. \tag{2.24}$$

Now, formulas (1.5) and (1.12) of §1 give

$$\frac{x_3}{R} = -\sin\Theta\cos\Psi, \text{ where } \cos\Theta = \frac{M_0}{L_0} \text{ by §1, (1.11);}$$

hence, on V,

$$\left\langle R_x, \Psi_x \right\rangle = 0, \qquad R^2\left\langle \Psi_x, \Psi_x \right\rangle = 1 + \frac{\cot^2\Theta}{\sin^2\Psi},$$

$$R^2\Delta\Psi = \cot\Psi\left(1 - \frac{\cot^2\Theta}{\sin^2\Psi}\right).$$

The expression for $R^2\Delta$ acting on functions of (R, Ψ) follows. Substituted into (2.24), it gives

$$\Delta_0 = \left(1 + \frac{\cot^2\Theta}{\sin^2\Psi}\right)\frac{\partial^2}{\partial\Psi^2} + \cot\Psi\left(1 - \frac{\cot^2\Theta}{\sin^2\Psi}\right)\frac{\partial}{\partial\Psi} \tag{2.25}$$

on these functions; hence, from (2.6),

$$\begin{aligned}\chi^{-1/2}\Delta_0(\beta\chi^{1/2}) &= \sqrt{\sin\Psi}\,\Delta_0\left(\frac{\beta}{\sqrt{\sin\Psi}}\right)\\ &= -2L_0\frac{d\tau}{d\Psi}F\beta + \tfrac{1}{4}\beta,\end{aligned} \tag{2.26}$$

by a trivial calculation. Thus (2.12) follows from (2.22).

Proof of 2°). By lemma 2.1, β locally depends only on the variables (R, Ψ). By this lemma and (2.12), β_0 depends only on R; by (2.5), β_0 is not identically zero.

For $r = 0$, 2°) is evident. Proceeding by induction on r, we can assume that $\beta_0, \ldots, \beta_{r-1}$ are polynomials in τ whose coefficients are functions of R and that 2°) is true when r is replaced by $r - 1$. Then

$$c_L = L_0^2 - \frac{5}{4v^2} + \frac{2L_0 L_{r+1}}{v^{r+1}} \mod \frac{1}{v^{r+2}}, \text{ where } L_{r+1} \in \mathbf{C};$$

thus equation $(2.13)_1$ may be written

$$\frac{1}{v} D_L \beta + \left(\frac{5}{4v^2} - \frac{2L_0 L_{r+1}}{v^{r+1}}\right)\beta = 0 \mod \frac{1}{v^{r+2}}. \tag{2.27}$$

By the induction hypothesis, this equation holds $\mathrm{mod}(1/v^{r+1})$; thus, by (2.12), it may be written

$$d\beta_r = (F\beta_{r-1})\,d\tau + L_{r+1}\beta_0\,d\Psi \text{ for } R = \text{const.}$$

Then, since F is an operator with polynomial coefficients, β_r is the sum of a polynomial in $\tau = \cot \Psi$ and the function $L_{r+1}\beta_0\Psi$, where $\beta_0 \neq 0$. But β_r is a function of Ψ with period 2π. Hence

$$L_{r+1} = 0, \qquad \beta_r \text{ is a polynomial in } \tau;$$

thus (2.27) may be written

$$d\beta = \frac{1}{v}(F\beta)\,d\tau \mod \frac{1}{v^{r+1}} \text{ for } R = \text{const.}$$

If 3°) is true, (2.15) follows.

Proof of 3°). The condition that (2.16) satisfy (2.17) $\mathrm{mod}(1/v^r)$ may be written

$$f_0 = \text{const.}, \qquad \frac{df_s}{d\tau} = Ff_{s-1} \text{ for } s = 1, \ldots, r - 1.$$

Thus f_s is a polynomial of degree $3s$ in τ containing an arbitrary constant of integration. Then f is well defined up to multiplication by a formal number with vanishing phase.

Let us choose the constants of integration to be real. Then the f_r are real.

There exists a unique choice of the constants of integration of the f_{2s-1} such that the f_r have the parity of r.

Proof of (2.19). Let $\bar{f}$ be the complex conjugate of f. Since v is purely imaginary and F is real, (2.17) implies

$$\frac{d\bar{f}}{d\tau} = -\frac{1}{v}F\bar{f}, \qquad \frac{d}{d\tau}(f\bar{f}) = \frac{1}{v}(\bar{f}Ff - fF\bar{f}),$$

that is, by the definition of F,

$$\frac{d}{d\tau}(f\bar{f}) = \frac{1}{v}\left[\bar{f}\frac{d}{d\tau}\left(F_2\frac{df}{d\tau}\right) - f\frac{d}{d\tau}\left(F_2\frac{d\bar{f}}{d\tau}\right)\right] = \frac{1}{v}\left[\frac{d}{d\tau}F_2\left(\bar{f}\frac{df}{d\tau} - f\frac{d\bar{f}}{d\tau}\right)\right];$$

thus

$$f\bar{f} - \frac{1}{v}F_2\left(\bar{f}\frac{df}{d\tau} - f\frac{d\bar{f}}{d\tau}\right) = \sum_{s\in\mathbf{N}}\frac{c_s}{v^s}, \text{ where } c_0 = 1,$$

that is, since the f_r are real, for all $s \in \mathbf{N}$

$$c_{2s+1} = 0,$$

$$\sum_{s'=0}^{2s}(-1)^{s'}f_{2s-s'}f_{s'} = 2F_2\sum_{s'=0}^{2s-1}(-1)^{s'}f_{s'}\frac{d}{d\tau}f_{2s-1-s'} + c_{2s}$$

If $s > 0$, this formula expresses f_{2s} using $f_1, \ldots, f_{2s-1}$ and c_{2s}, which is cancelled by an appropriate choice of the constant of integration of f_{2s}.

Proof of $(2.21)_2$. By (2.12) and (2.17),

$$D_L f = -\frac{5}{4v}f,$$

that is, by the definition (2.9) of D_L,

$$\left[a_{L^2}^{+}\left(v, x, \frac{1}{v}\frac{\partial}{\partial x}\right) - L_0^2 + \frac{5}{4v^2}\right](\sqrt{\chi}fe^{v\varphi_{R_0}}) = 0,$$

or, by the expression (2.3) for $a_{L^2}^{+}$ and the expression (2.6) for χ, since Δ_0 acts on the restrictions of functions to spheres R = const. and since φ_R can be expressed by (3.4) of §1, where Ω only depends on R,

$$\left(\frac{1}{v^2}\Delta_0 - L_0^2 - \frac{1}{4v^2}\right)U'(v, x) = 0$$

where U' is defined by (2.20); U' is homogeneous of degree 0 in x; $(2.21)_2$ follows from this by the definition (2.24) of Δ_0.

Proof of $(2.21)_1$. From the definition (2.16) of f and the expression (2.10) for D_M it follows that

$$D_M f = 0,$$

or, by the definition (2.8)–(2.9) of D_M,

$$(a_M^+ - M_0)(\sqrt{\chi}\, f e^{\nu\varphi_{R_0}}) = 0.$$

In the beginning of section 2, we obtained

$$a_M^+ = \frac{1}{\nu}\left(x_1 \frac{\partial}{\partial x_2} - x_2 \frac{\partial}{\partial x_1}\right).$$

Thus a_M^+ annihilates functions of the single variable R. Then by the expression (2.6) for χ and the expression (3.4) of §1 for φ_{R_0}, the preceding relation is equivalent to $(2.21)_1$.

LEMMA 2.3. There exists an operator D, acting on formal functions g defined on $\Gamma[L_0, M_0]\backslash\Sigma$, such that

$$D_H[f(\nu, \Psi)g(\nu, R)] = f(\nu, \Psi)Dg(\nu, R). \tag{2.28}$$

Locally,

$$D = \sum_{s \in \mathbf{N}} \frac{1}{\nu^s} D_s\left(R, \frac{d}{dR}\right), \tag{2.29}$$

where $D_s(R, d/dR)$ is a differential operator defined on $\Gamma[L_0, M_0]\backslash\Sigma$ and

$$D_0 = RH_Q \frac{d}{dR}. \tag{2.30}$$

Proof. Since D_H commutes with D_L, which is expressed by (2.12),

$$\frac{d\tau}{d\Psi}\left(\frac{\partial}{\partial\tau} - \frac{1}{\nu}F\right) D_H[f(\nu, \Psi)g(\nu, R)]$$
$$= D_H\left[\frac{d\tau}{d\Psi}\left(\frac{\partial}{\partial\tau} - \frac{1}{\nu}F\right) f(\nu, \Psi)g(\nu, R)\right] = 0.$$

Thus by lemma 2.2,3°), $D_H[fg]$ is multiplication of f by a formal function, which is defined on $\Gamma[L_0, M_0]\backslash\Sigma$ and denoted Dg.

Theorem 4 of II,§3 proves that D_H has the form

$$D_H = \sum_{s\in\mathbf{N}} \frac{1}{v^s} D_{H,s}\left(R, \Psi, \frac{\partial}{\partial R}, \frac{\partial}{\partial \Psi}\right),$$

where $D_{H,s}$ is a differential operator. It follows that D has the form (2.29).

Since the characteristics of H satisfy (2.14) of §1, the same theorem proves that

$$D_H(fg)\,dt = d(fg) \bmod \frac{1}{v} \quad \text{for } dR = RH_Q\,dt, \quad d\Psi = H_L\,dt.$$

But

$$f = 1 \bmod \frac{1}{v};$$

thus

$$D_H g = RH_Q \frac{dg}{dR} \bmod \frac{1}{v}.$$

This relation is equivalent to (2.30).

THEOREM 2. 1°) *Problem* $(2.1)_r$, *defined at the beginning of section 2, is solvable if and only if*

$$c_L = L_0^2 - \frac{5}{4v^2}, \qquad c_M = M_0.$$

2°) ***Problem*** $(2.1)_r$ ***is equivalent to problem*** $(2.31)_r$, *which is stated as follows: To define a formal function* $\bmod(1/v^{r+1})$ *on* $\Gamma[L_0, M_0]\backslash\Sigma$,

$$g(v) = \sum_{s\in\mathbf{N}} \frac{1}{v^s} g_s, \text{ where } g_0 = 1, \quad g_s : \Gamma[L_0, M_0]\backslash\Sigma \to \mathbf{C},$$

such that

$$\begin{aligned} &Dg = 0 \bmod(1/v^{r+1}), \\ &(H_Q)^{3r} g \text{ is regular}, \bmod(1/v^{r+1}), \text{ on } \Gamma[L_0, M_0]. \end{aligned} \qquad (2.31)_r$$

Any formal function satisfying condition $(2.31)_r$ *is,* $\bmod(1/v^{r+1})$, *the product of* g *and a formal number with vanishing phase.*

The condition that g *be a solution to* $(2.31)_r$ *evidently is equivalent to the following:*

g is a solution of problem (2.31)$_{r-1}$,

$$RH_Q dg_r/dR + \Sigma_{s=1}^{r} D_s g_{r-s} = 0 \text{ on } \Gamma[L_0, M_0]\backslash\Sigma, \tag{2.32$_r$}$$

$(H_Q)^{3r} g_r$ *is regular on* $\Gamma[L_0, M_0]$.

On $\Gamma[L_0, M_0]\backslash\Sigma$, *define the formal function*

$$U''(v) = \frac{ge^{v\Omega}}{\sqrt{R^3 H_Q}},$$

where Ω *is defined by* (2.8) *of* §1. *Define* U' *by remark* 2 *and lemma* 2.2. *Then* $U'(v)U''(v)$ *is the expression* U_{R_0} *of a solution* U *of problem* (2.1)$_r$. *Every solution of the system* (2.1)$_r$ *defined on* $V[L_0, M_0]$ *is*, $\mathrm{mod}(1/v^{r+1})$, *the product of that solution and a formal number.*

3°) *Suppose that* g *is a solution of problem* (2.31)$_{r-1}$. *Then there exists a function* $g_r : \check{\Gamma}[L_0, M_0]\backslash\check{\Sigma} \to \mathbf{C}$ *satisfying* (2.32)$_r$, *where* Γ *is replaced by its universal covering space* $\check{\Gamma}$. *The function* g_r *is defined up to an additive constant; moreover,* g_r *is defined on* $\Gamma[L_0, M_0]\backslash\Sigma$ *if and only if the equivalent problems* (2.1)$_r$ *and* (2.31)$_r$ *are solvable.*

Proof of 1°). Use lemmas 2.1,2°) and 2.2,2°).

Proof of 2°). Use (2.6), (2.8), lemmas 2.1,2°), 2.2,2°), and 2.3, and theorems 5 and 7.2 of II,§3.

Proof of 3°). Use the above theorems.

3. A Special Case

In order to supplement theorem 2, choose

$$\begin{aligned} H[L, M, Q, R] &= \frac{1}{2}\left\{P^2 - \frac{1}{R^2}K[R, M]\right\} \\ &= \frac{1}{2R^2}\{L^2 + Q^2 - K[R, M]\} \end{aligned} \tag{3.1}$$

as in §1,4, and choose K to be *an affine function of* M. Then condition (1.1) is satisfied. By (4.5) of §1, the expression in R_0 of the operator associated to H is

$$a = \frac{1}{2v^2}\Delta - \frac{1}{2R^2}K\left[R, \frac{1}{v}\left(x_1\frac{\partial}{\partial x_2} - x_2\frac{\partial}{\partial x_1}\right)\right]. \tag{3.2}$$

Moreover,

$$L_0^2 + Q^2 - K[R, M_0] = 0 \text{ on the curve } \Gamma[L_0, M_0]. \tag{3.3}$$

Σ is the set of points on this curve where $Q = 0$.

LEMMA 3. We have

$$D = \frac{Q}{R}\left[\frac{d}{dR} - \frac{1}{v} G,\right] \tag{3.4}$$

where G acts on functions

$$g:\Gamma[L_0, M_0]\backslash\Sigma \to \mathbf{C}$$

and is given by

$$Gg = G_1 g + \frac{d}{dR}\left[G_2 \frac{dg}{dR}\right], \tag{3.5}$$

G_1 and G_2 being functions defined on $\Gamma[L_0, M_0]\backslash\Sigma$ given by

$$G_1(R, Q) = \frac{G_3(R)}{Q^5}, \qquad G_2(R, Q) = -\frac{1}{2}\frac{R}{Q}, \tag{3.6}$$

where

$$G_3(R) = -\frac{5}{32}R(K_R)^2 + \frac{1}{8}(K - L_0^2)(K_R + RK_{R^2}).$$

Proof. The operator D may be described as follows with the aid of II,§3, theorem 4: in formula (4.5) of this theorem, $s = 2$ and

$$\left[\exp\frac{1}{2}\left\langle\frac{\partial}{\partial x}, \frac{\partial}{\partial p}\right\rangle\right]H^{(2)}(x, p) = \left[\exp\frac{1}{2}\left\langle\frac{\partial}{\partial x}, \frac{\partial}{\partial p}\right\rangle\right]\frac{1}{2}P^2 = \frac{1}{2}P^2; \tag{3.7}$$

the equations (2.14) of §1 giving the characteristics imply that

$$dR = \frac{Q}{R}dt, \qquad d\Psi = \frac{L_0}{R^2}dt; \tag{3.8}$$

then, by the definition (2.9) of D_H, when β only depends on the coordinates (R, Ψ) of V, this theorem gives

$$D_H \beta dt = d\beta + \frac{1}{2v}\chi^{-1/2}\Delta(\beta\chi^{1/2})dt$$

for dt, dR, and $d\Psi$ satisfying (3.8); in other words,

$$D_H\beta = \frac{\partial\beta}{\partial R}\frac{Q}{R} + \frac{\partial\beta}{\partial\Psi}\frac{L_0}{R^2} + \frac{1}{2v}\chi^{-1/2}\Delta(\beta\chi^{1/2}). \tag{3.9}$$

By (2.6)

$$\chi = [QR\sin\Psi\sin\Theta]^{-1}, \qquad \Theta = \text{const.};$$

therefore, by the definition (2.24) of Δ_0, which acts on the restrictions of functions to spheres $R = \text{const.}$,

$$\chi^{-1/2}\Delta(\beta\chi^{1/2}) = \frac{1}{R^2}\sqrt{\sin\Psi}\,\Delta_0\left(\frac{\beta}{\sqrt{\sin\Psi}}\right) + \sqrt{QR}\left(\frac{\partial^2}{\partial R^2} + \frac{2}{R}\frac{\partial}{\partial R}\right)\left(\frac{\beta}{\sqrt{QR}}\right).$$

Replacing Δ_0 by its expression (2.26) and substituting the result into (3.9) yields

$$D_H\beta = \frac{Q}{R}\frac{\partial\beta}{\partial R} + \frac{\sqrt{QR}}{2v}\left(\frac{\partial^2}{\partial R^2} + \frac{2}{R}\frac{\partial}{\partial R}\right)\left(\frac{\beta}{\sqrt{QR}}\right) + \frac{L_0}{R^2}\frac{d\tau}{d\Psi}\left(\frac{\partial}{\partial\tau} - \frac{1}{v}F\right)\beta + \frac{1}{8v}\frac{\beta}{R^2}.$$

Choose

$$\beta(v, R, \Psi) = f(v, \Psi)g(v, R),$$

where f is the formal function of Ψ defined by lemma 2.2 and g is a formal function defined on $\Gamma[L_0, M_0]\backslash\Sigma$. In accordance with lemma 2.3, we obtain

$$D_H(fg) = f\,Dg$$

with

$$Dg = \frac{Q}{R}\left[\frac{dg}{dR} + \frac{1}{2v}\frac{1}{\sqrt{QR}}\left(R^2\frac{d^2}{dR^2} + 2R\frac{d}{dR}\right)\left(\frac{g}{\sqrt{QR}}\right) + \frac{1}{8v}\frac{g}{QR}\right].$$

By a trivial calculation, we deduce the expression (3.4) for D, with G expressed by (3.5) and G_1 and G_2 given by

$$G_1 = \frac{1}{4}\frac{d}{dR}\left[\frac{R}{Q^2}\frac{dQ}{dR}\right] + \frac{1}{8}\frac{R}{Q^3}\left(\frac{dQ}{dR}\right)^2, \qquad G_2 = -\frac{1}{2}\frac{R}{Q}.$$

Now by the equation (3.3) for $\Gamma[L_0, M_0]$,

$$Q^2 = K[R, M_0] - L_0^2;$$

hence the expression (3.6) for G_1 follows.

The preceding lemma permits a statement of the properties of problem $(2.3)_r$, to which problem $(2.1)_r$ has been reduced by theorem 2.

Definition 3. A function $g:\Gamma[L_0, M_0]\backslash\Sigma \to \mathbf{C}$ is said to be *even* or *odd* when

$$g(Q, R) = \pm g(-Q, R) \qquad \forall(\pm Q, R) \in \Gamma[L_0, M_0]\backslash\Sigma.$$

THEOREM 3.1. (Complement to theorem 2) *Assume* (3.1).

1°) *If problem* $(2.31)_r$ *has a solution* g *such that* $g_0 = 1$, *then it has a unique solution such that*

$$g_0 = 1, \qquad g_s \text{ is real and has the parity of } s \ (s \leqslant r),$$

and

$$g\bar{g} = 1 - \frac{1}{2\nu}\frac{R}{Q}\left(\bar{g}\frac{dg}{dR} - g\frac{d\bar{g}}{dR}\right) \bmod \frac{1}{\nu^{r+1}}. \tag{3.10}$$

Formula (3.10) *means that, for* $2 \leqslant 2s \leqslant r$,

$$\sum_{s'=0}^{2s} (-1)^{s'} g_{2s-s'} g_{s'} = -\frac{R}{Q}\sum_{s'=0}^{2s-1} (-1)^{s'} g_{s'} \frac{d}{dR} g_{2s-1-s'}; \tag{3.11}$$

thus it expresses g_{2s} *using* $g_1, \ldots, g_{2s-1}$.

2°) *If problem* $(2.31)_{2s-1}$ *has a solution, then problem* $(2.31)_{2s}$ *has a solution.*

Remark 3. The formal function U'' defined on $\Gamma[L_0, M_0]\backslash\Sigma$ by theorem 2,2°) is evidently given by

$$U''(\nu) = \frac{g(\nu)}{\sqrt{QR}} e^{\nu\Omega}. \tag{3.12}$$

It satisfies

$$\frac{1}{\nu^2}\Delta U''(\nu, R) = \frac{1}{R^2}\left\{K[R, M_0] - L_0^2 - \frac{1}{4\nu^2}\right\} U''(\nu, R), \tag{3.13}$$

where Δ is the laplacian, specifically $(d^2/dR^2) + (2/R)(d/dR)$.

Proof of 1°). Assume that 1°) has been proved when $r - 1$ is substituted for r and that problem $(2.3)_r$ is solvable. Then, by theorem 2,2°) and lemma 3, g_r is defined up to a constant of integration by the conditions

$$\frac{dg_r}{dR} = Gg_{r-1} \text{ on } \Gamma[L_0, M_0]\backslash\Sigma, \qquad Q^{3r}g_r \text{ is regular on } \Gamma[L_0, M_0]. \tag{3.14$_r$}$$

Now, G changes parity. If r is odd, then a convenient choice of the constant of integration makes g_r real and odd. If $r = 2s$ is even, then g_{2s} is even and can be chosen to be real.

We have

$$\frac{dg}{dR} = \frac{1}{v}Gg \bmod \frac{1}{v^{r+1}}, \qquad Gg = G_1 g + \frac{d}{dR}\left[G_2 \frac{dg}{dR}\right].$$

By a trivial calculation analogous to the proof of (2.19), we deduce (3.10), up to the addition of a formal number $\Sigma_s(c_s/v^s)$, which is canceled by conveniently choosing the constants of integration of the g_{2s}.

Proof of 2°). Assume that $r = 2s$ is even and that problem $(2.31)_{2s-1}$ is solvable. By theorem 2,3°), problem $(3.14)_{2s}$ has a solution g_{2s} when Γ is replaced by its universal covering space $\check{\Gamma}$. The proof of (3.11) remains valid; (3.11) proves that g_{2s} is defined on $\Gamma[L_0, M_0]\backslash\Sigma$; thus g_{2s} is a solution of problem $(3.14)_{2s}$.

Proof of remark 3. By the expression (3.2) for a and theorem 2,2°),

$$\left\{\frac{1}{v^2}\Delta - \frac{1}{R^2}K\left[R, \frac{1}{v}\left(x_1\frac{\partial}{\partial x_2} - x_2\frac{\partial}{\partial x_1}\right)\right]\right\}[U'(v, x)U''(v, x)] = 0,$$

where U' is homogeneous in x of degree 0 and U'' only depends on R. This implies

$$\Delta(U' \cdot U'') = U' \cdot \Delta U'' + U'' \cdot \Delta U';$$

hence (3.13) follows by $(2.12)_2$.

Notation. Let $[R_1, R_2] \subset \mathbf{R} \subset \mathbf{C}$ be the set of values taken by R on $\Gamma[L_0, M_0]\backslash\Sigma$. Let ω be a simply connected neighborhood of the real closed segment $[R_1, R_2]$ in the complex plane $\mathbf{C}$; $R \in \mathbf{C}$.

THEOREM 3.2. *Assume that K is holomorphic in ω. Then Q, defined by* (3.3), *is holomorphic in* $\omega\backslash[R_1, R_2]$.

1°) *If problem* $(2.31)_{2s}$ *has a solution* g, *then, for all* $r \leqslant 2s$,

$$Q^{3r}g_r(Q, R) = (-Q)^{3r}g_r(-Q, R)$$

is the value of a function of R *that is holomorphic in* ω. *In particular,* g_{2s} *is a function of* R *that is meromorphic in* ω, *with poles* R_1 *and* R_2.

2°) *Problem* $(2.31)_{2s+1}$ *is solvable* [*and thus problem* $(2.31)_{2s+2}$ *also*] *if and only if the primitive of* $(Gg_{2s})dR$ *is defined* (*that is, uniform*) *in* $\omega\backslash[R_1, R_2]$.

Proof. Suppose 1°) is true. [1°) is evident for $s = 0$]. Then, by (3.14), the function $g_{2s+1}\colon \check{\Gamma}[L_0, M_0]\backslash\check{\Sigma} \to \mathbf{C}$ is obviously the restriction of the primitive of $(Gg_{2s})dR$ to the edges of the cut $[R_1, R_2]$ in $\mathbf{C}$, which is defined on the universal covering space of $\omega\backslash[R_1, R_2]$. Obviously, g_{2s+1} is defined on $\Gamma[L_0, M_0]\backslash\Sigma$ if and only if this primitive is defined on $\omega\backslash[R_1, R_2]$.

Assume that this condition is satisfied.

With a convenient choice of the constant of integration, this primitive g_{2s+1} is an odd meromorphic function of Q in a neighborhood of R_1 and R_2. Like Q, it takes opposite values on the edges of the cut $[R_1, R_2]$.

For R near R_1 or R_2, Q is near 0 and g_{2s} is an even function of Q having a pole of order $6s$ at $Q = 0$. From (3.3), (3.5), (3.6), and (3.14),

$$\frac{dg_{2s+1}}{dQ} = \frac{dR}{dQ}Gg_{2s} = 2\frac{G_3}{K_R}g_{2s}\frac{1}{Q^4} - \frac{1}{4}\frac{d}{dQ}\left[\frac{RK_R}{Q^2}\frac{dg_{2s}}{dQ}\right], \text{ where } K_R \neq 0.$$

Thus g_{2s+1} is locally an odd function of Q having a pole of order $3(2s + 1)$ at $Q = 0$.

Consequently $Q^{3(2s+1)}g_{2s+1}$ is holomorphic in ω.

g_{2s+2} is defined by (3.11). Then $Q^{3(2s+2)}g_{2s+2}$ is holomorphic in ω. Consequently 1°) holds when s is replaced by $s + 1$.

4. The Schrödinger–Klein-Gordon Case

In order to establish, for any r, the solvability of problem $(2.1)_r$, which we have reduced to problem $(2.31)_r$, an appropriate assumption is clearly necessary.

Assume that K is a second-degree polynomial:

$$K[R, M] = -R^2A(M) + 2RB(M) - C(M).$$

In other words, *the expression of a in R_0 is the Schrödinger–Klein-Gordon operator* (§1,4) and *A, B, and C are affine functions of M.*

Then in theorem 3.2, $\omega = \mathbf{C}$. In the definition (3.5) of the operator G, G_3 is a polynomial in R of degree 3 and G_1 and G_2 are functions of R holomorphic in $\mathbf{C}\backslash[R_1, R_2]$ and at infinity, where G_1 vanishes to second order. If g is holomorphic in $\mathbf{C}\backslash[R_1, R_2]$ and at infinity, then so is the primitive of $(Gg)dR$. Then all the g_r exist and are holomorphic in $\mathbf{C}\backslash[R_1, R_2]$ and at infinity. Since g_r is holomorphic at infinity and since $Q^{3r}g_r$ is holomorphic in $\mathbf{C}$, $Q^{3r}g_r$ is a polynomial in R of degree $3r$. Thus we have proven the following two theorems.

THEOREM 4.1. (Existence and uniqueness) *Assume that the expression of a in R_0 is the Schrödinger–Klein-Gordon operator* (*example* 4 *of* §1).

1°) *The condition that the lagrangian system*

$$aU = (a_{L^2} - c_L)U = (a_M - c_M)U = 0, \tag{4.1}$$

where c_L and c_M are two formal numbers such that

$$c_L - L_0^2 = c_M - M_0 = 0 \mod \frac{1}{\nu^2}, \tag{4.2}$$

have a solution defined on a **compact** *lagrangian manifold V is the following:*

i. $c_L = L_0^2 + (1/4\nu^2)$, $c_M = M_0$;

ii. *V is one of the lagrangian tori* $V[L_0, M_0] = T(l, m, n)$ *defined by theorem* 4.1 *of* §1.

2°) *There exists a lagrangian solution U of* (4.1) *defined on such a torus V having lagrangian amplitude*

$\beta_0 = 1.$

Any lagrangian solution of (4.1) *defined on V is the product of U and a formal number with vanishing phase.*

Remark 4.1. The projection of V on X is

$V_X : R_1 \leqslant |x| \leqslant R_2, \qquad M_0|x| \leqslant L_0\sqrt{x_1^2 + x_2^2},$

where R_1 and R_2 are the two roots of the equation

$A_0R^2 - 2B_0R + C_0 + L_0^2 = 0$

(A_0, B_0, and C_0 are the values of A, B, and C at M_0.)

THEOREM 4.2. (Structure) *There exists a unique solution U of* (4.1) *defined on such a torus V having the following structure: Its expression U_{R_0} in the frame R_0* (§1,1) *has the form*

$$U_{R_0}(\nu) = U'(\nu)U''(\nu). \tag{4.3}$$

Using the local coordinates $x \in V_X$ on V, $U'(\nu)$ is a formal function of x that is homogeneous of degree 0 *and satisfies*

$$\begin{aligned} \frac{1}{\nu}\left(x_1\frac{\partial}{\partial x_2} - x_2\frac{\partial}{\partial x_1}\right)U'(\nu, x) &= M_0 U'(\nu, x), \\ \frac{1}{\nu^2}\Delta U'(\nu, x) &= \frac{1}{R^2}\left(L_0^2 + \frac{1}{4\nu^2}\right)U'(\nu, x). \end{aligned} \tag{4.4}$$

$U''(\nu)$ is a formal function of R satisfying

$$\frac{1}{\nu^2}\Delta U''(\nu, x) = \frac{1}{R^2}\left\{K[R, M_0] - L_0^2 - \frac{1}{4\nu^2}\right\}U''(\nu, x). \tag{4.5}$$

U' and U'' are defined by the formulas

$$\begin{aligned} U'(\nu, x) &= \frac{f(\nu, \tau)}{\sqrt{\sin\Psi}}e^{\nu(L_0\Psi + M_0\Phi)}, \\ U''(\nu, x) &= \frac{g(\nu, Q, R)}{\sqrt{QR}}e^{\nu\Omega}, \end{aligned} \tag{4.6}$$

where $\tau = \cot\Psi$ *and* Ω *is the function of R defined by* (2.8) *of* §1; $\arg\sin\Psi$ *and* $\arg Q$ *have jumps of* $+\pi$ *at the points* $\Psi = 0 \bmod \pi$ *on* $\mathbf{R}$ *and* $Q = 0$ *on* $\Gamma[L_0, M_0]$, *which are oriented in the directions* $d\Psi > 0$ *and* $Q\,dR > 0$. *Moreover,*

$$f(\nu) = \sum_{r\in\mathbf{N}} \frac{1}{\nu^r} f_r \quad \textit{and} \quad g(\nu) = \sum_{r\in\mathbf{N}} \frac{1}{\nu^r} g_r$$

are formal functions with vanishing phase defined on $\mathbf{R}$ *and on* $\Gamma[L_0, M_0]\backslash\Sigma$, *respectively, by the following set of properties:*

$$\frac{df}{d\tau} = \frac{1}{\nu}Ff, \qquad \frac{dg}{dR} = \frac{1}{\nu}Gg, \tag{4.7}$$

the differential operators F and G being given by

$$Ff = F_1 f + \frac{d}{d\tau}\left[F_2 \frac{df}{d\tau}\right], \qquad Gg = G_1 g + \frac{d}{dR}\left[G_2 \frac{dg}{dR}\right] \tag{4.8}$$

where F_1 and F_2 are the even polynomials in τ of degree 2 and 4 defined by lemma 2.2,1°); $Q^5 G_1$ and QG_2 are the polynomials in R of degree 3 and 1 defined by (3.6); $f_0 = 1$, $g_0 = 1$; the functions f_r and g_r are real and have the parity of r;

$$f\bar{f} = 1 + \frac{1}{v}F_2\left(\bar{f}\frac{df}{d\tau} - f\frac{d\bar{f}}{d\tau}\right), \qquad g\bar{g} = 1 + \frac{1}{v}G_2\left(\bar{g}\frac{dg}{dR} - g\frac{d\bar{g}}{dR}\right). \tag{4.9}$$

These formulas (4.9) give the even functions f_{2s} and g_{2s} in terms of $f_1, \ldots, f_{2s-1}$ and $g_1, \ldots, g_{2s-1}$. The odd functions f_{2s+1} and g_{2s+1} are defined by the quadratures

$$df_{2s+1} = (Ff_{2s})d\tau, \qquad dg_{2s+1} = (Gg_{2s})dR.$$

f_r is a polynomial in τ of degree 3r. $Q^{3r} g_r$ is a polynomial in R of degree 3r.

Remark 4.2. Thus $g_{2s+1} - G_2 dg_{2s}/dR$ is the odd function on $\Gamma[L_0, M_0]\backslash\Sigma$ that is the primitive of the differential form $G_1 g_{2s} dR$, namely, of a form $\Pi'(R)Q^{-6s-5}dR$, where Π' is a polynomial of degree $6s + 3$. By the preceding theorem, this primitive shall be $\Pi(R)Q^{-6s-3}$, where Π is a polynomial of degree $6s + 3$. Let us give a more direct proof of this essential fact.

LEMMA 4. For each $s \in \mathbf{N}$, the differentiation

$$\frac{d}{dR}\frac{\Pi(R)}{Q^{2s+1}} = \frac{\Pi'(R)}{Q^{2s+3}} \tag{4.10}$$

(Q^2 a polynomial in R of degree 2, discriminant $Q \neq 0$) defines an automorphism $\Pi \mapsto \Pi'$ of the vector space of polynomials of degree $2s + 1$.

Proof. Let Π be a polynomial of degree $2s + 1$; $\Pi(R)Q^{-2s-1}$ is holomorphic at infinity; thus its derivative has a double zero at infinity. Consequently (4.10) defines a polynomial Π' of degree $2s + 1$. Hence the mapping

$$\Pi \mapsto \Pi'$$

is an endomorphism of a finite-dimensional vector space. Now it is evidently a monomorphism; thus it is an isomorphism.

Conclusion

This §3 *is concerned with finding, for* dim $X = 3$, *a lagrangian system, with one lagrangian unknown, having a solution defined on a* **compact** *lagrangian manifold unique up to a multiplicative factor. In this* §3, *we have found a system of this type: the system used in wave mechanics for studying atoms (or ions) with a unique and spinless electron.*

Remark 4.3. Replacing the Schrödinger–Klein-Gordon hamiltonian by the *harmonic oscillator* hamiltonian for $\mathbf{E}^3$,

$$H(x, p) = \frac{1}{2\mu}[P^2 + AR^2 - 2B],$$

that is, choosing

$$K[R, M] = -AR^4 + 2BR^2$$

where A and B are affine functions of M, one obtains another lagrangian system (2.1) belonging to the *same type.*

This can be shown by computations similar to those performed above, where R^2 plays the role that R played above.

The energy levels are still those of wave mechanics.

Remark 4.4. Of course, the study of the harmonic oscillator in $\mathbf{R}$ is simpler; in $\mathbf{E}^3$, for A and B independent of M, it gives new lagrangian solutions for the operator associated to the harmonic oscillator, but again the same energy levels.

§4. The Schrödinger–Klein-Gordon Equation

0. Introduction

The following classical boundary-value problem will be called *problem* (0.1): To find nonzero square-integrable functions

$$u : \mathbf{E}^3 \to \mathbf{C}$$

that have square-integrable gradients and that are solutions of the differential equation

$$au = 0, \tag{0.1}$$

where a is the *differential operator* associated to the hamiltonian

$$H[L, M, Q, R] = \frac{1}{2}\left\{P^2 - \frac{1}{R^2}K[R, M]\right\}, \tag{0.2}$$

K being an affine function of M [compare (3.1) and (3.2) of §3]. This operator is defined by substituting $\nu_0 = i/\hbar$ for ν in the lagrangian operator associated to H [(4.5) of §1]. Hence it is

$$a = \frac{1}{2\nu_0^2}\Delta - \frac{1}{2R^2}K\left[R, \frac{1}{\nu_0}\left(x_1\frac{\partial}{\partial x_2} - x_2\frac{\partial}{\partial x_1}\right)\right]. \tag{0.3}$$

In section 2, we assume (compare §3,4)

$$K[R, M] = -R^2A(M) + 2RB(M) - C(M), \tag{0.4}$$

where A, B, and C are affine functions of M. Then a is the Schrödinger–Klein-Gordon operator.

In §4, we briefly recall the solution of the classical problem (0.1) in order to observe two facts:

i. There are formal analogies between its solution and that of the lagrangian problem (2.1) of §3, which is to solve

$$aU = (a_{L^2} - c_L)U = (a_M - c_M)U = 0 \text{ on compact } V.$$

ii. The condition for the existence of a solution to this lagrangian problem [or to this problem mod$(1/\nu^2)$, namely, to (3.1) of §1] is the same as that for the classical boundary-value problem (0.1) in the Schrödinger–Klein-Gordon case, that is, under assumption (0.4).

This assumption is essential.

Remark 0. For a suitable choice of A, B, and C, the Schrödinger–Klein-Gordon equation is the Schrödinger or Klein-Gordon equation (4.21) or (4.22) in §1, where the $\mathscr{H}^2$ terms have been omitted. It is customary to treat the terms of these equations that are linear in $\mathscr{H}$ by "perturbation" theory. We do not do this, but we rigorously solve problem (0.1) in order to show that assertion (ii) is rigorous.

1. Study of Problem (0.1) without Assumption (0.4)

Review of the properties of spherical harmonics $u'_{l,m}$. The set of polynomials in $x \in \mathbf{E}^3$ that are harmonic and homogeneous of degree l is a vector space over $\mathbf{C}$ of dimension $2l + 1$. A basis consists of the polynomials

$$R^l u'_{l,m}, \qquad l \text{ and } m \text{ integers}, \quad |m| \leqslant l,$$

defined by the following system, where $u'_{l,m}$ is homogeneous of degree 0 and thus satisfies $R^2 \Delta u' = \Delta_0 u'$ [see (2.4) and (2.24) of §3]:

$$\Delta u'_{l,m} + \frac{1}{R^2} l(l+1) u'_{l,m} = 0, \qquad \left(x_1 \frac{\partial}{\partial x_2} - x_2 \frac{\partial}{\partial x_1}\right) u'_{l,m} = imu'_{l,m}. \quad (1.1)$$

Let $\mathbf{S}^2$ be the unit sphere in $\mathbf{E}^3$ and let σ be the usual measure on $\mathbf{S}^2$. Then

$$\int_{\mathbf{S}^2} \left|\frac{\partial}{\partial x} u'_{l,m}\right|^2 \sigma = l(l+1) \int_{\mathbf{S}^2} |u'_{l,m}|^2 \sigma$$

and for all $(l_1, m_1) \neq (l_2, m_2)$

$$\int_{\mathbf{S}^2} \left\langle \frac{\partial}{\partial x} u'_{l_1,m_1}, \frac{\partial}{\partial x} u'_{l_2,m_2} \right\rangle \sigma = \int_{\mathbf{S}^2} u'_{l_1,m_1} \bar{u}'_{l_2,m_2} \sigma = 0,$$

where $\langle \cdot, \cdot \rangle$ is the (sesquilinear) scalar product.

The restrictions of the $u'_{l,m}$ to $\mathbf{S}^2$ form a complete system of functions on $\mathbf{S}^2$: every square-integrable function $u : \mathbf{E}^3 \to \mathbf{C}$ with square-integrable gradient has a unique expansion

$$u = \sum_{l,m} u'_{l,m} u''_{l,m},$$

where the $u''_{l,m}$ are functions only of the variable $R > 0$, such that

$$\int_{\mathbf{E}^3} |u|^2 d^3x = \sum_{l,m} \int_{\mathbf{S}^2} |u'_{l,m}| \sigma \int_0^{+\infty} R^2 |u''_{l,m}|^2 \, dR < \infty$$

and

$$\int_{\mathbf{E}^3} \left|\frac{\partial}{\partial x} u\right|^2 d^3x = \sum_{l,m} \int_{\mathbf{S}^2} \left|\frac{\partial}{\partial x} u'_{l,m}\right|^2 \sigma \cdot \int_0^{+\infty} |u''_{l,m}|^2 \, dR$$
$$+ \sum_{l,m} \int_{\mathbf{S}^2} |u'_{l,m}|^2 \sigma \cdot \int_0^{+\infty} R^2 \left|\frac{d}{dR} u''_{l,m}\right|^2 dR < \infty.$$

Resolution of problem (0.1). By (0.3) and (1.1), the condition that u be a solution of problem (0.1) is formulated as

$$\left(\frac{d^2}{dR^2} + \frac{2}{R}\frac{d}{dR}\right)u''_{l,m}(R) + \frac{1}{R^2}\left\{\frac{1}{\hbar^2}K[R, \hbar m] - l(l+1)\right\}u''_{l,m}(R) = 0 \tag{1.2}$$

for all l, m. By writing $Ru''_{l,m} = v$, we may evidently state that result as follows.

THEOREM 1.1. *Problem* (0.1) *is equivalent to problem* (1.3), *which is to find integers l, m and nonzero functions v:*$]0, +\infty[\to \mathbf{R}$ *such that*

$$|m| \leqslant l,$$

$$\frac{d^2 v}{dR^2} + \frac{1}{R^2}\left\{\frac{1}{\hbar^2}K[R, \hbar m] - l(l+1)\right\}v = 0 \tag{1.3}$$

and

$$\int_0^{+\infty}\left(1 + \frac{1}{R^2}\right)v^2\,dR < \infty, \qquad \int_0^{+\infty}\left(\frac{dv}{dR}\right)^2 dR < \infty. \tag{1.4}$$

Remark 1.1. Thus problem (0.1) is solvable if and only if the following problem is solvable: To find two integers l, m and a nonzero square-integrable function $u:\mathbf{E}^3 \to \mathbf{C}$ with square-integrable gradient such that

$$|m| \leqslant l$$

and

$$au = 0, \qquad \Delta_0 u + l(l+1)u = 0, \qquad \left(x_1\frac{\partial}{\partial x_2} - x_2\frac{\partial}{\partial x_1}\right)u = imu. \tag{1.5}$$

Recall that system (1.5) plays an essential role in physics.

Remark 1.2. Equations (1.1) and (1.2) and the system (1.5) are obtained formally by replacing

$$v, \quad U', \quad U'', \quad U$$

by

$$i/\hbar, \quad u'_{l,m}, \quad u''_{l,m}, \quad u$$

in equations (2.21) and (3.13) and in the system (2.1) of §3, where account is taken of theorem 3 of §1 and theorem 2 of §3, which require

$$L_0 = \hbar\left(l + \frac{1}{2}\right), \qquad M_0 = \hbar m, \qquad c_L = L_0^2 - \frac{5}{4\nu^2}, \qquad c_M = M_0,$$

that is, by (2.3) of §3,

$$\begin{aligned} a_{L^2}^+ - c_L &= \frac{1}{\nu^2}\Delta_0 - \hbar^2\left[l + \frac{1}{2} - \frac{i}{2\nu\hbar}\right]\left[l + \frac{1}{2} + \frac{i}{2\nu\hbar}\right] \\ &= -\hbar^2[\Delta_0 + l(l+1)] \text{ for } \nu = i/\hbar, \end{aligned}$$

$$a_M^+ - c_M = \frac{1}{\nu}\left(x_1\frac{\partial}{\partial x_2} - x_2\frac{\partial}{\partial x_1}\right) - \hbar m.$$

Assume that K is a holomorphic function of R at the origin. Let

$$C_0 = -K[0, \hbar m], \qquad \gamma = \sqrt{(l + \tfrac{1}{2})^2 + C_0\hbar^{-2}}. \tag{1.6}$$

Apply *Fuchs's theorem*, that is, the theory of *regular points* of analytic ordinary differential equations: see, for instance, [19], section 10.3.

Fuchs's theorem constructs two independent solutions of equation (1.3). If 2γ is not an integer, the respective quotients of these solutions with $R^{\pm\gamma+1/2}$ are holomorphic at the origin. If γ is purely imaginary, then every nonzero solution of (1.3) makes the integrals (1.4) diverge at the origin. The same is true if $\gamma = 0$ (see Fuchs). Assume $\gamma > 0$. The solution v of (1.3) that is a product of $R^{\gamma+1/2}$ and a holomorphic function at the origin makes the integrals (1.4) converge at the origin, while the other solutions of (1.3) make them diverge (see Fuchs). Thus, since problem (1.3) is equivalent to problem (0.1), the following theorem holds.

THEOREM 1.2. *If K is holomorphic at the origin, then problem* (0.1) *is solvable if and only if the following problem is solvable: To find two integers l and m such that*

$$|m| \leqslant l, \qquad \gamma > 0,$$

and such that the nonzero solution v of (1.3), *whose quotient with $R^{\gamma+1/2}$ is holomorphic at the origin, satisfies*

$$\int_1^{+\infty} v^2\, dR < \infty, \qquad \int_1^{\infty} \left(\frac{dv}{dR}\right)^2 dR < \infty. \tag{1.7}$$

2. The Schrödinger–Klein-Gordon Case

In this case, that is, under assumption (0.4), the preceding theorem can be supplemented. Let

$$M_0 = \hbar m, \qquad A_0 = A(M_0), \qquad B_0 = B(M_0), \qquad C_0 = C(M_0),$$

$$\alpha = \sqrt{A_0}\hbar^{-1}, \qquad \beta = B_0\hbar^{-2}. \tag{2.1}$$

Equation (1.3) becomes the *confluent hypergeometric equation*:

$$\frac{d^2v}{dR^2} + \left[-\alpha^2 + \frac{2\beta}{R} - \frac{\gamma^2 - 1/4}{R^2}\right]v = 0, \tag{2.2}$$

where α^2, β, and $\gamma \in \mathbf{R}$, $\gamma > 0$. If $\alpha^2 > 0$, then choose $\alpha > 0$.

Let v be the nonzero solution of (2.2) such that $vR^{-\gamma-1/2}$ is an entire function of R; it is defined up to a multiplicative constant (Fuchs).

LEMMA 2. The convergence of the integrals (1.7) is equivalent to the following condition:

$$\frac{\beta}{\alpha} + \frac{1}{2} - \gamma \text{ is a positive integer.} \tag{2.3}$$

Proof. Express v by

$$v(R) = R^{\gamma+1/2}e^{-\alpha R}\sum_{s\in\mathbf{N}} c_s R^s, \qquad c_s \in \mathbf{C}. \tag{2.4}$$

Substituting into (2.2), we obtain (Fuchs) the recurrence formula defining the c_s in terms of c_0:

$$s(s + 2\gamma)c_s = 2[\alpha(s + \gamma - \tfrac{1}{2}) - \beta]c_{s-1}. \tag{2.5}$$

Condition (2.3) holds if and only if

$$\sum_{s\in\mathbf{N}} c_s R^s \text{ is a polynomial.}$$

It implies $\alpha^2 > 0$; thus $\alpha > 0$ and thus the integrals (1.7) converge.

Now let us prove that (1.7) does not hold when (2.3) is not satisfied.

Case when $\alpha > 0$: If $\Sigma_s c_s R^s$ is not a polynomial, then, for an appropriate choice of the sign of c_0,

$$c_s > 0 \text{ for } s \text{ near } +\infty.$$

Let $\varepsilon \in]0, \alpha[$. From (2.5) we have

$$sc_s > 2\varepsilon c_{s-1} \text{ for } s \text{ near } +\infty;$$

thus $\Sigma_s c_s R^s > c'e^{2\varepsilon R}$ for R near $+\infty$, where $c' = \text{const.} > 0$,

and the integral $(1.7)_1$ diverges.

Case when $\alpha^2 < 0$: The classical integral expression for v gives an asymptotic expression for v which is also classical: for R near $+\infty$,

$$v(R) = ce^{\alpha R}R^{-\beta/\alpha} + \bar{c}e^{-\alpha R}R^{\beta/\alpha} + \cdots,$$

where c and $\bar{c}$ are constants and α is purely imaginary. Hence the integral $(1.7)_1$ diverges. (See Whittaker and Watson, [19], chapter XVI, "The Confluent Hypergeometric Function: Asymptotic Expansion.")

Case when $\alpha = 0, \beta < 0$: By (2.5),

$$v(R) = R^{\gamma+1/2} \sum_{s\in\mathbf{N}} c_s R^s, \text{ where } c_s > 0 \qquad \forall s;$$

thus the integral $(1.7)_1$ diverges.

Case when $\alpha = 0, \beta > 0$: $R^{-\gamma-1/2}v$ satisfies a differential equation with linear coefficients. Laplace's expression for its solution gives

$$v(R) = \sqrt{R}\int_T t^{-2\gamma-1}e^{\sqrt{2\beta R}(t-t^{-1})}dt,$$

where T is the boundary of a half-strip in $\mathbf{C}$ containing the cut $]-\infty, 0[$. By replacing T by a path on which $\mathrm{Re}(t - t^{-1}) \leqslant 0$ ($=0$ only for $t = i$) and assuming that R is near $+\infty$, we obtain the asymptotic value

$$v(R) = R^{1/4}[ce^{2i\sqrt{2\beta R}} + \bar{c}e^{-2i\sqrt{2\beta R}}] + \cdots.$$

Thus integral $(1.7)_1$ diverges.

Theorem 1.2 and lemma 2, where α and β are defined by (2.1) and γ by (1.6), prove the following.

THEOREM 2. *When a is the Schrödinger–Klein-Gordon operator, then, for each of the following problems, namely,*

1°) *the classical problem* (0.1) *that we have just reviewed,*
2°) *the lagrangian problem* (2.1) *in* §3,

$$aU = (a_{L^2} - c_L)U = (a_M - c_M)U = 0 \textit{ on compact } V,$$

3°) *the same problem* $\mathrm{mod}(1/v^2)$, *that is, problem* (3.1) *in* §1,

the condition that there exist a solution is the same: the existence of a triple of integers (l, m, n) *satisfying condition* (4.11) *of* §1, *theorem* 4.

Remark 2. A comparison of theorem 1.2 and theorem 3.1, 1°) of §1 proves that the preceding theorem does not hold for every operator a associated to a hamiltonian H of the form (0.2).

Conclusion

Although the classical boundary-value problems and the lagrangian problems are completely independent, they define the same energy levels for the Schrödinger and Klein-Gordon equations.

As a matter of fact the experimental values of the energy levels agree with those of the Dirac equation, which is studied in the following chapter.

IV Dirac Equation with the Zeeman Effect

Introduction

In §1, we solve, $\mathrm{mod}(1/v^2)$, a homogeneous lagrangian problem in several unknowns. That is one of the simplest problems to which theorem 3 of II,§4 can be applied. The resolution of this problem introduces the quadruple of quantum numbers that arises in the study of the Dirac equation.

In §2, we use lagrangian analysis to reduce the Dirac equation $\mathrm{mod}(1/v^2)$ to the simpler system solved in §1. This reduction is analogous to the reduction theorem 2.2 of II,§4. Thus we prove that *the lagrangian solutions of the Dirac equation (one-electron atoms in a magnetic field) defined* $\mathrm{mod}(1/v^2)$ *on compact manifolds have energy levels that are exactly those for which the classical solution of this equation exists* (taking into account the Zeeman effect and even the Paschen-Back effect).

§1. A Lagrangian Problem in Two Unknowns

In this §1, we give an application of II,§4,theorem 3.

1. Choice of Operators Commuting $\mathrm{mod}(1/v^3)$

As in III,§1,1, we let

$$X = X^* = \mathbf{E}^3, \qquad x, p \in \mathbf{E}^3,$$
$$R(x) = |x|, \quad P(p) = |p|, \quad Q(x,p) = \langle p, x\rangle, \quad L(x,p) = |x \wedge p|, \tag{1.1}$$

so that

$$L^2 + Q^2 = P^2R^2. \tag{1.2}$$

Let $(M_1, M_2, M_3 = M)$ be the components of $x \wedge p$.

By III,§1,(1.3), the following formulas define three functions $H^{(1)}$, $H^{(2)}$, and $H^{(3)}$ that are pairwise in involution (see III,§1,2):

$$H^{(1)}(x,p) = H[L(x,p), M(x,p), Q(x,p), R(x)],$$
$$H^{(2)}(x,p) = L^2(x,p), \qquad H^{(3)}(x,p) = M(x,p). \tag{1.3}$$

To use theorem 3 of II,§4, we must choose three $\mu \times \mu$ matrices $J^{(1)}$, $J^{(2)}$, and $J^{(3)}$ that are functions of $(x,p) \in \mathbf{E}^3 \oplus \mathbf{E}^3$. These must be such that the matrices of lagrangian operators associated to the three matrices

$$H^{(k)}E + \frac{1}{v}J^{(k)} \qquad (E \text{ the } \mu \times \mu \text{ identity matrix}) \tag{1.4}$$

commute $\mathrm{mod}(1/v^3)$.

Since the $H^{(k)}$ are in involution, it follows by remark 3 of II,§4 that this commutivity condition is equivalent to the condition that for all i, k,

$$(H^{(i)}, J^{(k)}) - (H^{(k)}, J^{(i)}) + J^{(i)}J^{(k)} - J^{(k)}J^{(i)} = 0, \tag{1.5$_{i,k}$}$$

where $(\cdot, \cdot)$ is the Poisson bracket.

We choose $\mu = 2$. Then the values of the $J^{(k)}$ are matrices acting on vectors in the space $\mathbf{C}^2$, which we provide with a hermitian structure. We choose the $iJ^{(k)}$ to be self-adjoint. (The vectors in $\mathbf{C}^2$ are called *spinors*.)

Notation. Let $\sigma_0 = 1$ be the 2×2 identity matrix. Every 2×2 self-adjoint matrix has the form $x_0\sigma_0 + \sigma$, where x_0 is a real number and σ is *a self-adjoint matrix with zero trace*:

$$\sigma = \sigma[x_1, x_2, x_3] = \begin{pmatrix} x_3 & x_1 - ix_2 \\ x_1 + ix_2 & -x_3 \end{pmatrix}$$

$$= x_1\sigma_1 + x_2\sigma_2 + x_3\sigma_3, \tag{1.6}$$

where

$(x_1, x_2, x_3) \in \mathbf{E}^3$

and

$$\sigma_1 = \begin{pmatrix} 0 & 1 \\ 1 & 0 \end{pmatrix}, \qquad \sigma_2 = \begin{pmatrix} 0 & -i \\ i & 0 \end{pmatrix}, \qquad \sigma_3 = \begin{pmatrix} 1 & 0 \\ 0 & -1 \end{pmatrix}$$

are the *Pauli matrices*, which satisfy the relations

$$\sigma_k^2 = 1, \qquad \sigma_i\sigma_k = -\sigma_k\sigma_i, \qquad \sigma_1\sigma_2\sigma_3 = i. \tag{1.7}$$

It follows that

$$\sigma^2[x] = |x|^2. \tag{1.8}$$

Remark 1.1. Recall the consequences of (1.8): Let u_2 be a measure-preserving automorphism of $\mathbf{C}^2$, that is, $u_2 \in SU(2)$; $u_2\sigma[x]u_2^{-1}$ is self-adjoint and has zero trace; thus it is of the form $\sigma[y]$. Let $y = ux$. Then

$$\sigma[ux] = u_2\sigma[x]u_2^{-1}.$$

By (1.8), u preserves $|x|^2$. More precisely, u is a rotation in $\mathbf{E}^3$, that is, $u \in SO(3)$, whence we obtain a morphism

$$SU(2) \ni u_2 \mapsto u \in SO(3),$$

which proves that $SU(2)$ is the covering group of $SO(3)$ of order 2. It is the universal covering group of $SO(3)$ (see Steenrod [17], p. 115).

LEMMA 1. Assume

$$J^{(1)} = if\sigma_3 + ig\sigma[x \wedge p], \tag{1.9}$$

is a datum, where

$$f(x, p) = f[L(x, p), M(x, p), Q(x, p), R(x)], \qquad g = g[L, M, Q, R].$$

Then (1.5) is satisfied by choosing

$$J^{(2)} = 0, \qquad J^{(3)} = \frac{i}{2}\sigma_3. \tag{1.10}$$

Remark 1.2. Relations (1.5) evidently remain valid when arbitrary complex numbers (that is, their products with the identity matrix denoted by $\sigma_0 = 1$) are added to the $H^{(i)}$ and $J^{(k)}$.

Proof of $(1.5)_{1,2}$. $(H^{(2)}, J^{(1)}) = 0$ because, by III,§1,1, L is in involution with Q, R, and $M = M_3$, and similarly with M_1 and M_2.

Proof of $(1.5)_{1,3}$. Since M is in involution with L, Q, and R,

$$(H^{(3)}, J^{(1)}) = ig(M, \sigma[x \wedge p]) = ig\sigma[-M_2, M_1, 0] \tag{1.11}$$

because an immediate and classical calculation gives

$$(M, M_1) = -M_2, \qquad (M, M_2) = M_1.$$

Moreover, (1.7) implies

$$\sigma_3\sigma[x \wedge p] - \sigma[x \wedge p]\sigma_3 = 2i\sigma[-M_2, M_1, 0];$$

whence, by the definitions (1.9) and (1.10) of $J^{(1)}$ and $J^{(3)}$,

$$J^{(3)}J^{(1)} - J^{(1)}J^{(3)} = -ig\sigma[-M_2, M_1, 0]; \tag{1.12}$$

$(1.5)_{1,3}$ results from (1.11) and (1.12).

The proof of $(1.5)_{2,3}$ is evident.

2. Resolution of a Lagrangian Problem in Two Unknowns

Notation. Let

$$a_{f,g}, \quad a_{L^2}, \quad a_M$$

be the 2×2 matrices of lagrangian operators associated to the 2×2 matrices

$$H[L, M, Q, R] + \frac{i}{v} f[L, M, Q, R]\sigma_3 + \frac{i}{v} g[L, M, Q, R]\sigma[x \wedge p],$$
$$L^2, \quad M. \tag{2.1}$$

Let us study the *solutions* $U = (U_1, U_2)$ of the system

$$\begin{aligned} a_{f,g}U &= \left(a_{L^2} - L_0^2 - \frac{2i}{v} L_0 l'\right) U \\ &= \left(a_M - M_0 - \frac{i}{v} m' + \frac{i}{2v}\sigma_3\right) U = 0 \bmod \frac{1}{v^2}, \end{aligned} \tag{2.2}$$

where U_1 and U_2 are two lagrangian functions *defined on a* ***compact*** *lagrangian manifold* V, l' and m' are real numbers near 0 *whose squares are negligible*, and L_0 and M_0 are two real numbers. (The interest of this system is shown by lemma 1 and remark 1.2.)

As in III,§1, let $V[L_0, M_0]$ denote the lagrangian manifold in $\mathbf{E}^3 \oplus \mathbf{E}^3$ given by the equations

$$V[L_0, M_0]: H[L_0, M_0, Q, R] = L(x, p) - L_0 = M(x, p) - M_0 = 0. \tag{2.3}$$

When this manifold is *compact*, it is a *torus* whose points have the coordinates

$$t \bmod c[L_0, M_0], \qquad \Psi \bmod 2\pi, \qquad \Phi \bmod 2\pi.$$

$\Gamma[L_0, M_0]$, t, and $c[L_0, M_0]$ are defined in III,§1 by (2.5), (2.6), and (2.7), respectively. Assume that on $V[L_0, M_0]$

$$(f - \tfrac{1}{2}H_M)c[L, M] \text{ and } Lgc[L, M]$$

are near zero and have *negligible squares*. For $V[L_0, M_0]$, and hence $\Gamma[L_0, M_0]$, compact, let

$$\begin{aligned} f_0 &= \pi N_M[L_0, M_0] + \int_{\Gamma[L_0, M_0]} f[L_0, M_0, Q, R]\, dt, \\ g_0 &= L_0 \int_{\Gamma[L_0, M_0]} g[L_0, M_0, Q, R]\, dt, \end{aligned} \tag{2.4}$$
$$2\pi\varepsilon[L_0, M_0] = \sqrt{f_0^2 + 2\frac{M_0}{L_0} f_0 g_0 + g_0^2} > 0,$$

where $|M_0| < L_0$. We shall see that f_0, g_0, and ε have negligible squares.

THEOREM 2. *The **compact** manifolds V in $\mathbf{E}^3 \oplus \mathbf{E}^3$ on which solutions U of the system* (2.2) *are defined are the tori $V[L_0, M_0]$ such that*

$$\begin{gathered} L_0 = \hbar(l + \tfrac{1}{2} - l'), \qquad M_0 = \hbar(m - m'), \\ \hbar(l + \tfrac{1}{2}) + N[\hbar(l + \tfrac{1}{2}), \hbar m] = \hbar n \pm \hbar\varepsilon[\hbar(l + \tfrac{1}{2}), \hbar m], \end{gathered} \tag{2.5}$$

where $l, m - \frac{1}{2}$, and n are three integers satisfying the inequalities

$$|m| \leqslant l + \tfrac{1}{2}, \qquad l < n; \tag{2.6}$$

l' and m' are numbers near 0 *that are arbitrary unless $m = l + \frac{1}{2}$. Then they must be chosen so that*

$$|M_0| < L_0. \tag{2.7}$$

Remark 2.1. U will be denoted by U_+ or U_- depending on whether $(2.5)_3$ holds with the choice $+$ or $-$. The lagrangian amplitude $\beta = (\beta_1, \beta_2)$ of $U = U_\pm$ will be denoted by $\beta_\pm$. It satisfies the relations

$$\begin{aligned} \beta_\pm(t + c[L_0, M_0], \Psi, \Phi) &= c_{\pm 1}\beta_\pm(t, \Psi, \Phi), \\ \beta_\pm(t, \Psi + 2\pi, \Phi) &= c_{\pm 2}\beta_\pm(t, \Psi, \Phi), \\ \beta_\pm(t, \Psi, \Phi + 2\pi) &= c_{\pm 3}\beta_\pm(t, \Psi, \Phi). \end{aligned} \tag{2.8}$$

The $c_{\pm k}$ are complex numbers with modulus 1 given by

$$\begin{gathered} c_{\pm 1} = e^{2\pi i(l'N_L[L_0,M_0] + m'N_M[L_0,M_0] \mp \varepsilon[L_0,M_0])}, \\ c_{\pm 2} = e^{2\pi i l'}, \qquad c_{\pm 3} = -e^{2\pi i m'}. \end{gathered} \tag{2.9}$$

Proof. By lemma 1 and remark 1.2, theorem 3 of II,§4 can be applied. Then the $\check{V}$ on which a solution U of the system (2.2) is defined are the covering spaces of the manifolds V given by the equations (2.3). The compact V are the tori $V[L_0, M_0]$. Let us determine when U is defined on $V[L_0, M_0]$, that is, when

$$\left(\frac{\eta}{d^3x}\right)^{1/2} e^{\nu_0 \varphi_{R_0}} \beta : V[L_0, M_0] \to \mathbf{C}^2, \tag{2.10}$$

where R_0 is the frame we are using,

$$\arg \eta = 0, \qquad \arg d^3x = m_{R_0},$$

and m_{R_0} and φ_{R_0} are the Maslov index and the phase of $V[L_0, M_0]$ in R_0.

By (3.5) and (3.4) of III,§1,

$$m_{R_0} = \left[\frac{1}{\pi}\Psi\right] - \left[\frac{1}{\pi}\arctan\frac{H_Q}{H_R}\right], \text{ where } [\cdot] = \text{integer part,}$$
$$\varphi_{R_0} = \Omega + L_0\Psi + M_0\Phi, \tag{2.11}$$

where Ω is defined in III,§1 by (2.8).

We now finish applying theorem 3 of II,§4. By the remarks 3.2 and 3.3 of III,§1, the characteristic vectors κ, κ_{L^2}, and κ_M of the hamiltonians H, L^2, and M are

$$\begin{aligned} \kappa: dt &= 1, & d\Psi &= H_L, & d\Phi &= H_M; \\ \kappa_{L^2}: dt &= 0, & d\Psi &= 2L_0, & d\Phi &= 0; \\ \kappa_M: dt &= 0, & d\Psi &= 0, & d\Phi &= 1. \end{aligned}$$

Then by this theorem the system (2.2) is equivalent to the condition

$$\left(\frac{\partial}{\partial t} + H_L\frac{\partial}{\partial\Psi} + H_M\frac{\partial}{\partial\Phi}\right)\beta + if\sigma_3\beta + ig\sigma[x \wedge p]\beta = 0,$$

$$\frac{\partial\beta}{\partial\Psi} - il'\beta = 0, \qquad \frac{\partial\beta}{\partial\Phi} - im'\beta + \frac{i}{2}\sigma_3\beta = 0.$$

In other words, it is equivalent to the pfaffian system

$$\begin{aligned} d\beta + i\{f\sigma_3\,dt + g\sigma[x \wedge p]\,dt - l'(d\Psi - H_L\,dt) \\ - (m' - \tfrac{1}{2}\sigma_3)(d\Phi - H_M\,dt)\}\beta = 0, \end{aligned} \tag{2.12}$$

which is guaranteed to be *completely integrable* by *this theorem*. By (1.5) of III,§1,

$$x \wedge p = L_0 J_3;$$

thus by $(1.12)_3$ and (1.11) of III,§1,

$$\sigma[x \wedge p] = L_0[\sigma_1\cos\Phi\sin\Theta + \sigma_2\sin\Phi\sin\Theta + \sigma_3\cos\Theta],$$

where

$$\cos\Theta = \frac{M_0}{L_0}. \tag{2.13}$$

Then, by (1.7),

$$\begin{aligned}\sigma[x \wedge p] &= L_0\sigma_1(\cos\Phi + i\sigma_3 \sin\Phi)\sin\Theta + L_0\sigma_3\cos\Theta \\ &= L_0\sigma_1 e^{i\Phi\sigma_3}\sin\Theta + L_0\sigma_3\cos\Theta \\ &= L_0 e^{-i\Phi\sigma_3/2}(\sigma_1\sin\Theta + \sigma_3\cos\Theta)e^{i\Phi\sigma_3/2}.\end{aligned}$$

Moreover, the functions of $[L, M, t]$, λ, and μ, which are defined by (2.10) of III,§1, satisfy

$$\lambda_t = -H_L, \qquad \mu_t = -H_M \tag{2.14}$$

by (2.3) of III,§2. Let

$$\beta = e^{-i\sigma_3\Phi/2 + il'(\Psi+\lambda) + im'(\Phi+\mu)}\gamma. \tag{2.15}$$

Then the system (2.12) may be written

$$d\gamma + i[(f - \tfrac{1}{2}H_M)\sigma_3 + L_0 g(\sigma_1\sin\Theta + \sigma_3\cos\Theta)]\gamma\,dt = 0. \tag{2.16}$$

It has now become *evident* that this *system is completely integrable*: here γ is independent of Φ and Ψ and is only a function of t.

Condition (2.10), which is that U be defined on $V[L_0, M_0]$, is then formulated as

$$\left(\frac{\eta}{d^3x}\right)^{1/2} e^{\nu_0\Omega + \nu_0 L_0\Psi + \nu_0 M_0\Phi - i\sigma_3\Phi/2 + il'(\Psi+\lambda) + im'(\Phi+\mu)}\gamma : V \to \mathbf{C}^2.$$

Now Ω, λ, and μ only depend on t and, by (1.4) and (1.5) of III,§2, increase by

$$\begin{aligned}&\Delta_t\Omega = 2\pi N[L_0, M_0], \qquad \Delta_t\lambda = 2\pi N_L[L_0, M_0], \\ &\Delta_t\mu = 2\pi N_M[L_0, M_0]\end{aligned} \tag{2.17}$$

when t increases by $c[L_0, M_0]$; N is defined by (2.9) of III,§1.

Let $\nu_0 = i/\hbar$ ($\hbar > 0$); then, by (2.11), condition (2.10) that U be lagrangian on V may be formulated as follows, since $e^{-\pi i\sigma_3} = -1$: There exists $\varepsilon \in \mathbf{R}$ such that

$$\gamma(t + c[L_0, M_0]) = e^{2\pi i\varepsilon}\gamma(t), \tag{2.18}$$

$$\begin{aligned}\frac{1}{\hbar}N[L_0, M_0] + l'N_L + m'N_M + \tfrac{1}{2} + \varepsilon &= \frac{1}{\hbar}L_0 + l' - \tfrac{1}{2} \\ &= \frac{1}{\hbar}M_0 + m' - \tfrac{1}{2} = 0 \bmod 1.\end{aligned} \tag{2.19}$$

The search for solutions γ of (2.16) satisfying (2.18) is a classical problem:

$$(f - \tfrac{1}{2}H_M)\sigma_3 + L_0 g(\sigma_1 \sin\Theta + \sigma_3 \cos\Theta)$$

is a *self-adjoint matrix* with *zero trace*; it is a *periodic* function of t; the period is $c[L_0, M_0]$; then the 2×2 matrix $u(t)$ defined by

$$du + i[(f - \tfrac{1}{2}H_M)\sigma_3 + L_0 g(\sigma_1 \sin\Theta + \sigma_3 \cos\Theta)]u\,dt = 0, \qquad u(0) = 1, \tag{2.20}$$

is a unitary matrix with determinant 1, that is, $u \in SU(2)$; it satisfies

$$u(t + c[L_0, M_0]) = u(t)u(c[L_0, M_0]);$$

the general solution of (2.16) is

$$\gamma(t) = u(t)\delta, \text{ where } \delta \in \mathbf{C}^2;$$

γ satisfies (2.18) if and only if δ is one of the eigenvectors of the matrix $u(c[L_0, M_0]) \in SU(2)$. Let

$$e^{\pm 2\pi i \varepsilon[L_0, M_0]} \qquad (0 \leqslant \varepsilon[L_0, M_0] \leqslant \tfrac{1}{2}) \tag{2.21}$$

be the eigenvalues of $u(c[L_0, M_0])$. Then, in (2.18),

$$\varepsilon = \mp\varepsilon[L_0, M_0].$$

Thus, *there exists a lagrangian solution* U *of* (2.2) *defined on the torus* $V[L_0, M_0]$ *if and only if there exist three integers* $l, m - \frac{1}{2}$, and n such that

$$\frac{1}{\hbar}L_0 + l' + \frac{1}{\hbar}N[L_0, M_0] + l'N_L + m'N_M = n \pm \varepsilon[L_0, M_0],$$

$$\frac{1}{\hbar}L_0 + l' - \tfrac{1}{2} = l, \qquad \frac{1}{\hbar}M_0 + m' = m, \text{ where } |M_0| < L_0. \tag{2.22}$$

This assertion *does not assume* that l', m', $(f - \frac{1}{2}H_M)c[L, M]$, and $Lgc[L, M]$ have negligible squares on $V[L_0, M_0]$.

U and β will be denoted by $U_\pm$ and $\beta_\pm$ depending on whether (2.22) holds with $\pm\varepsilon[L_0, M_0]$. By (2.15), (2.17), and (2.18), where $\varepsilon = \mp\varepsilon[L_0, M_0]$, $\beta_\pm$ satisfies (2.8), where the $c_{\pm k}$ have the values (2.9).

Assume that the squares of l' and m' and the squares of the derivatives of $\hbar\varepsilon$ are negligible. Then (2.22) reduces to (2.5). The integers l, $m - \frac{1}{2}$, and n must satisfy *condition* $(2.5)_3$, *which is independent of* l' and m'. Since l and n are integers, since ε is near 0, and since $N > 0$, $(2.5)_3$ implies

$l < n$. Since l' and m' are near 0, $(2.5)_1$ and $(2.5)_2$ imply $|m| \leqslant l + \frac{1}{2}$, which completes the proof of (2.6).

It remains to prove that (2.4) is an approximate expression for $\varepsilon[L_0, M_0]$. Since we have assumed that, on $V[L_0, M_0]$,

$$(f - \tfrac{1}{2}H_M)c[L, M] \text{ and } Lgc[L, M]$$

are near zero and have negligible squares, the solution of (2.20), modulo these squares, is

$$u = 1 - i\left[\int_0^t (f + \tfrac{1}{2}\mu_t)\,dt\,\sigma_3 + L_0 \int_0^t g\,dt\,(\sigma_1 \sin\Theta + \sigma_3 \cos\Theta)\right.$$

by $(2.14)_2$. In other words, choosing the right-hand side to be in $SU(2)$, like u,

$$u = e^{-i[\int_0^t (f + \mu_t/2)\,dt\,\sigma_3 + L_0 \int_0^t g\,dt\,(\sigma_1 \sin\Theta + \sigma_3 \cos\Theta)}.$$

Thus, by (2.17), definition (2.21) of $\varepsilon[L_0, M_0]$, and definition (2.4) of f_0 and g_0, the eigenvalues $\pm 2\pi\varepsilon[L_0, M_0]$ are approximately the eigenvalues of the self-adjoint matrix with zero trace given by

$$f_0\sigma_3 + g_0(\sigma_1 \sin\Theta + \sigma_3 \cos\Theta).$$

Now, by (1.7) and (2.13), its square is

$$(f_0 + g_0 \cos\Theta)^2 + g_0^2 \sin^2\Theta = f_0^2 + 2\frac{M_0}{L_0} f_0 g_0 + g_0^2;$$

hence the approximate expression $(2.4)_2$ for $\varepsilon[L_0, M_0]$ follows.

Remark 2.2. The quantum numbers

$$l, \quad m, \quad n, \quad \pm 1$$

become those used in the classical solutions of the Dirac equation by imposing a condition

- stricter than (2.6)
- less strict than the condition that, on account of (2.7), would result from the choice $l' = m' = 0$, namely,

$$|m| \leqslant l - \tfrac{1}{2}, \qquad l < n. \tag{2.6)*$$

This condition is as follows.

Definition 2. In III,§1, we solved a system in one unknown, analogous to the system (2.2) when l' and m' are chosen to be zero. In III,§1, the lagrangian amplitude of the unknown is necessarily constant. It is in conformity with the nature of the amplitude α and of the phase $v_0\varphi_R$ in physics: α must vary more slowly than $e^{v_0\varphi_R}$. Let us make the present situation as similar as possible: consider a *quadruple*

$$l, \quad m, \quad n, \quad \pm 1$$

satisfying the condition $(2.5)_3$ *for the existence of a solution of* (2.2) *that is lagrangian on* $V[L_0, M_0]$; this condition is independent of (l', m'); require that at (l', m') the function

$$(l', m') \mapsto |c_{\pm 1} - 1|^2 + |c_{\pm 2} - 1|^2 + |c_{\pm 3} + 1|^2 \in \mathbf{R}$$

takes on values close to its minimum; in other words, by (2.9) and since l'^2 and m'^2 are negligible in comparison with l' and m', at (l', m') the function

$$(l', m') \mapsto l'^2 + m'^2 + [l'N_L + m'N_M \mp \varepsilon]^2$$

takes on values close to its minimum; if such a choice, namely,

$$(l', m') \text{ near } \pm \frac{\varepsilon}{1 + N_L^2 + N_M^2}(N_L, N_M),$$

satisfies the condition (2.7), then and only then will we say that the quadruple $(l, m, n, \pm 1)$ is *admissible.*

In §2 (Dirac equation), the following case arises.

Example 2. (N_L, N_M) is near $(-1, 0)$; thus (l', m') is near $(\mp \varepsilon/2, 0)$. By (2.7), the admissible quadruples $(l, m, n, \pm 1)$ are those that satisfy condition $(2.5)_3$ and the condition

$$0 \leqslant l < n, \qquad |m| \leqslant l \pm \tfrac{1}{2} \tag{2.23}$$

(the signs correspond).

It is customary in quantum mechanics to let

$$j = l \pm \tfrac{1}{2}$$

and to use *quadruples of quantum numbers*

$$(l, m, n, j = l \pm \tfrac{1}{2}).$$

They are admissible if they satisfy $(2.5)_3$ *and*

$$0 \leqslant l < n, \qquad |m| \leqslant j. \tag{2.24}$$

Then for each admissible choice of (l', m') there correspond solutions U of (2.2) that are equal up to a proportionality factor. This factor is a constant $\in \mathbf{C}$.

§2. The Dirac Equation

0. Summary

The Dirac equation is a system in 4 unknowns. Theorems 2.1 and 2.2 of II,§4 do not apply in this case because the zeros of $\det a_0^0$ are double zeros.

Theorem 1, whose statement resembles that of theorem 2.2 of II,§4, reduces, $\operatorname{mod}(1/v^2)$, this system in 4 unknowns to a *self-adjoint* system in 2 unknowns.

By suppressing terms that are negligible in view of the order of magnitude of the magnetic field, this system is transformed in section 2 into the *reduced Dirac equation*, which is a system of the form solved in §1.

In section 3, we observe that *the energy levels defined by this reduced Dirac equation in lagrangian analysis are those defined by the classical resolution* of the Dirac equation, even when the magnetic field is strong enough to produce the *Paschen-Back effect.*

But the probability of the presence of an electron obtained in section 4 differs from that obtained in wave mechanics; it is connected with the first quantum theory.

1. Reduction of the Dirac Equation in Lagrangian Analysis

Suppose we are given two infinitely differentiable mappings and a constant:

$$A:\mathbf{E}^3\backslash\{0\} \to \mathbf{R}_+; \qquad B:\mathbf{E}^3 \to \mathbf{E}^3; \qquad C \in \mathbf{R}_+.$$

Let a' and a'' be the two 2×2 matrices of lagrangian operators whose expressions in $Z(3) = \mathbf{E}^3 \oplus \mathbf{E}^3$ are

$$a' = A(x) + \sigma\left[\frac{1}{v}\frac{\partial}{\partial x} + B(x)\right], \qquad a'' = A(x) - \sigma\left[\frac{1}{v}\frac{\partial}{\partial x} + B(x)\right],$$

where σ is the 2×2 matrix defined by (1.6) of §1; a' and a'' are self-adjoint. They are associated to the self-adjoint matrices

$A(x) + \sigma[p + B(x)]$ and $A(x) - \sigma[p + B(x)]$.

The Dirac equation is the system

$$a'U' = CU'', \qquad a''U'' = CU', \tag{1.1}$$

in which the unknowns U' and U'' are vectors. Let us require that the two components of each of these vectors be *lagrangian functions, the four of them defined on a single* ***compact*** *lagrangian manifold V in $Z(3) = \mathbf{E}^3 \oplus \mathbf{E}^3$.*

The Dirac equation (1.1) is evidently equivalent to

i. the system

$$[C^2 - a'' \circ a']U' = 0 \tag{1.2}$$

in the unknown U',

ii. the system

$$[C^2 - a' \circ a'']U'' = 0 \tag{1.3}$$

in the unknown U''.

The calculation of $C^2 - a'' \circ a'$ and $C^2 - a' \circ a''$ is easy and standard. In order to state the result, let

$$\begin{aligned} H(x, p) &= \frac{1}{2}|p + B(x)|^2 + \frac{C^2}{2} - \frac{1}{2}A^2(x), \\ J'(x, p) &= \frac{i}{2}\sigma\left[\frac{\partial}{\partial x} \wedge B(x)\right] + \frac{1}{2}\sigma[A_x], \\ J'' &= \frac{i}{2}\sigma\left[\frac{\partial}{\partial x} \wedge B\right] - \frac{1}{2}\sigma[A_x], \end{aligned} \tag{1.4}$$

where

$$\frac{\partial}{\partial x} \wedge B = \operatorname{curl} B;$$

$\frac{1}{2}[C^2 - a'' \circ a']$ is the matrix associated to the 2×2 matrix

$$H + \frac{1}{\nu}J';$$

$\frac{1}{2}[C^2 - a' \circ a'']$ is the matrix associated to the 2×2 matrix

$$H + \frac{1}{v} J''.$$

Remark 1.1. These matrices are *not self-adjoint*; but one is the adjoint of the other.

Notation. Let W be the hypersurface in $\mathbf{E}^3 \oplus \mathbf{E}^3$ given by the equation

$$W: H(x, p) = 0. \tag{1.5}$$

Let β' and β'' be the lagrangian amplitudes of U' and U'':

$$\beta' : V \to \mathbf{C}^2, \qquad \beta'' : V \to \mathbf{C}^2.$$

By theorem 4 of II,§3, equation (1.2) is equivalent, $\mathrm{mod}(1/v^2)$, to the conditions:

$$V \subset W$$

$$\frac{d\beta'}{dt} + J'\beta' = 0, \tag{1.6}$$

where d/dt is the Lie derivative $\mathscr{L}_\kappa$ in the direction of the characteristic vector κ of V. Similarly, (1.3) is equivalent, $\mathrm{mod}(1/v^2)$, to the conditions: $V \subset W$

$$\frac{d\beta''}{dt} + J''\beta'' = 0. \tag{1.7}$$

By (1.1), β' and β'' satisfy the two equivalent relations

$$\{A(x) + \sigma[p + B(x)]\}\beta' = C\beta'', \qquad \{A - \sigma[p + B]\}\beta'' = C\beta' \tag{1.8}$$

on V. The equivalence of (1.2) and (1.3) proves the equivalence of the equations (1.6) and (1.7). Relation (1.8) evidently transforms the solutions of one of these equations into the solutions of the other.

The definition (1.5) of W means that there exists a function

$$\rho : W \to \mathbf{R}_+$$

such that

$$A = C \cosh 2\rho, \qquad |p + B| = C \sinh 2\rho. \tag{1.9}$$

Let

$$\tau = \frac{1}{C \sinh 2\rho}\sigma[p + B], \qquad \text{so that } \tau^2 = 1 \tag{1.10}$$

by (1.8) of §1. It follows that

$$Ce^{\pm 2\rho\tau} = A \pm \sigma[p + B].$$

Relation (1.8) connecting β' and β'' is then written

$$\beta'' = e^{2\rho\tau}\beta'.$$

It means that there exists a function

$$\beta : V \to \mathbf{C}^2$$

such that

$$\beta' = e^{-\rho\tau}\beta, \qquad \beta'' = e^{\rho\tau}\beta.$$

The equivalent relations (1.6) and (1.7) may then be written

$$\frac{d}{dt} + J\beta = 0, \tag{1.11}$$

where the 2×2 matrix J is given by

$$J = e^{\rho\tau}\frac{d}{dt}(e^{-\rho\tau}) + e^{\rho\tau}J'e^{-\rho\tau} = e^{-\rho\tau}\frac{d}{dt}(e^{\rho\tau}) + e^{-\rho\tau}J''e^{\rho\tau}. \tag{1.12}$$

Let us show how this definition of J is equivalent to (1.14), which makes the following theorem evident.

THEOREM 1. *Solving the systems* (1.2) *and* (1.3) (*which are equivalent to the Dirac equation*) $\operatorname{mod}(1/v^2)$ *on* V *is equivalent to solving the system*

$$aU = 0 \ \operatorname{mod}\frac{1}{v^2} \tag{1.13}$$

in two unknowns, where a *is the matrix of lagrangian operators associated to the* 2×2 *matrix*

$$H + \frac{1}{v}J.$$

H is defined by $(1.4)_1$ *and J by*

$$J(x, p) = \frac{i}{2}\sigma\left[\frac{\partial}{\partial x} \wedge B\right] + \frac{i}{2}\frac{1}{A + C}\sigma[(p + B) \wedge A_x]. \tag{1.14}$$

Remark 1.2. $\frac{1}{v}J$, and thus a, are *self-adjoint*, which shows by remark 1.1 the advantage of substituting (1.13) for (1.2)–(1.3).

Proof of (1.14). By the definitions (1.4) and (1.12) of J', J'', and J, it suffices to prove the following formulas, where $\langle\cdot,\cdot\rangle$ is the scalar product in $\mathbf{E}^3$ and $\operatorname{curl} B = \partial/\partial x \wedge B$:

$$\begin{aligned}\frac{C}{2}\left\{e^{\rho\tau}\frac{d}{dt}(e^{-\rho\tau}) + e^{-\rho\tau}\frac{d}{dt}(e^{\rho\tau})\right\} = &-\frac{i}{2}(A - C)\sigma[\operatorname{curl} B] \\ &-\frac{i}{2}\frac{A}{A + C}\sigma[(p + B) \wedge A_x] \\ &+\frac{i}{2}\langle p + B, \operatorname{curl} B\rangle\tau\tanh\rho\end{aligned} \tag{1.15}$$

$$\begin{aligned}&\frac{iC}{4}\{e^{\rho\tau}\sigma[\operatorname{curl} B]e^{-\rho\tau} + e^{-\rho\tau}\sigma[\operatorname{curl} B]e^{\rho\tau}\} \\ &= \frac{i}{2}A\sigma[\operatorname{curl} B] - \frac{i}{2}\langle p + B, \operatorname{curl} B\rangle\tau\tanh\rho,\end{aligned} \tag{1.16}$$

$$\frac{C}{4}\{e^{\rho\tau}\sigma[A_x]e^{-\rho\tau} - e^{-\rho\tau}\sigma[A_x]e^{\rho\tau}\} = \frac{i}{2}\sigma[(p + B) \wedge A_x]. \tag{1.17}$$

Remark 1.3. The Pauli relations [(1.7) of §1] may be written

$$\sigma[x]\sigma[y] = \langle x, y\rangle + i\sigma[x \wedge y] \qquad \forall x, y \in \mathbf{E}^3. \tag{1.18}$$

It follows, in particular, that

$$\sigma[x]\sigma[y] + \sigma[y]\sigma[x] = 2\langle x, y\rangle, \tag{1.19}$$

$$\sigma[x]\sigma[y] - \sigma[y]\sigma[x] = 2i\sigma[x \wedge y]. \tag{1.20}$$

Proof of (1.15). Since $\tau^2 = 1$ and ρ is a number,

$$e^{\pm\rho\tau} = \cosh\rho \pm \tau\sinh\rho.$$

Hence

$$e^{\mp\rho\tau}\frac{d}{dt}(e^{\pm\rho\tau}) = e^{\mp\rho\tau}\left[\frac{d\rho}{dt}\sinh\rho \pm \tau\frac{d\rho}{dt}\cosh\rho \pm \frac{d\tau}{dt}\sinh\rho\right]$$
$$= \pm\tau\frac{d\rho}{dt} \pm e^{\mp\rho\tau}\frac{d\tau}{dt}\sinh\rho,$$

therefore

$$\frac{1}{2}\left[e^{\rho\tau}\frac{d}{dt}(e^{-\rho\tau}) + e^{-\rho\tau}\frac{d}{dt}(e^{\rho\tau})\right] = -\tau\frac{d\tau}{dt}\sinh^2\rho. \tag{1.21}$$

In order to calculate $\tau(d\tau/dt)$, differentiate the definition (1.10) of τ:

$$C\frac{d\tau}{dt}\sinh 2\rho + 2C\tau\frac{d\rho}{dt}\cosh 2\rho = \sigma\left[\frac{d}{dt}(p + B)\right].$$

It follows by (1.9) and (1.10) that

$$C^2\tau\frac{d\tau}{dt}\sinh^2 2\rho = -\frac{1}{2}\frac{dA^2}{dt} + \sigma[p + B]\sigma\left[\frac{d}{dt}(p + B)\right],$$

whence by (1.18) and considering that $A^2 = C^2 + |p + B|^2$ on W,

$$C^2\tau\frac{d\tau}{dt}\sinh^2 2\rho = i\sigma\left[(p + B) \wedge \frac{d}{dt}(p + B)\right]. \tag{1.22}$$

By the definition of d/dt [(3.10) of II,§3],

$$\frac{dx}{dt} = H_p(x, p), \qquad \frac{dp}{dt} = -H_x(x, p);$$

hence, by some easy calculations,

$$\frac{d}{dt}(p + B) = -(p + B) \wedge \operatorname{curl} B + AA_x,$$

$$(p + B) \wedge \frac{d}{dt}(p + B) = |p + B|^2 \operatorname{curl} B - \langle p + B, \operatorname{curl} B\rangle(p + B)$$
$$+ A(p + B) \wedge A_x.$$

Therefore, since

$$2C\cosh^2\rho = C[\cosh 2\rho + 1] = A + C, \qquad |p + B|^2 = A^2 - C^2,$$

(1.22) may be written

$$2C\tau\frac{d\tau}{dt}\sinh^2\rho = i(A - C)\sigma[\operatorname{curl} B] - i\langle p + B, \operatorname{curl} B\rangle\tau\tanh\rho$$
$$+ i\frac{A}{A + C}\sigma[(p + B) \wedge A_x]$$

by (1.10); (1.15) follows by (1.21).

Proof of (1.16) *and* (1.17). Let $y \in \mathbf{E}^3$;

$$e^{\pm\rho\tau}\sigma[y]e^{\mp\rho\tau} = \sigma[y]\cosh^2\rho \pm \tfrac{1}{2}\{\tau\sigma[y] - \sigma[y]\tau\}\sinh 2\rho - \tau\sigma[y]\tau\sinh^2\rho. \quad (1.23)$$

Now by the definition (1.10) of τ and the commutation formula (1.20),

$$\frac{C}{2}\{\tau\sigma[y] - \sigma[y]\tau\}\sinh 2\rho = i\sigma[(p + B) \wedge y]. \quad (1.24)$$

By (1.10) and (1.19),

$$\tau\sigma[y] = \frac{2}{C\sinh 2\rho}\langle p + B, y\rangle - \sigma[y]\tau,$$

whence

$$\tau\sigma[y]\tau = \frac{2}{C\sinh 2\rho}\langle p + B, y\rangle\tau - \sigma[y]. \quad (1.25)$$

By (1.24) and (1.25), (1.23) may be written

$$Ce^{\pm\rho\tau}\sigma[y]e^{\mp\rho\tau} = A\sigma[y] - \langle p + B, y\rangle\tau\tanh\rho \pm i\sigma[(p + B) \wedge y],$$

whence the two formulas

$$\frac{C}{2}\{e^{\rho\tau}\sigma[y]e^{-\rho\tau} + e^{-\rho\tau}\sigma[y]e^{\rho\tau}\} = A\sigma[y] - \langle p + B, y\rangle\tau\tanh\rho,$$

$$\frac{C}{2}\{e^{\rho\tau}\sigma[y]e^{-\rho\tau} - e^{-\rho\tau}\sigma[y]e^{\rho\tau}\} = i\sigma[(p + B) \wedge y],$$

which prove (1.16) and (1.17), respectively.

2. The Reduced Dirac Equation for a One-Electron Atom in a Constant Magnetic Field

Let us choose A, B, and C such that the function H defined by $(1.4)_1$ is the hamiltonian of such an electron: H has to be the function defined by the formulas (4.15) and (4.18) of III,§1, up to the factor μ. Thus,

A is a function of R, the function $R \mapsto RA(R)$ is *affine*,

$B = \frac{1}{2}(-bx_2, bx_1, 0)$, where $b \in \mathbf{R}$.

In the expression $(1.4)_1$ for H, neglect B^2, as is done in (4.20) of III,§1. In the expression (1.14) for J, similarly neglect the term

$$\frac{i}{2}\frac{1}{A + C}\sigma[B \wedge A_x] \text{ in comparison with } \frac{i}{2}\left[\frac{\partial}{\partial x} \wedge B\right],$$

because, on V, $xA_x/(A + C)$ will be negligible in comparison with 1. Then the expressions (1.4) and (1.14) for H and J reduce to

$$H(x, p) = \tfrac{1}{2}[P^2 + bM + C^2 - A^2(R)], \tag{2.1}$$

where b and $C \in \mathbf{R}$ and $R \mapsto RA(R)$ is affine,

$$J(x, p) = \frac{i}{2}b\sigma_3 - \frac{i}{2}\frac{A_R}{R(A + C)}\sigma[x \wedge p]. \tag{2.2}$$

The matrix of lagrangian operators associated to the matrix

$$H + \frac{1}{v}J,$$

where H and J are defined by (2.1) and (2.2), will be denoted by a_r and called the *reduced Dirac operator*.

Remark 2. Thus H, defined by (2.1), becomes the function defined by formula (4.20) of III,§1, up to the factor μ. It follows that

$$A = \frac{1}{c}\left(E + \frac{\varepsilon^2 Z}{R}\right), \qquad b = \frac{\varepsilon\mathscr{H}}{c}, \qquad C = \mu c,$$

where

- c is the speed of light,
- μ and ε are the mass and charge of the electron,
- Z is the atomic number of the nucleus,
- E is the energy level of the atom, which is close to μc^2,
- $\mathscr{H}$ is the intensity of the magnetic field.

We replace the study of the Dirac equation by the study of the system

$$a_r U = \left(a_{L^2} - L_0^2 - \frac{2i}{v} L_0 l'\right) U = \left(a_M - M_0 - \frac{i}{v} m' + \frac{i}{2v} \sigma_3\right) U = 0, \tag{2.3}$$

where the vector U has two components, which are lagrangian functions defined on a compact lagrangian manifold V in $\mathbf{E}^3 \oplus \mathbf{E}^3$, and where l' and m' are real and have negligible squares.

Theorem 2 of §1, can be applied to system (2.3); its statement becomes more precise by means of the following two lemmas.

LEMMA 2.1. An approximate value for the function ε defined by $(2.4)_3$ of §1 is

$$\varepsilon[L_0, M_0] = \frac{1}{2}\sqrt{(1 + N_L)^2 + 2\frac{M_0}{L_0}(1 + N_L)N_M + N_M^2}, \tag{2.4}$$

where

$$N_L = N_L[L_0, M_0], \qquad N_M = N_M[L_0, M_0].$$

Proof. Let us explicitly give the values of f_0 and g_0, defined by $(2.4)_1$ and $(2.4)_2$ in §1. By (2.1),

$$H[L, M, Q, R] = \frac{L^2 + Q^2}{2R^2} + \frac{b}{2}M + \tfrac{1}{2}C^2 - \tfrac{1}{2}A^2(R), \tag{2.5}$$

whence

$$H_L = \frac{L}{R^2}, \qquad H_M = \frac{b}{2};$$

now by (2.3) of III,§2,

$$\lambda_t = -H_L, \qquad \mu_t = -H_M;$$

by (1.5) of III,§2, we have, writing Γ for $\Gamma[L_0, M_0]$,

$$\int_\Gamma \lambda_t \, dt = \Delta_t \lambda = 2\pi N_L, \qquad \int_\Gamma \mu_t \, dt = \Delta_t \mu = 2\pi N_M;$$

hence,

$$\int_\Gamma \frac{L_0}{R^2} dt = -2\pi N_L, \qquad \int_\Gamma \frac{b}{2} dt = -2\pi N_M; \tag{2.6}$$

the last formula and formula $(2.4)_1$ of §1, where $f = b/2$ by the choice (2.2) of J, give

$$f_0 = -\pi N_M. \tag{2.7}$$

By the choice (2.2) of J and the definition §1, $(2.4)_2$ of g_0,

$$g_0 = -\frac{L_0}{2}\int_\Gamma \frac{A_R}{R(A + C)}dt; \tag{2.8}$$

now by (2.14) of III,§1 and (2.5), we have, on Γ,

$$dt = \frac{dR}{RH_Q} > 0, \qquad H_Q = \frac{Q}{R^2}; \qquad \text{thus } dt = \frac{R\,dR}{Q} > 0; \tag{2.9}$$

hence relations (2.8) and $(2.6)_1$ may be written

$$g_0 = -\frac{1}{2}\int_\Gamma \frac{L_0}{Q}\frac{A_R dR}{A + C}, \qquad 2\pi N_L = -\int_\Gamma \frac{L_0}{Q}\frac{dR}{R}. \tag{2.10}$$

Let us calculate an approximate value for g_0, namely, its value for $b = 0$. By (2.5), in that case the equation of Γ is

$$\Gamma: Q^2 = A'(R)A''(R) - L_0^2,$$

where

$$A'(R) = RA(R) + CR, \qquad A''(R) = RA(R) - CR; \tag{2.11}$$

A' and A'' are affine functions; by remark 2, A' is increasing and $A'(R) > 0$ for $R > 0$; it follows by an easy calculation using residues that

$$\int_\Gamma \frac{L_0}{Q}\frac{A'_R}{A'}dR = 2\pi; \tag{2.12}$$

now by $(2.11)_1$,

$$\frac{A'_R}{A'} = \frac{1}{R} + \frac{A_R}{A + C};$$

then the relations (2.10) and (2.12) prove that

$$g_0 = -\pi(1 + N_L). \tag{2.13}$$

Formulas (2.7) and (2.13) and formula $(2.4)_3$ of §1 prove the lemma.

The assumptions made in §1,2, including those of §1, example 2, are satisfied, as it is proved by the following lemma.

LEMMA 2.2. The values of the physical quantities defining A, B, and C (remark 2) are such that for the energy levels E used in section 3,

$$f_0 = -\pi N_M, \quad g_0 = -\pi(1 + N_L), \quad N_{L^2}, \quad N_{LM}, \quad N_{M^2}$$

are small in comparison with 1.

Proof. $N[\cdot,\cdot]$ is expressed by (4.8) of III,§1, where A_0, B_0, and C_0 are defined by identifying formulas (4.6) and (4.20) in III,§1:

$$\frac{1}{\hbar}N[L_0, M_0] = \alpha Z\left[\frac{\mu c^2(\mu c^2 + 2\beta\mathscr{H}M_0/\hbar)}{E^2} - 1\right]^{-1/2} - \left[\left(\frac{L_0}{\hbar}\right)^2 - \alpha^2 Z^2\right]^{1/2}, \tag{2.14}$$

where Z, c, μ, ε, E, and $\mathscr{H}$ are the physical quantities defined by remark 2,

$$\alpha = \frac{\varepsilon^2}{\hbar c} \simeq \frac{1}{137} \text{ is the dimensionless } \textit{fine-structure constant} \tag{2.15}$$

($\simeq$ means of the order of magnitude of), and

$$\beta = \frac{\varepsilon\hbar}{2\mu c} \text{ is the } \textit{Bohr magneton.} \tag{2.16}$$

In section 3, we choose

- $0 < (\mu c^2/E) - 1 \simeq \alpha^2 Z^2/(2n^2)$, where n is an integer,
- the magnetic energy $\beta\mathscr{H}$ to be very small in comparison with μc^2, and
- $2M_0/\hbar$ to have integer values which are not large,
- $L_0/\hbar > \frac{1}{2}$.

The lemma follows.

3. The Energy Levels

Notation. Formulas (4.23) and (4.25) of III,§1 have already defined and used the function F given by

$$F(n, k) = \frac{1}{\sqrt{1 + \left(\dfrac{\alpha Z}{n - k + \sqrt{k^2 - \alpha^2 Z^2}}\right)^2}}. \tag{3.1}$$

Note the sign of its derivatives:

$$F_n > 0, \qquad F_k > 0.$$

Using F, it is possible to give the relation

$$\hbar k + N[\hbar k, \hbar m] = \hbar n, \tag{3.2}$$

where N is expressed by (2.14), the form

$$E^2 = \mu c^2[\mu c^2 + 2\beta\mathscr{H}m]F^2(n, k).$$

In other words, to the degree of approximation used here, and assuming $E > 0$,

$$E = \mu c^2 F(n, k) + \beta\mathscr{H}m. \tag{3.3}$$

In these formulas, α is the *fine-structure constant* (2.15) and β is the *Bohr magneton* (2.16).

THEOREM 3. *The energy levels E for which system* (2.3) (*where l' and m' are real and have negligible squares*) *has* ***admissible*** *lagrangian solutions* (*definition* 2 *of* §1) *on a* ***compact*** *lagrangian manifold are defined by the quadruples of quantum numbers*

$$j = l \pm \tfrac{1}{2}, \quad m, \quad n \tag{3.4}$$

such that

$$l,\ m - \tfrac{1}{2},\ \text{and } n \text{ are integers} \tag{3.5}$$

and

$$|m| \leqslant j, \qquad 0 \leqslant l < n. \tag{3.6}$$

Up to some negligible quantities, E is expressed as a function of these quantum numbers as follows:

$$E = \mu c^2 F\left(n, l + \frac{1}{2}\right) \pm \frac{\mu c^2}{2}\sqrt{F_k^2 + \frac{4m}{2l+1}F_k\frac{\beta\mathscr{H}}{\mu c^2} + \frac{\beta^2\mathscr{H}^2}{\mu^2 c^4}} + \beta\mathscr{H}m. \tag{3.7}$$

The $\pm$ sign is the same in (3.4) *and* (3.7).

Remark 3.1. For $\mathscr{H} = 0$, since $F_k > 0$, (3.7) reduces to

$$E = \mu c^2 F(n, k), \tag{3.8}$$

where

$$k = j + \tfrac{1}{2} \in \mathbf{Z}. \tag{3.9}$$

Proof. By theorem 2 of §1, the energy levels E such that the system (2.3) has solutions defined on compact lagrangian manifolds are those that satisfy condition (2.5) of §1. From the approximate value of $\varepsilon[L_0, M_0]$ given by lemma 2.1, these values of E are approximately those satisfying the condition

$$\hbar(l + \tfrac{1}{2}) + N[\hbar(l + \tfrac{1}{2}), \hbar m]$$
$$= \hbar n \pm \frac{\hbar}{2}\sqrt{(1 + N_L)^2 + \frac{4m}{2l + 1}(1 + N_L)N_M + N_M^2}, \tag{3.10}$$

$|m| \leqslant l + \tfrac{1}{2}, l < n;$ $\quad l, m - \tfrac{1}{2}, n$ integers; $\quad 1 + N_L$ and N_M small.

Then by the equivalence of the relations (3.2) and (3.3), a crude approximation for E is

$$E \simeq \mu c^2 F(n, l + \tfrac{1}{2}) \simeq \mu c^2\left[1 - \frac{\alpha^2 Z^2}{2n^2}\right], \tag{3.11}$$

which is sufficient to justify the use of lemma 2.2.

From example 2 of §1, admissible solutions of the system (2.3) correspond to a choice of l, m, n, and E satisfying (3.10) if and only if

$$|m| \leqslant l \pm \tfrac{1}{2},$$

which amounts to (3.6) and (3.4), where the $\pm$ sign is the same in (3.4) and (3.10).

Let us now express E as a function of (l, m, n).

Since (3.2) is equivalent to (3.3), on the one hand (3.10) may be written

$$E = \mu c^2 F\left(n \pm \tfrac{1}{2}\sqrt{(1 + N_L)^2 + \frac{4m}{2l + 1}(1 + N_L)N_M + N_M^2}, l + \tfrac{1}{2}\right)$$
$$+ \beta\mathscr{H} m; \tag{3.12}$$

on the other hand, the following two relations are equivalent (for all dk, dm, dn):

$$(1 + N_L)dk + N_M dm = dn, \qquad \mu c^2(F_n dn + F_k dk) + \beta\mathscr{H}\, dm = 0.$$

This equivalence means that

$$\mu c^2 F_n = -\frac{\mu c^2 F_k}{1 + N_L} = -\frac{\beta \mathscr{H}}{N_M},$$

whence since $F_n > 0$,

$$F_n \sqrt{(1 + N_L)^2 + \frac{4m}{2l + 1}(1 + N_L)N_M + N_M^2}$$

$$= \sqrt{F_k^2 + \frac{4m}{2l + 1} F_k \frac{\beta \mathscr{H}}{\mu c^2} + \frac{\beta^2 \mathscr{H}^2}{\mu^2 c^4}}.$$

Thus (3.12) is equivalent to (3.7).

Remark 3.2. The energy levels obtained by theorem 3 are exactly those obtained by the classical theory of the Dirac atom: see, for example, Bethe and Salpeter [2]. For $\mathscr{H} = 0$, their formula (14.29), p. 68, can be identified with our formula (3.8) and, for $\mathscr{H} \neq 0$, their formulas (46.12), (46.13), and (46.15), p. 211, can be identified with our formula (3.7): see table for the correspondence between their notation and ours.

Table

Bethe-Salpeter [2]	Leray (Remark 2 and Section 3)
$E_0 = mc^2$	μc^2
E_+	$\mu c^2 F(n, l + 1)$
E_-	$\mu c^2 F(n, l)$
$\frac{1}{2}(E_+ + E_-)$	$\mu c^2 F(n, l + \frac{1}{2})$
$\Delta E = E_+ - E_-$	$\mu c^2 F_k$
μ_0	β
$\xi = \mathscr{H}\mu_0/\Delta E$	$\beta\mathscr{H}/(\mu c^2 F_k)$
$E' = \Delta E\left[\xi m \pm \frac{1}{2}\sqrt{1 + \xi\frac{4m}{2l + 1} + \xi^2}\right]$	$\beta\mathscr{H}m \pm \frac{\mu c^2}{2}\sqrt{F_k^2 + \frac{4m}{2l + 1}F_k\frac{\beta\mathscr{H}}{\mu c^2} + \frac{\beta^2\mathscr{H}^2}{u^2 c^4}}$

Remark 3.3. When $\beta\mathscr{H}$ is small in comparison with $\mu c^2 F_k$, formula (3.7) evidently may be written

$$E = \mu c^2 F(n, k) + \beta \mathscr{H} gm, \quad \text{where } g = (j + \tfrac{1}{2})(l + \tfrac{1}{2});$$

g is *Landé's factor* [see Bethe-Salpeter [2], (46.4) and (46.6) p. 209].

4. Crude Interpretation of the Spin in Lagrangian Analysis

The crude approximation consists of neglecting $\beta\mathscr{H}/(\mu c^2)$ in comparison with α^2; α^2 itself, and hence F_k, in comparison with 1.

Remark 4.1. Wave mechanics suceeds in evaluating the energy levels of the *helium atom* at the cost of this *crude approximation*: see a summary of these very laborious calculations in Bethe-Salpeter [2], chapter II.

In the equation $H = 0$, this approximation identifies the hamiltonian H of the relativistic electron [(4.20) of III,§1] with the nonrelativistic hamiltonian [(4.19) of III,§1], where $\mathscr{H} = 0$. In this second hamiltonian, E denotes the nonrelativistic energy.

The crude statement of theorem 3 is the following.

THEOREM 4.1. *Let* $(j = l \pm \frac{1}{2}, l, m, n)$ *be a quadruple of quantum numbers defining a system* (2.3) *that has* ***admissible*** *lagrangian solutions on a* ***compact*** *lagrangian manifold* V; *it is a quadruple satisfying* (3.5)–(3.6). *The crude expressions for the nonrelativistic energy* E *and for* L_0 *and* M_0 *as functions of this quadruple are*

$$E = -\mu c^2 \frac{\alpha^2 Z^2}{2n^2}, \qquad L_0 = \hbar(l + \tfrac{1}{2}), \qquad M_0 = \hbar m. \tag{4.1}$$

The crude equations for V *are*

$$V\colon H = 0, \textit{ where } H = \frac{P^2}{2\mu} - E - \frac{\varepsilon^2 Z}{R}, \qquad L = \hbar(l + \tfrac{1}{2}), \qquad M = \hbar m. \tag{4.2}$$

Remark 4.2. In the case

$$|m| = j = l + \tfrac{1}{2},$$

we have, by (4.2) and the definition of L and M (III,§1,1), that, on V,

$$|M| = L; \text{ thus } x_3 = p_3 = 0;$$

whence

$\dim V = 2$, *contrary* to the definition that $\dim V = 3$.

In order to give a meaning to this result, other than by reintroducing nonzero l' and m', one perhaps has to use the notion of a lagrangian distribution and to refer to some of the papers of Voros [25], [26].

Notation. Let $SO(3)$ be the group of rotations $\rho : \mathbf{E}^3 \to \mathbf{E}^3$ leaving the origin invariant. Let it act on $\mathbf{E}^3 \oplus \mathbf{E}^3$ as follows:

$$\rho(x, p) = (\rho x, \rho p), \text{ where } (x, p) \in \mathbf{E}^3 \oplus \mathbf{E}^3.$$

Let $SO(2)$ be the subgroup of rotations leaving I_3 (III,§1,1) that is, the direction of the magnetic field, invariant. $SO(2)$ evidently leaves V invariant. More precisely, by sections 1 and 2 of I,§1,

$$V = SO(2) \times T^2,$$

where T^2 is a 2-dimensional torus, every point of which is left invariant by $SO(2)$. Recall (remark 1.1 of §1) that *the universal covering group of* $SO(3)$ *is* $SU(2)$, *which is a covering group of order* 2. Let $SU(1)$ be the covering group of $SO(2)$ of order 2. Then we have the commutative diagram

$$\begin{array}{ccc} SU(1) & \hookrightarrow & SU(2) \\ \downarrow & & \downarrow \\ SO(2) & \hookrightarrow & SO(3) \end{array}$$

where the vertical arrows represent natural projections of covering groups; the horizontal arrows are inclusions. Let

$$V_2 = SU(1) \times T^2.$$

V_2 *is a covering space of* V *of order* 2; $SU(1)$ acts on V_2, leaving each point of T^2 invariant.

Definition 4. Let V be a lagrangian manifold, φ_{R_0} its phase in a 2-frame R_0, and β a function $\check{V} \to \mathbf{C}$. By theorem 2.2 of II,§2,

$$\check{U}_{R_0} = \left(\frac{\eta}{d^l x}\right)^{1/2} \beta e^{\nu\varphi_{R_0}}$$

is the expression in R_0 of a lagrangian function $\check{U}$ defined mod$(1/\nu)$ on $\check{V}$. The definition of a lagrangian function on V (definition 3.2 of II,§2) may be generalized as follows: Let $\tilde{V}$ be any one of the *covering spaces of* V; $\check{U}$ is said to be *lagrangian on* $\tilde{V}$ when the restriction of $\check{U}_{R_0}$ to v_0.

$$\check{U}^0_{R_0} = \left(\frac{\eta}{d^l x}\right)^{1/2} \beta e^{\nu_0 \varphi_{R_0}},$$

is a mapping $\tilde{V}\backslash\tilde{\Sigma}_{R_0} \to \mathbf{C}$.

The calculations that establish the conditions (3.11) in III,§1 prove the following theorem.

THEOREM 4.2. *The conditions* (3.4)–(3.6), *which are satisfied by the quadruples of quantum numbers defining solutions of* (2.3), *express the existence of a function with constant lagrangian amplitude defined* mod$(1/\nu)$ *on* V_2 *but not on* V.

Conclusion. Let V be the lagrangian manifold given by equation (4.2). Let $\check{U}$ be a function with constant lagrangian amplitude defined mod$(1/\nu)$ on $\check{V}$. *The following is a crude interpretation of the spin*:

- $\check{U}$ *is lagrangian on* V *in the spin* 0 *case* (*Schrödinger*)*;*
- $\check{U}$ *is lagrangian on* V_2 (and not on V) *in the spin* $\frac{1}{2}$ *case* (*Dirac*).

5. The Probability of the Presence of the Electron

Recall that x is the position and p the momentum of an electron; thus

$(x, p) \in \mathbf{E}^3 \oplus \mathbf{E}^3 = X \oplus X^*$.

With the conventions of remark 2, let

$$R_1 = \frac{\hbar^2}{\mu\varepsilon^2},$$

which is the length called the *radius of the Bohr atom* in the first quantum theory. Let

$$x' = \frac{Zx}{R_1}, \qquad R' = \frac{ZR}{R_1},$$

where Z is the atomic number of the nucleus.

THEOREM 5. *The position* x *of the electron with quantum numbers* ($j = l \pm \frac{1}{2}, l, m, n$) *stays in the projection* V_X *of* $V[L_0, M_0]$ *onto* X*;* V_X *is a subset of* X *lying between two spheres and outside of a cone of revolution around the magnetic field; the center of these spheres and the vertex of this cone coincide with the nucleus of the atom.*

Assume that on $V[L_0, M_0] \subset \mathbf{E}^3 \oplus \mathbf{E}^3$ *the probability of the presence of the electron is proportional to the invariant measure* η_V [*see the con-*

clusion of the introduction to chapter III]. *Then on* V_X *the probability of the presence of the electron is crudely*

$$\frac{1}{2\pi^3}\frac{l+\frac{1}{2}}{n^2}\frac{d^3x'}{\sqrt{-R'^2+2n^2R'-(l+\frac{1}{2})^2n^2}\sqrt{(l+\frac{1}{2})^2(x_1'^2+x_2'^2)-m^2R'^2}}. \tag{5.1}$$

A crude definition of V_X *is that* V_X *is the subset of* X *where the above radicals are all real.*

Remark 5.1. If $|m| = l + \frac{1}{2}$, then V_X is the disk where

$$x_3 = 0, \qquad R'^2 - 2n^2R' + (l+\tfrac{1}{2})^2n^2 \leqslant 0.$$

The probability of the presence of the electron on this disk is

$$\frac{1}{2\pi^2}\frac{1}{n^2}\frac{d^2x'}{\sqrt{-R'^2+2n^2R'-(l+\frac{1}{2})^2n^2}}.$$

Proof. On $V[L_0, M_0]$ the probability of the presence of the electron is, by definition,

$$\frac{1}{4\pi^2}\left[\int_\Gamma dt\right]^{-1} dt \wedge d\Psi \wedge d\Phi, \text{ since } \eta_V = dt \wedge d\Psi \wedge d\Phi.$$

By (1.5), (1.11), and $(1.12)_1$ of III,§1, the projection V_X of $V[L_0, M_0]$ onto X is the set of points x in X for which

$$\left|\frac{x_3}{R}\right| \leqslant \sin\Theta, \text{ where } \cos\Theta = \frac{M_0}{L_0}, \tag{5.2}$$

and the second-degree equation in Q,

$$H[L_0, M_0, Q, R] = 0, \tag{5.3}$$

has real roots.

Thus V_X is indeed a subset of X lying between two spheres and outside a cone of revolution whose centers and vertex are at 0.

Formulas (1.5), (1.6), and (1.12) of III,§1 show that every point x inside V_X is the projection of 4 points (t, Ψ, Φ) in $V[L_0, M_0]$. By (1.16) and (3.6) of III,§1 and by (2.9),

$$d^3x = -QR\sin\Psi\sin\Theta\, dt \wedge d\Psi \wedge d\Phi, \qquad dt = \frac{R\,dR}{Q} > 0,$$

where $|Q|$ is defined by giving x. Then on V_X the probability of the presence of the electron is

$$\frac{1}{\pi^2}\left[\int_\Gamma \frac{R\,dR}{Q}\right]^{-1} \frac{d^3x}{|Q|R|\sin\Psi|\sin\Theta}. \tag{5.4}$$

By $(1.12)_1$ of III,§1,

$$x_3 = R\cos\Psi\sin\Theta;$$

therefore, from $(5.2)_2$,

$$R|\sin\Psi|\sin\Theta = \frac{1}{L_0}[L_0^2(x_1^2 + x_2^2) - M_0^2R^2]^{1/2}, \text{where } \frac{M_0}{L_0} = \frac{m}{l+\frac{1}{2}}. \tag{5.5}$$

The definition of Q by equations (5.3) and (4.2), where E has the value (4.1) and $P^2 = (Q^2 + L_0^2)/R^2$, crudely gives

$$Q \simeq \frac{\hbar}{n}[-R'^2 + 2n^2R' - (l+\tfrac{1}{2})^2n^2]^{1/2}. \tag{5.6}$$

It follows that

$$\int_\Gamma \frac{R\,dR}{Q} \simeq 2\pi\frac{R_1^2}{Z^2}\frac{n^3}{\hbar}. \tag{5.7}$$

Formulas (5.5), (5.6), and (5.7) give the probability (5.4) its crude expression (5.1).

Now, V_X, being the subset of X where (5.5) and Q are real, is crudely defined by the condition that the radicals entering into (5.1) be real.

Remark 5.2. *The probability of the presence* of the electron defined by theorem 5 *is extremely different from that defined by wave mechanics,* but *closely connected to the first quantum mechanics*: the electron stays in a compact subset V_X of X that is the union of the trajectories used by the first quantum mechanics; the probability of its presence results as simply as possible from the equations characterizing V_X.

Conclusion

This survey has elucidated *Maslov's quantization.* On the one hand, theorem 3 of IV,§2 proves that *it is related to classical wave mechanics* and preserves its essential results; on the other hand, remark 5.2 of IV, §2 proves that *it is related to the first quantum theory* and preserves the simplicity of its calculations. Chapters I and II have shown how Maslov's quantification replaces the Bohr-Sommerfeld quantization by a *quantization that mathematical motivations justify.*

We considered only one-electron atoms (or ions). Let us review very briefly how quantum mechanics has studied several-electrons atoms. It has been by using the Schrödinger equation.

i. There are rich theoretical results (by T. Kato, B. Simon, and many other mathematicians); but they do not lead to numerical results.
ii. There are some numerical results (by E. A. Hylleraas, C. L. Pekeris, and many other physicists) that strikingly agree with experimental results; but their mathematical accuracy is not completely established.

[In its present state Maslov's quantification can reobtain neither (i) nor (ii).]

iii. There is also a very crude method that foresees the behavior of the *plasmas*; it pays attention to only one of the electrons, to which is applied the theory of one-electron atoms. *V. P. Maslov* asserts that such a procedure becomes more efficient when his quantification is used.

That last fact and the fact that (i), (ii), and (iii) together do not at all constitute a complete theory show how also *physics* called for a coherent elucidation of Maslov's quantification and for the required tools.

Bibliography

1. Cited Publications

[1] Arnold, V. I.
On a characteristic class intervening in quantum conditions, *Funct. Anal. Appl.* 1:1–14, 1967 (in Russian and English translation).

[2] Bethe, H., and Salpeter, E.
Quantum Mechanics of One- and Two-Electron Atoms, New York: Springer-Verlag, 1957.

[3] Buslaev, V. C.
Quantization and the W.K.B. method, *Trudy Mat. Inst. Steklov* 110:5–28, 1970 (in Russian).

[4] Cartan, E.
La théorie des groupes finis et continus et la géometrie différentielle, traitées par la méthode du repère mobile, Paris: Gauthier-Villars, 1937 (chapter XI).

[5] Cartan, E.
Leçons sur les invariants intégraux, Paris: Hermann, 1922.

[6] Choquet-Bruhat, Y.
Géométrie différentielle et systèmes extérieurs, Paris: *Dunod*, 1968.

[7] De Paris, J. C.
Problème de Cauchy à données singulières pour un opérateur différentiel bien décomposable, *J. math. pures appl.* 51:465–488, 1972.

[8] Gårding, L., Kotake, T., and Leray, J.
Uniformisation et développement asymptotique de la solution du problème de Cauchy linéaire à données holomorphes; analogie avec la théorie des ondes asymptotiques et approchées, *Bull. soc. math. France* 92:263–361, 1964 (introduction and chapter 7).

[9] Lefschetz, S.
Algebraic topology, *Amer. Math. Soc. Colloq. Publications* 27, 1942 (chapter III, §5).

[10] Maslov, V. P.
Perturbation Theory and Asymptotic Methods, Moscow: M.G.U., 1965 (in Russian and English translation).

[11] Maslov, V. P.
Théorie des perturbations et méthodes asymptotiques, suivie de deux notes complémentaires de V. I. Arnold, et V. C. Buslaev, Paris: Dunod, 1972 (translated by J. Lascoux and R. Senor).

[12] Morse, M., and Cairns, S.
Critical Point Theory in Global Analysis and Differential Geometry, New York: Academic Press, 1969 (§4, reduction theorem).

[13] Schwartz, L.
Théorie des distributions, Paris: Hermann, 1951.

[14] Segal, I. E.
Foundations of the theory of dynamical systems of infinitely many degrees of freedom (I), *Mat. Fys. Medd. Danske Vid. Selsk.* 31 (12):1–39, 1959.

[15] Shale, D.
Linear symmetries of free boson fields, *Trans. Amer. Math. Soc.* 103:149–167, 1962.

[16] Souriau, J. M.
Construction explicite de l'indice de Maslov. Applications. *4th International Colloquium on Group Theoretical Methods in Physics, University of Nijmegen*, 1975.

[17] Steenrod, N.
The Topology of Fiber Bundles, Princeton: Princeton University Press, 1951 (§2, §15).

[18] Weil, A.
Sur certains groupes d'opérateurs unitaires, *Acta math.* 111:143–211, 1964.

[19] Whittaker, E. T., and Watson, G. N.
Modern Analysis, Cambridge: Cambridge University Press, 1927.

2. Related Publications

[20] Crumeyrolle, A.
Revêtements spinoriels du groupe symplectique et indices de Maslov, *C. R. Acad. Sci. Paris* 280: 1753–1756, 1975; Alqèbre de Clifford symplectique èt revêtement spinoriel du groupe symplectique, *C. R. Acad. Sci. Paris* 280:1689–1692, 1975.

[21] Crumeyrolle, A.
Algèbre de Clifford symplectique, revêtements du groupe symplectique, indices de Maslov et spineurs symplectiques, *J. math. pures appl.* 56:205–230, 1977.

[22] Dazord, P.
La classe de Maslov-Arnold; L'opérateur canonique de Maslov, *Seminaire de Geometrie; Université Claude Bernard* [Lyon I], 1975/76; Une interprétation géométrique de la classe de Maslov-Arnold, *J. math. pures appl.* 56:231–250, 1977.

[23] Guillemin, V., and Sternberg, S.
Geometric Asymptotics, Providence; American Mathematical Society, 1977.

[24] Malgrange, B.
Intégrales asymptotiques et monodromie, *Ann. Sci. Ecole Norm. Sup.* 7 (4):405–430, 1974.

[25] Voros, A.
The W.K.B.-Maslov method for non-separable systems, *Colloques internationaux du C.N.R.S.*, 237: *Géométrie symplectique et physique mathématique.* Aix-en-Provence, 1974.

[26] Voros, A.
Asymptotic $\hbar$-expansions of stationary quantum states, *Ann. Inst. Henri Poincaré* 26A: 343–370, 1977.

3. Preliminary Publications

[27] Leray, J.
Solutions asymptotiques des équations aux dérivées partielles (une adaptation du traité de V. P. Maslov), *Convegno internazionale: metodi valutativa nella fisica matematica*; *Accad. Naz. dei Lincei, Roma, 1972.*

[28] Leray, J.
Complement à la théorie d'Arnold de l'indice de Maslov, *Convegno di geometrica simplettica et fisica matematica, Istituto di Alta Matematica, Roma, 1973.*

[29] Leray, J.
Solutions asymptotiques et groupe symplectique, *Colloque sur les opérateurs de Fourier intégraux et les équations aux dérivées partielles, Nice*; Lecture Notes, Springer-Verlag, 1974.

[30] Leray, J.
Solutions asymptotiques et physique mathématique, *Colloques internationaux du C.N.R.S.*, 237: *Géométrie symplectique et physique mathématique*; *Aix-en-Provence, 1974.*

[31] Leray, J.
Solutions asymptotiques de l'équation de Dirac, *Trends in Applications of Pure Mathematics to Mechanics, Conference at the University of Lecce*, London: Pitman, 1975.

[32] Leray, J.
Analyse lagrangienne et mécanique quantique, *Séminaire du Collège de France, 1976–1977*, and *R.C.P. 25, Strasbourg, 1978.*

[33] Leray, J.
The meaning of Maslov's asymptotic method: the need of Planck's constant in mathematics, *Bulletin of the American Mathematical Society*, to appear, 1981. (Symposium on the Mathematical Heritage of Henri Poincaré, April 1980).

The present publication is a revised version of [32], which encompasses [27]–[31], and [33] constitutes a comment of the present publication.